NATIONAL
GEOGRAPHIC

OCEAN

OCEAN

| A GLOBAL ODYSSEY

SYLVIA A. EARLE

NATIONAL GEOGRAPHIC
WASHINGTON, D.C.

The shallow inshore waters of
South Africa's Table Mountain host
jellyfish and curious fur seals.
PAGE 1: A lemonpeel angelfish
populates French Polynesian reefs.
PAGES 2-3: An undersea garden of
soft corals adorns a healthy reef.
PAGES 6-7: As blue as the sea, blue
button jellyfish (Porpita porpita), a
global High Seas drifter, is actually
a colony of hydroids joined as one.

CONTENTS

PREFACE | HOPE FOR THE OCEAN

Aturning point for ocean protection came in 2006 when President George W. Bush designated 362,598 square kilometers of ocean around the Northwestern Hawaiian Islands as the Papahānaumokuākea Marine National Monument, the largest place in the world where ocean wildlife was safeguarded. At the designation ceremony on June 15, President Bush provided the rationale for doing so:

The vibrant beauty of the oceans is a blessing to our country. And it's a blessing to the world. The oceans contain countless natural treasures. They carry much of our trade; they provide food and recreation for billions of people. We have a responsibility, a solemn responsibility, to be good stewards of the oceans and the creatures who inhabit them . . . Our duty is to use the land and seas wisely, or sometimes not use them at all. Good stewardship of the environment is not just a personal responsibility, it is a public value . . . We're just beginning to appreciate what the seas have to offer humanity. The waters of this new national monument will be a living laboratory that offers new opportunities to discover new life, that helps us to better manage our ocean ecosystems, and allows us to pursue advances in science.

Ten years later, President Barack Obama quadrupled the size of the monument and made these comments at the September 15, 2016, Our Ocean Conference in Washington, D.C.:

We cannot truly protect our planet without protecting our ocean . . . Our children prove every day that they care deeply about this planet. Their right to inherit a healthy planet is a sacred responsibility for all of us. And how we treat our oceans is a big part of that burden . . . It is the dominant feature of our planet. It's why we share a blue marble, as opposed to a brown or gray one . . . Dangerous changes in our climate, caused mainly by human activity; dead zones in our ocean, caused mainly by pollution that we create here on land; unsustainable fishing practices; unprotected marine areas, in which rare species and entire ecosystems are at risk—all those things are happening now. They've been happening for a long time. So if we're going to leave our children with oceans like the ones that were left to us,

then we're going to have to act. And we're going to have to act boldly. The ocean's health is our health . . . Probably the most important thing that you can do on this planet Earth is to make sure that you're making it better for future generations.

Many countries are committing to the United Nations goal of protecting 30 percent of the ocean, including the High Seas, by 2030. Presidents Michelle Bachelet and Sebastián Piñera have led Chile to commit nearly half of the country's waters for protection. As Piñera observed recently, "Oceans don't need us for their survival, we need them for ours." The small Pacific island nation of Palau has set a high bar by highly protecting 80 percent of the ocean under its jurisdiction and carefully managing the rest, commencing January 1, 2020. In a December 2020 speech, Tommy E. Remengesau, Jr., president of Palau, remarked:

The ocean is our past, our present, and our future. We do not have to choose between ocean protection and production; we can have both for a healthy, prosperous, and equitable tomorrow if we properly manage our impacts upon it.

Understanding the vital importance of the ocean to everything we care about is the key to taking action. With knowing comes caring, and with caring, there is hope that we will find an enduring place for ourselves within the mostly blue systems that sustain us.

ABOVE: A healthy coral reef in Hawaii's French Frigate Shoals. OPPOSITE: Palau president Thomas E. Remengesau and Sylvia Earle "high five" designation of Palau's marine sanctuary. BELOW LEFT: First Lady Laura Bush and guests including Sylvia Earle watch as President George W. Bush establishes Hawaii's Papahānaumokuākea Marine National Monument in June 2006. BELOW RIGHT: A decade later, President Barack Obama and Sylvia Earle visit the monument after he expanded its boundaries to 1,508,870 square kilometers.

FOREWORD

> "MOONLIGHT RIPPLED ON THE OCEAN FLOOR,
> CREATING PHANTOM SHAPES ALONG THE REEF."
>
> —SYLVIA A. EARLE, *NATIONAL GEOGRAPHIC* MAGAZINE, AUGUST 1971

Fifty years ago, oceanographer Sylvia Earle captivated the world, sharing the mysteries of life deep below the ocean's surface. When she was denied a position with a crew of men testing an undersea habitat, she accepted the challenge of leading an all-woman team as part of the program and lived underwater for two straight weeks—the first of many times she broke barriers and dared to explore new frontiers.

She's been called Her Deepness, A Living Legend, A Hero for the Planet. Through her science and environmental stewardship, Sylvia has worked to protect Earth's blue heart and all of its treasures, becoming a global icon and champion of the natural world. Through her powerful storytelling, she has connected millions of people around the globe to the splendor and awe of the sea. And amid escalating threats to the ocean, from pollution and overfishing to climate change, Sylvia has shown the world how profoundly valuable the ocean is and why we must safeguard it.

In *National Geographic Ocean,* Sylvia offers a riveting and in-depth look at this remarkable natural wonder—its beauty and complexity, drama and despair—and chronicles her lifelong commitment to protecting it. The book also spotlights National Geographic's unparalleled legacy of exploring, honoring, and conserving the ocean—from leading some of the most significant underwater discoveries in history to helping protect millions of square kilometers of open sea in the face of urgent threats. With unrivaled insight, Sylvia shares stories of fellow National Geographic legends, including Jacques Cousteau, Bob Ballard, James Cameron, Jonatha Giddens, Enric Sala, Katy Croff Bell, and many others.

Today, in our roles at National Geographic, we have the great privilege of working with innovators and solution seekers across all seven continents who are dedicated to illuminating and protecting the wonder of our world. Few have embodied that mission as fully as Sylvia.

National Geographic Ocean is a compelling and important read by a masterful storyteller and conservation champion. It gives everyone the opportunity to experience the many wonders of the sea through the eyes of some of the greatest explorers of our time. By reading her story, we hope you will be inspired to be like Sylvia Earle—and become a champion for the planet.

Jill Tiefenthaler
Chief Executive Officer, National Geographic Society

Gary Knell
President and Chief Executive Officer,
National Geographic Society, 2014–2018
Chairman, National Geographic Partners, 2018–2021

RIGHT: A molluscan "sea butterfly" Clione flutters through Arctic waters in Baffin Bay. OPPOSITE: A sharpear enope firefly squid lights its deep-ocean home with its own bioluminescence.

INTRODUCTION

A child might ask, "What is water?" "Where do waves come from?" "How deep is the ocean?" "What lives in the sea?" Such questions seem simple, but the answers are not.

In this volume, I have tried, with support from the National Geographic Society and legions of global experts, to provide answers to these questions and more by bringing together the latest insights about the nature of the ocean, its origin, its present state, and how its future and that of humankind are now inextricably bound together. The content is shaped in some measure by my lifetime of personal experiences on, around, and especially under the sea, primarily as a scientist but with opportunities to view the ocean as a government official, the founder of engineering companies, a participant in dozens of scientific and conservation organizations, a child of the 20th

century, and a parent—and grandparent of some who may greet the 22nd century. Most important, I am a witness, a participant in the greatest era of ocean discovery—and ocean decline—in the history of humankind.

Twenty years ago, in 2001, I authored the *Atlas of the Ocean: The Deep Frontier,* the National Geographic Society's first stand-alone ocean atlas, complementing the Society's iconic series *Atlas of the World,* then in its 9th edition. The first *Atlas of the World,* published in 1963, portrayed the ocean with the best data then available, that is, a mostly featureless wash of blue adjacent to highly detailed maps of the land. Knowledge of the planet's undersea terrain was just beginning to come into focus, largely through the efforts of a young woman scientist at Columbia University, Marie Tharp, who meticulously plotted data gathered by men who worked at sea from

Author Sylvia Earle, at home on a reef in the *Aquarius* underwater laboratory near Key Largo, Florida

research vessels that did not allow women aboard.

It was headline news the following year when I joined the 1962–64 International Indian Ocean Expedition as the sole botanist and only woman for a six-week voyage of exploration and research. The *Mombasa Daily Times* reported, "Sylvia Sails Away With Seventy Men, but She Expects No Problems." Actually, I did have problems, those shared by oceanographers then and even now: how to meaningfully explore, from the deck of a rolling ship, the dynamic, living liquid realm below that averages about 4,000 meters in depth. The scuba gear we had on board enabled us to dive to about 30 meters, but the hooks, nets, and grabs made blind sampling possible all the way to the bottom. I likened it to flying over New York City dangling a net from far above the clouds, deciphering the nature of what was below from captured pieces of cement, fragments of trees, and maybe a taxi, a mailbox, or some unwary pedestrians.

By 2001, I was among the many women scientists who were not only at sea but also leading expeditions, living underwater, designing, building, and piloting submersibles, starting and leading tech companies, and commanding research ships. Maps of the ocean had progressed significantly, portrayed in detail in the first ocean atlas in striking shades of blue adjacent to landforms shown in black, a deliberate effort to graphically emphasize the ocean as Earth's dominant feature. National Geographic Books' then director of maps, Carl Mehler, worked closely with marine geologist Linda Glover and then executive editor Barbara Brownell Grogan to produce and win approval for publishing the maps in this revolutionary form, a style that continues in this and many other volumes.

For me, social change was moving in the right direction, and so was ocean exploration. At a National Academy of Sciences meeting in 2000, geophysicist Robert Ballard quipped, "The next generation will explore more of the Earth than all preceding generations combined." I was at sea most of that year, leading the 1998–2003 NGS-Goldman Foundation-NOAA–supported Sustainable

In deep waters off the Philippines, myriad tiny shrimp larvae encased in egg sacs swarm the mother who has just produced them.

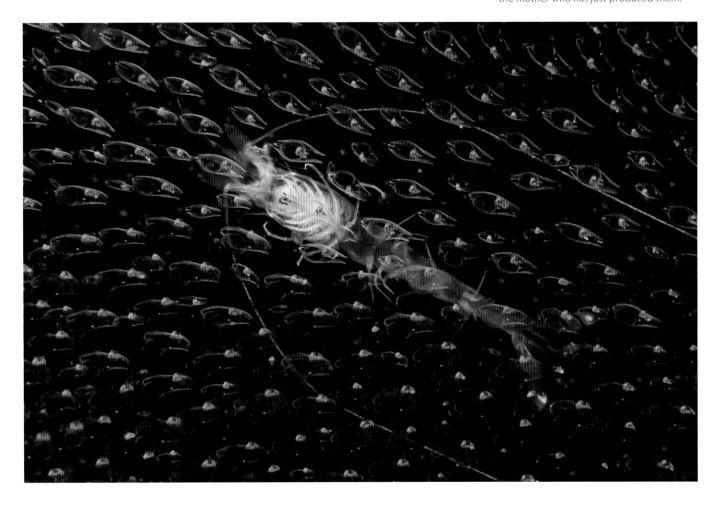

Seas Expeditions, using one-person submersibles that I helped design to be so simple to operate that "even a scientist can do it." Frequently I relied on a satellite phone to discuss with Grogan editorial details of the book. At the last possible moment, we were able to include news of the unexpected discovery of the "Lost City," an 18-story-high limestone formation surrounding a warm hydrothermal vent 900 meters deep in the North Atlantic Ocean, a revolutionary find that inspired new thinking about the origin of life and the ocean's geological processes.

New discoveries about the ocean were occurring so fast that as the 2001 *Atlas of the Ocean* was about to go to press, I joked that it would be out of date before the ink on the pages was dry. Straightaway, the book team began gathering updates for what we hoped would be a new version. Seven years later, in 2008, *National Geographic Ocean: An Illustrated Atlas* went to press, a hefty volume that Glover and I co-authored, with Grogan providing vital editorial oversight. Two years earlier, I had teased a Google executive, John Hanke, about the absence of something big and blue on their popular digital Google Earth platform. "You should call it Google Dirt," I suggested. Remarkably, Hanke invited me to help do something about the "missing piece." We assembled a team of 30 international experts, including Glover, who, with years of experience in the Office of the Oceanographer of the Navy, navigated the right channels to obtain release of previously undisclosed seafloor data. The creative Google engineers, working with the Navy, NOAA, the company Deep Ocean Exploration and Research (DOER), and various new data sources, released the first publicly available digital portrayal of the entire world in 2009. The 2008 ocean atlas benefited by producing for National Geographic the first significant updates to its seafloor maps in 30 years.

More than new maps, the 2008 atlas reflected the impact of the digital revolution in terms of photography, data gathering, storage, and important new ways of data transmission. Live views from the deep sea were being transmitted from undersea robots and submersibles to surface ships via satellite to viewers globally. Global information systems (GIS) technologies, with a history of integrating, visualizing, and analyzing terrestrial data, began to be applied to the ocean, notably by the California-based company Environmental Systems Research Institute (ESRI). Data for ESRI's ArcGIS Ocean Basemap was under way, with the goal of portraying a global visualization of the ocean digitally, in three dimensions. The 10-year Census of Marine Life had already found thousands of new kinds of ocean organisms, and Craig Venter, renowned for

> "KNOWLEDGE OF THE OCEAN IS MORE THAN A MATTER OF CURIOSITY. OUR VERY SURVIVAL MAY HINGE UPON IT."
>
> —PRESIDENT JOHN F. KENNEDY (1961)

sequencing the human genome, had discovered millions of new genes while undertaking another ambitious project: sequencing the ocean's genome.

But exhilaration about new discoveries was more than matched by the painful realization that I was witnessing the ocean's rapid decline. NOAA declared the Caribbean monk seal, a charismatic creature still swimming in the Gulf of Mexico when I was a teenager, to be officially extinct in 2008. Mangrove forests and seagrass meadows I knew and loved as a child were being paved over in Florida. I mourned the transformation of a favorite 500-year-old mound of brain coral in the Bahamas into what appeared to be a massive snowball. It, like reef corals globally, had succumbed to warming seas, changing ocean chemistry, and aggressive fishing. In the Galápagos and Cocos Islands, where I once felt a tinge of fear because of the abundance of sharks, I now was afraid because of their absence. Sharks, like grouper, snapper, cod, tuna, and even colorful parrotfish, were becoming rare except in restaurants.

In 2000, a fraction of one percent of the ocean was safeguarded with full protection, but there was compel-

Soft mushroom coral (*Anthomastus ritteri*)
snare plankton deep in the Pacific Ocean.

ling evidence that damaged systems could rebound if the pressures were removed. A turning point came in 2006 when President George W. Bush designated the waters around the northwestern Hawaiian Islands to 92 kilometers from shore as the world's largest protected area, the Papahānaumokuākea Marine National Monument. I was thrilled in the fall of 2008 when Chris Anderson, curator of TED (Technology, Entertainment, Design), called to tell me that I had won the TED Prize and to say that if I made a wish "big enough to change the world," the "TEDsters" would help make it happen. The one thing I felt could have the largest impact would be to embrace the wild ocean with care. It would take enhanced communications, new submersibles, and a new way of thinking about the value of the ocean, but most important, it would need a network of protected areas, "Hope Spots," large enough to restore and protect the ocean, Earth's blue heart.

Science fiction writer Isaac Asimov observed: "Science gathers knowledge faster than society gathers wisdom." Nowhere is this more obvious than in the way the living ocean is regarded in the 21st century. The ocean is in trou-

ble, and therefore, so are we. But old habits, existing laws, and vested economic and political interests fuel business as usual. Nonetheless, in 2016 nations came together at the World Conservation Congress in Hawaii with a resolution to protect at least 30 percent of the ocean by 2030. Hope for success surged when President Barack Obama at the meeting expanded the Papahānaumokuākea Marine National Monument to the very edge of U.S. jurisdiction, making it once again Earth's largest protected area.

How much of the ocean must be protected to secure a habitable planet? Many support maintaining at least half of the world, land and sea, in a natural state to protect diversity and healthy planetary functions. I am among those who regard Earth, all of it and all of us, as a Hope Spot, where we now have the knowledge that can enable us to find an enduring place for ourselves within the natural systems that make not only our livelihoods but our very lives possible. The next 10 years can be a time of renewal and recovery, a time to develop a peaceful relationship with the ocean, with nature, and among ourselves.

I THE

PART ONE

LIVING OCEAN

1

ORIGIN OF THE OCEAN

Chapter One

"IF SOME ALIEN CALLED ME UP . . . 'HELLO, THIS IS ALPHA, AND WE WANT TO KNOW WHAT KIND OF LIFE YOU HAVE,' I'D SAY, 'WATERBASED.'" —CHRISTOPHER MCKAY, NASA SCIENTIST

On December 24, 1968, astronaut William Anders glanced out of the window of the Apollo 8 spacecraft as it emerged from the far side of the moon and saw what appeared to be a glistening drop of water suspended against a vast desert of darkness.

"Oh, my God," he said. "Look at that picture over there! There's the Earth coming up. Wow, is that pretty." Later, he remarked, "To me it was strange that we had come all the way to the moon to study the moon, and what we really discovered was the Earth."

"A pale blue dot," astrophysicist Carl Sagan called it, looking at an image of Earth taken by the Voyager 1 spacecraft from the edge of the solar system. Now we know, with a shiver of awareness, what our long-ago ancestors could not know: Nowhere else in the universe is there a planet blessed with a built-in life support system just right for the likes of us, and the key to making it so is the ocean.

Even deep within the Earth, between 410 and 660 kilometers down, abundant water exists, bound to hydrous minerals. Water can exist without life but life cannot exist without water. Ninety-seven percent of Earth's water available to life is ocean, and most of the rest of Earth's water is frozen in glaciers and polar ice. All lakes, rivers, streams, and groundwater combined make up only a bit more than 0.7 percent of this planet's available water. The ocean covers more than two-thirds of Earth's surface with an average depth of about 4 kilometers, a maximum of nearly 11 kilometers. That precious fluid stands between us and oblivion.

Where did it come from, that miraculous substance that morphs into vaporous clouds and takes the form of lacy snowflakes, delicate crystals, and solid blocks of ice? Mysteries about the origin of Earth's water are being resolved by analyzing ancient rocks and considering new evidence about the widespread occurrence of water within our solar system and far beyond.

It is water plus the rare just-right distance from our sun-star (not too hot, not too cold), the ideal size (not too big, not too small), and Earth's physical and chemical composition, including a magnetic metal heart, that set our planet apart as a place fit for life as we know it—after billions of years of fine-tuning.

RIGHT: Earthrise—a mostly blue planet—is captured by astronaut William Anders aboard the Apollo 8 spacecraft in 1968. PAGES 16-17: A kaleidoscopic swirl of bigeye jacks off Darwin Island, Galápagos. PAGES 18-19: An ocher starfish, jade anemones, and barnacles share space in a tide pool in Washington's Olympic National Park.

The Earth Forms

During the formation of the solar system, masses of gas and dust—essentially "stardust"—aggregated into grains that gathered together as small planetoids that in turn collided, and from this swirling mass of material, the planets of our solar system were formed. The age of the Earth is thought to be 4.54 billion years, based on data derived from radioactive decay techniques applied to meteorites, from rocks brought back from the moon by astronauts, and by examination of zircon crystals embedded in ancient rocks found in Western Australia.

Earth began as a fiery ball of molten rock that gradually cooled over millions of years, forming a solid crust that is the foundation for the continents and seafloor. No one has yet seen, or even sent a probe into, the fiery interior of the Earth, but from measurements and models, it appears that there is a dense, hot inner iron core surrounded by a liquid metallic layer that releases enormous amounts of heat and, together with Earth's rotation, creates the planet's magnetic field.

The crust is made of solid rocks and minerals. The mantle, joined to the crust, is largely rock but also includes malleable areas of semisolid magma. Together, the crust and part of the upper mantle make up the lithosphere. In between the two is a layer where seismic waves change velocity known as the Mohorovicic discontinuity, or simply "the Moho." The depth of the Moho varies, as does the depth of the lithosphere.

"Where did we come from?" "What is our place in the universe?" "What new discoveries might there be over the horizon?" Or, as oceanographer Walter Munk observed in a 2015 article in *Science* magazine, "We ought to know something about what happens beneath us."

Only a few centuries ago, it was commonly believed that Earth was the center of the universe; that the sun, moon, and stars moved around our planet; and that plant and animal life as it then existed had remained unchanged throughout time. The continent of Antarctica was not discovered until two centuries ago, and as recently as the mid-20th century, Earth's landmasses were thought to be constant in their position within an ocean of fixed dimension.

It now seems obvious that the landmasses on opposite sides of the Atlantic Ocean were, in fact, once joined in a jigsaw puzzle–like fit as one supercontinent that astronomer Alfred Wegener called "Pangaea."

Only in recent decades has it been possible to acquire data that indicate that the lithosphere is composed of 15 or so enormous slabs, called tectonic plates. These plates glide over the Earth's surface at a stately geological pace. Over the ages, their dynamic movements and collision have caused the crust to buckle, lifting mountain ranges, building island chains, creating volcanoes, and generating earthquakes.

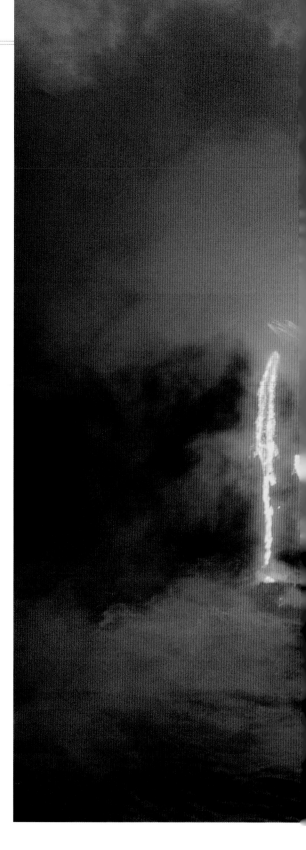

Spewing clouds of gas skyward, Hawaii's Kīlauea volcano sends molten rock into the sea, where it forms crusts and hollow lava tubes.

EARTH INSIDE OUT

Earth's hot heart: A core of molten iron-nickel is surrounded by liquid metal wrapped by a thick lower mantle. Above it floats the upper mantle and rocky crust, the lithosphere, composed of 15 or so tectonic plates.

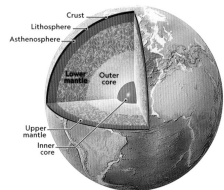

Crust
Lithosphere
Asthenosphere
Lower mantle
Outer core
Upper mantle
Inner core

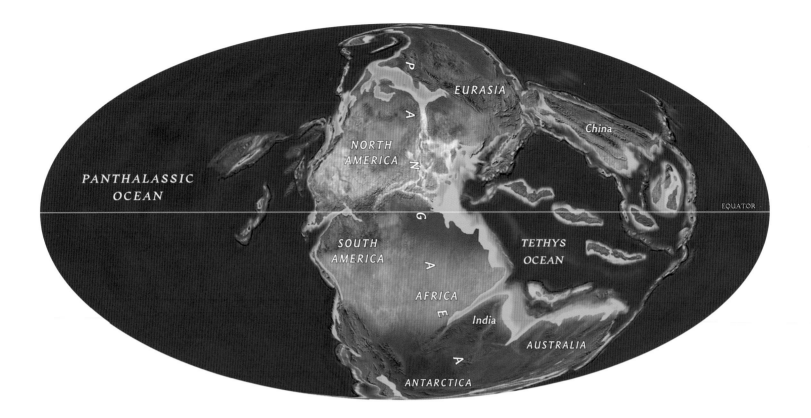

The origin of the continents is widely thought to have developed from rocky materials of lower density that were pushed to the surface by the movement of tectonic plates. Based on analysis of thousands of rocks from both ocean and continental crusts, some more than four billion years old, as reported in June 2015 *Nature Geoscience,* the lighter terrestrial crust rose above the ocean over three billion years ago. A study by geochemist Bruno Dhuime of the University of Bristol in England used an analysis of radioactive decay to derive this timeline. In 2014, a 4.4-billion-year-old zircon found in Australia seemed to confirm that what is now land may have been submerged for more than a billion years.

Once emerged, the amount of continental crust has been relatively constant, covering about 30 percent of the Earth's surface, with the ocean dominating the rest. Over the ages, continental masses have been in slow but constant motion, pushed together at times, then split apart by the restless, rifting, spreading, subducting seafloor, endlessly creating and destroying ocean crust. At least five times in Earth's history the continents have been brought together in one immense "supercontinent," then have been

TOP: Some 200 million years ago, shifting plates collided to form the supercontinent Pangaea, surrounded by ocean. ABOVE: As Indian and Eurasian plates slammed together, their crust crumpled upward, forming the Himalaya. BELOW: Geologists use this time line to sequence events and their relationships over Earth's 4.5-billion-year history.

GEOLOGIC TIME LINE

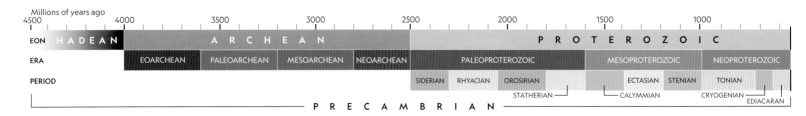

torn apart by relentless tectonic movements, a pattern that recurs about every 250 million years.

As the landmasses shift, so does the shape of the ocean. Continental shelves now border major landmasses in depths to about 200 meters, sloping along their margins to the great deep ocean basins beyond. Four large basins with an average depth of 3.7 kilometers occupy about three-quarters of the total area of the ocean. The Pacific is the largest, encompassing about a third of Earth's surface with an average depth of 4,000 meters. The Atlantic Ocean is about half as large as the Pacific with an average depth of 3,646 meters. And the slightly smaller Indian Ocean basin is also slightly deeper, with an average depth of 3,960 meters. The smallest ocean basin is at the top of the world beneath the Arctic Ocean, with a total area that is about 10 percent of the size of the Pacific Ocean with an average depth of 1,250 meters. The Southern Ocean does not have a basin but flows over the submerged terrain surrounding Antarctica.

All ocean basins feature volcanic ocean ridges, the most distinctive occurring in the Atlantic, a mighty mountain range that extends in a sinuous line from the edge of the Arctic to the Southern Ocean. Seamounts rise from within broad abyssal plains that occupy the greatest part of ocean basins. Rimming the edge in places are steep-sided trenches that plunge thousands of meters deeper than the adjacent plains.

Knowledge of the nature of Earth's deep ocean basins has increased rapidly in recent decades with the advent of techniques to determine accurate water depth measurement (bathymetry) over wide areas, to sample by drilling, and to observe and document directly with manned and robotic underwater vehicles. Nonetheless, it is surprising to some to learn that only about 15 percent of the ocean has been seen, explored, or even mapped with the same degree of accuracy as has been done for the moon, Mars, Jupiter, and the terrestrial parts of Earth.

When ocean bathymetry was portrayed digitally on Google Earth in 2009, it was based on the best publicly available data, and for the first time made knowledge of the basic configuration of the Earth beneath the sea readily accessible on digital devices around the globe. Limitations in detailed coverage came into sharp focus when it proved impossible to find a Malaysian Airlines passenger plane that was lost in the southern Indian Ocean in March 2014. Large areas of ocean, especially in the Southern Hemisphere, are less well mapped than the far side of the moon.

WHY IT MATTERS

Drilling Into the Depths

O ver cocktails with colleagues one evening in 1957, oceanographer Walter Munk conceived a plan to drill through the ocean's crust and continue into the mantle. The first Deep-Sea Drilling Program flourished from 1963 to 1983 and gave rise to global programs that continue today. Carving out cylindrical core samples deep into the ocean floor, to be read like the cross-section of a tree trunk, allows

scientists to decipher the past, to inform our future. Early data supported Alfred Wegener's theories of seafloor spreading, plate tectonics, and the great ancient continent Pangaea. Global drilling has uncovered such wonders as the makeup of the 80-million-year-old lost Pacific continent Zealandia, which tells us of volcanism and climate change; rare metal deposits from the Brothers Volcano submerged 2.4 kilometers below the Pacific surface; and deep-subsurface microbes that may survive on minuscule amounts of energy for millennia.

A research vessel is positioned in deep water for exploratory offshore drilling.

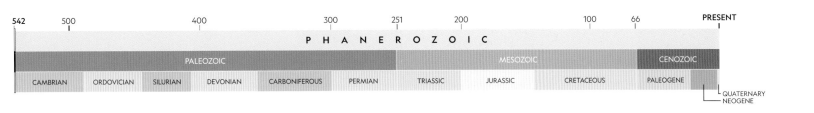

| 542 | 500 | | 400 | | 300 | 251 | 200 | | 100 | 66 | PRESENT |

Underwater Mountains in Motion

S everal discoveries that led to the understanding of planetary dynamics began quietly, not while at sea but later, through painstaking plotting of data gathered offshore. Oceanographers Maurice Ewing and Bruce Heezen at Columbia University's Lamont Geological Observatory used sonar "echo sounders" in the 1950s aboard research vessels to map the floor of the Atlantic Ocean, while Heezen's graduate student, Marie Tharp, analyzed the information back in Heezen's Columbia University laboratory. Gradually, the magnitude and meaning of a continuous mountain ridge with hundreds of peaks, some thousands of meters high, that ran like a giant backbone down the middle of the Atlantic Ocean began to come into focus. Based on their work, the National Geographic Society and the Geological Society of America began issuing a series of maps late in the 1950s that for the first time showed the seafloor as if the ocean had been drained away.

Beyond revealing the configuration of the Atlantic Ocean, Ewing, Heezen, and Tharp found an intriguing correlation between the Mid-Atlantic Ridge and the pattern of earthquakes in the Atlantic. By analyzing earthquake data worldwide, they proposed that ranges of mountains must extend throughout all of the oceans. During the International Geophysical Year (1957–58), data gathered by research vessels operating globally proved the existence of the largest feature on Earth's surface—more than 64,000 kilometers of nearly continuous mountain ranges in the depths of the Atlantic, Pacific, Indian, Arctic, and Southern Oceans.

But there was more. Marie Tharp noticed a recurrent pattern that appeared in the Mid-Atlantic Ridge, a cleft deeper and wider than the Grand Canyon, running down the center of the entire mountain range. Soon thereafter, ocean scientists Harry Hess and Robert Dietz independently proposed an explanation for this and other mysterious features of the ocean floor, and thereby launched a new field of science—plate tectonics—and a new understanding of the Earth's crust and its ocean basins. The clues were diverse: the continuous crack down the middle of the mid-ocean ridges; the deep trenches around the rim of the Pacific and elsewhere; the preponderance of volcanoes and earthquakes on one side of these deep trenches; the fact that sediments in the deep sea close to the center of the seafloor appeared to be younger than those along the edges of the continents; and the way the configuration of continents on one side of the Atlantic seemed to fit nicely against continents on the other, as if they had once been joined.

The evidence led to one explanation: The seafloor is spreading—with molten material rising from the mantle into the middle of the mid-ocean

ABOVE: At 2,700 meters deep along the Mid-Atlantic Ridge, the bright red Trachymedusa jellyfish uses two sets of tentacles to capture prey. OPPOSITE: Along Iceland's portion of the Mid-Atlantic Ridge, at Thingvellir National Park, a diver explores the Silfra canyon.

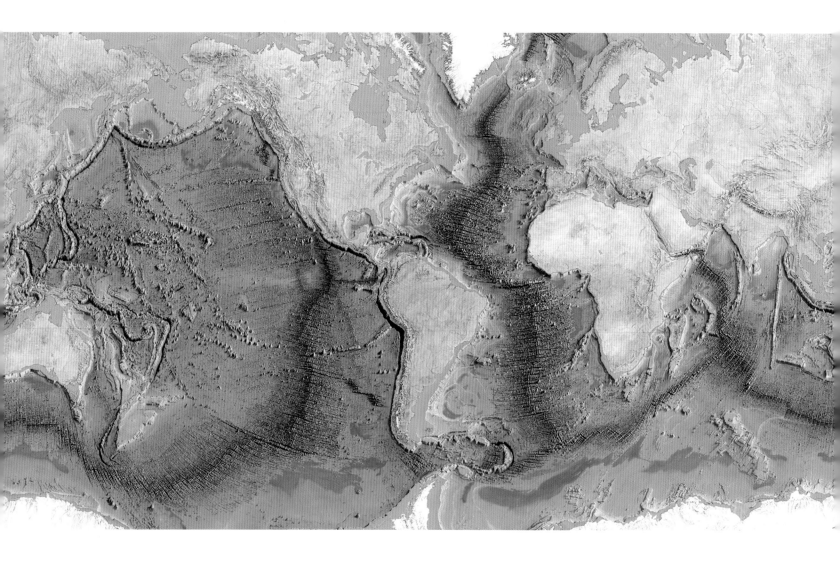

ABOVE: The first detailed map of the seafloor by oceanographers Bruce Heezen and Marie Tharp (pictured opposite) in 1957 revealed an Earth-encircling Mid-Ocean Ridge and paved the way for acceptance of the theory of seafloor spreading and continental drift. LEFT: Rocky terrain characterizes the Mid-Atlantic Ridge.

ridges, cooling, forming new seafloor, then moving away from both sides of the ridge, driven by currents of heat deep within the Earth. The Red Sea, the Gulf of California, and the great East African Rift Valley are now identified as new spreading centers, where briny lakes will eventually become new ocean basins and the land on either side will be pushed apart. The continents are drifting in response, riding along on the moving tectonic plates as the seafloor spreads apart. Where a moving oceanic plate plows into a thicker continental plate, the thinner ocean crust is subducted, or thrust down under the continent, pulling the seafloor into deep ocean trenches. The subducted plate heats and melts as it is pushed deeper and reabsorbs into the mantle, some of it returning explosively to the surface through volcanoes that form over the subducting plate. Many volcanic island arcs near deep trenches can be seen in the western Pacific. Where two continental plates collide as the spreading plates carry them, high plateaus and huge mountain chains are thrust up.

For reasons that are not well understood, the Earth's magnetic field periodically reverses itself, the North Pole becoming the South Pole and vice versa. When this happens, bits of iron embedded in cooling lava solidify into tiny magnets permanently aligned with the north magnetic pole at the time they become rock. This leaves a record of reversals of the magnetic poles. First in the Indian Ocean and then in the northern Atlantic, bands of magnetized rock perfectly matching in their widths, rock ages, and magnetic signature were discovered aligned on either side of seafloor-spreading ridges.

In the 1970s, a project called FAMOUS (French-American Mid-Ocean Undersea Study) involved scientists and several submersibles from both countries in a collaboration geophysicist Robert Ballard called "a combination of basic science and adventurous exploration." Ballard, one of the expedition's leaders, was among those who visually confirmed the phenomenon of seafloor spreading. He expressed amazement at the narrowness of the zone where fresh volcanic intrusion flowed up between the two plates, each one thousands of kilometers across, slowly drifting apart.

At the tip of Iceland's Reykjanes Peninsula, you can see the Mid-Atlantic Ridge rise above sea level and even walk across a bridge that spans the Eurasian plate on one side, the North American plate on the other. Ultra-clear water fills the cleft at Thingvellir National Park in southwest Iceland, a popular place for intrepid divers to witness the shallow-water extension of what is largely a deep-sea phenomenon.

VISIONARIES

Marie Tharp
Mapping Milestone

....

At first dismissed as "girl talk," Marie Tharp's plotting of the Mid-Atlantic Ridge soon revolutionized ocean cartography. In the 1950s few women joined research ships at sea, but that didn't stop the young geologist from gathering data of the ocean floor. With master's degrees in geology and mathematics at a time when women rarely studied earth sciences, Tharp worked as a research assistant at Columbia University's Lamont Geological Observatory. She collaborated with fellow geologist Bruce Heezen, who did go to sea and sent back sonar measurements of the North Atlantic depths. As Tharp translated the data by hand, mountains, canyons, and ridges including the Mid-Atlantic Ridge came into view. She rocked the world with the first detailed map of the ocean floor in 1957.

....

> "THE OBSCURE WE SEE EVENTUALLY. THE COMPLETELY APPARENT, IT SEEMS, TAKES LONGER."
>
> — EDWARD R. MURROW, AMERICAN JOURNALIST

Plate Tectonics: The Mechanics

You can't feel it any more than you can feel Earth spinning in its orbit around the sun. But it is happening beneath us in a continual global process, building mountains and island chains, expanding and shrinking the ocean, triggering volcanic eruptions, earthquakes, and tsunamis. It is plate tectonics.

Plate tectonics is simply this: The Earth's crust and upper part of the mantle make up the lithosphere, which is separated into some 15 massive plates. These move across Earth's surface, colliding, diving one below the other, spreading apart, and scraping past one another. In three key plate motions—convergent, transform, and divergent—they build and destroy landforms, change geologic boundaries, and launch awe-inspiring natural forces that change the face—and life—of Earth.

GEOLOGICAL FORCES

As massive plates drift across Earth's surface, colliding, ripping apart, and grazing past one another, their movements trigger seafloor spreading, volcanic eruptions, and earthquakes.

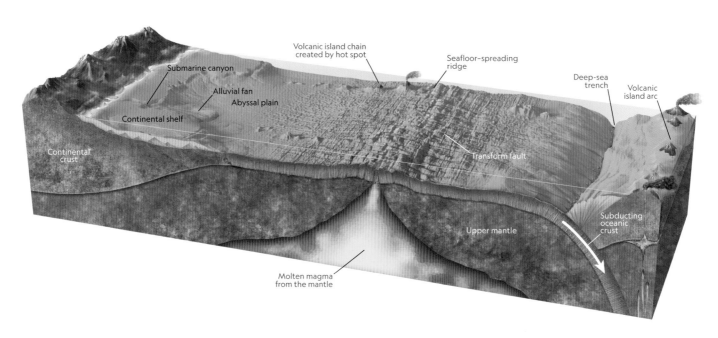

Volcanic island chain created by hot spot
Submarine canyon
Alluvial fan
Abyssal plain
Continental shelf
Continental crust
Seafloor-spreading ridge
Deep-sea trench
Volcanic island arc
Transform fault
Upper mantle
Subducting oceanic crust
Molten magma from the mantle

Convergent Boundaries

Convergent plate boundaries are crash zones. Two plates collide with such force that the edges of each plate may buckle, forming a massive mountain range—aka new continental crust. In places where one plate dives below the other, or subducts, a deep trench is formed.

Transform Boundaries

A transform boundary occurs when two plates are moving past each other. Friction strains the rocks at each boundary until they break, releasing energy as an earthquake. Most transform boundaries are created as two plates move away from the spreading center at a divergent boundary.

Divergent Boundaries

At divergent boundaries, plates pull apart, creating a split in the seafloor called a rift valley. Here geysers spew out superheated water and magma gushes from the mantle below, solidifying into basalt as it hits the cold seawater and creates new ocean crust.

Alfred Wegener
Earth's Puzzle Pieces

....

Like pieces of a puzzle, the eastern coasts of North and South America and the western coasts of Europe and Africa appear to have once fit together. As early as 1910, Alfred Wegener thought so, too. An astronomer who turned his interests to paleoclimatology, meteorology, and geophysics, he took a particular interest in these contours; he determined they could have been part of a single landmass he named Pangaea. He was ridiculed during his lifetime because he could not find any believable mechanism to support his theory. But he was right after all. In the 1960s, scientists found evidence of seafloor spreading and moving plates at the Atlantic Reykjanes Ridge off Iceland. Along the coast of New England they found fossil remains identical to those on Europe's west coast and similar remains along eastern Florida and western North Africa.

....

California's San Andreas transform fault is the birthplace of the historic magnitude 7.9 San Francisco earthquake in 1906.

Living in Hot Water

Among the most celebrated ocean discoveries of the 20th century are hydrothermal vents and the life associated with them, including microbes so different from all other forms of life that a new domain and a new kingdom, the Archaea, were designated. Lava erupting into near-freezing seawater at ridges spreading on the seafloor swiftly hardens and cracks. Cold seawater flows into deep crevices, where it comes into contact with molten rock and becomes hot enough to dissolve iron, manganese, silicon, and other minerals from the rock. This heated, mineral-laden water then rises out of the seafloor, sometimes as gently seeping warm water but also as hot, gushing geysers. As this superheated water pours back into the surrounding sea, minerals precipitate out into metallic crusts, or sometimes into tall chimney-like columns, fantastic shapes as high as multistoried buildings. Marine geochemist John Edmond calculated that all the water in the ocean circulates through deep hydrothermal vents about every 10 million years, and in so doing helps give the saltiness of the ocean its amazing consistency.

Though predicted for years, hydrothermal vents had not actually been seen in the deep sea until 1977, when geophysicists John Corliss and Robert Ballard and several dozen other researchers—with three ships and a submersible—explored a promising part of the Galápagos Rift 420 kilometers offshore from the Galápagos Islands. Crouched inside the research submersible *Alvin* at 2.4 kilometers under the sea, they were the first to personally witness glassy mounds of pillow lava and dark plumes of hot water laden with hydrogen sulfide gushing into the surrounding cold.

Hydrothermal vents now have been found globally: at the Gorda and Juan de Fuca Ridges on the northwestern coast of the United States, near Baja California, and in the Atlantic, Indian, Arctic, and Southern Oceans. They appear to mostly occur along active seafloor-spreading ridges, around fracture zones along those ridges, along subduction zones, and sometimes on isolated seamounts—in other words, in most places in the ocean where there is volcanism.

The serendipitous biological discoveries at the Galápagos site were revolutionary. No one expected to see a metropolis of strange-looking creatures flourishing in darkness, thousands of meters below the reach of life-giving sunlight. The communities of tubeworms, clams, and mussels were described as "lush oases in a sunless desert." Subsequent research proved that the animals derived sustenance from microbes that thrive on chemical energy through chemosynthesis—a process that bacteria and other microbial life appear to have mastered in the ocean for as much as a billion years prior to the advent of photosynthesis.

ABOVE: Small but active, Northwest Eifuku volcano vents in the Mariana Trench. OPPOSITE: A pale brachyuran crab clings to a giant tubeworm at a hydrothermal vent on the East Pacific Rise.

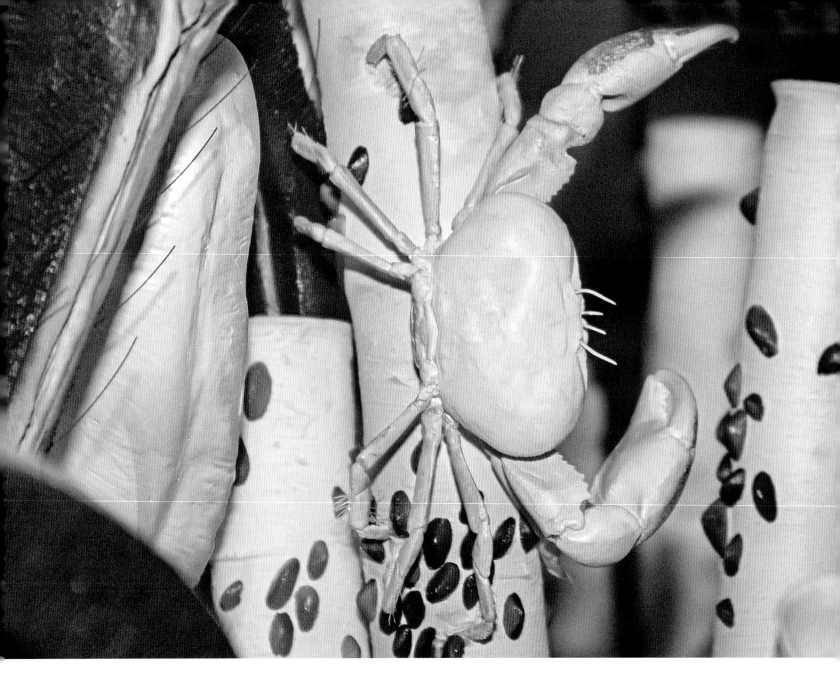

Holger Jannasch, a microbiologist at the Woods Hole Oceanographic Institution, pioneered studies on hydrothermal microbes, examining the pathways whereby some of them, taking advantage of sulfur compounds ejected from the vents and oxygen from the surrounding seawater, convert carbon dioxide into organic matter. Some organisms rely on symbiotic bacteria for sustenance, others filter microbes in the water for food, and predators dine on their neighbors.

Genome scientist Craig Venter intensively studied one microbe, *Methanococcus jannaschii,* beguiled by its weirdness. It lives at temperatures from 48°C to 94°C and at pressures 200 times as high as those at sea level; oxygen kills it; it generates sustenance from carbon dioxide, nitrogen, and hydrogen; and it produces methane. In a 1996 *Science* article, Venter remarked, "It's like something out of science fiction. Not so long ago, no one would have believed you if you'd told them such organisms existed on Earth." Those seeking life beyond Earth's atmosphere are heartened by the discovery of microbes that thrive under circumstances that may have prevailed on Earth in its formative years—and that may exist now on Mars, on Jupiter's moon Europa, or on other planets in this or other solar systems.

Landscape of the Ocean Floor

O n a map, the floor of the ocean might appear monotonous, but the reality is anything but. The land under the sea has higher mountains, broader plains, and deeper canyons than anything above the ocean's surface. When sea level was lower, rivers cut canyons into the now submerged continental shelves, rimmed in places by deep, steep trenches. Sediment from the rivers and the shells of myriad microscopic plankton blanket the seafloor across vast, flat abyssal plains. Thousands of isolated mountains, seamounts, are scattered across the plains, and the largest feature on Earth—more than 64,000 kilometers of almost continuous mountain ranges, the Mid-Ocean Ridge—laces the ocean depths of the Atlantic, Pacific, Indian, Arctic, and Southern Oceans. Running

A spotted ratfish cruises the deep seafloor off Monterey Bay, California.

down the middle of the ridge is a cleft where molten rock spills out on both sides, hardens, and moves away as the underlying plates spread apart.

The Mid-Ocean Ridge forms rough, new crust that cools and sinks deeper as it spreads away from the ridge, forming abyssal hills. Much of this terrain is covered by sediments derived from the shells of countless billions of planktonic organisms deposited over millions of years, combined with silt, sand, and mud eroded from the land into the sea. These sediments form broad, flat abyssal plains, masking the underlying terrain.

Vast areas of the seafloor are paved with manganese nodules: potato-size lumps of manganese, iron, copper, and other minerals. First discovered during the H.M.S. *Challenger* expedition of 1872–76, they were regarded mostly as a curiosity until the magnitude of their abundance and the value of their strategically important metals became known. Iron typically accounts for most of a nodule's composition. As much as 25 percent may be manganese, with cobalt and copper about 2 percent each and nickel 1.6 percent. Content and percentages vary widely from place to place, and so do the microbes that draw elements from the surrounding seawater and slowly build the lumpy formations. A nodule the size of a walnut may be a slowly growing living rock several thousand years old.

Fishermen trawling in deep water occasionally haul up chunks of icy material that fizzles and pops, then dissolves into a pool of gas and water. The glistening masses consist of methane frozen within a lattice-like structure of water molecules called clathrates, or gas hydrates by oil field geologists. Formed at depths greater than 300 meters and in near-freezing temperatures, gas hydrates are common in the deep sea along continental margins, and in latitudes near the poles. Gas hydrate layers mixed with sediment may exceed a thickness of 1,000 meters directly beneath the seafloor. Warming temperatures may cause a breakdown of the frozen methane, triggering underwater landslides and potentially releasing methane into the atmosphere.

A sea cucumber settles on the seabed of the Pacific's Clarion-Clipperton Zone, where mineral-rich manganese nodules are sought by miners.

DEEPER DIVE

The Ocean Landscape: It's Integrated

Rocky outcroppings, glorious coral formations, thick seafloor sediment, metallic nodules, and bedrock. These features are vital for shaping the geography of the ocean. In addition, each one brims with life and chemical change—vital for shaping the ocean's biologic and chemical processes. In rocky outcroppings from an ancient sea that once covered South Africa, researchers from England's University of Leeds found tiny tunnels, leading them to believe that shrimps and worms likely burrowed deep into seafloor sediment, surviving only on oxygen and nutrients there. In Pacific waters off Japan, coal beds more than two kilometers below the seafloor hold microbes possibly tens of millions of years old, whose longevity and efficient energy use may lead to revolutionary human applications.

Red shell coral

Above the seafloor, coral reef communities of living, soft-bodied polyps, cemented together by their limestone skeletons, create fantastic contours as they sustain and protect myriad creatures. Called "among the most spectacular features on the seafloor," manganese nodules form slowly around a shark's tooth or a whale's ear bone. Aided by microbial action, manganese, iron, and other metals extracted from the surrounding water may grow into fist-size lumps. Targeted for extraction for their valuable metal and mineral content, nodules have been protected from exploitation because mining them is expensive and difficult. New markets and improved deep-sea access are changing that, despite knowledge of the devastation mining causes to deep-sea ecosystems.

HOPE
SPOT

Galápagos Archipelago

A gem in the Eastern Pacific Seascape Hope Spot, which spans two million square kilometers of ocean from Central to South America, the Galápagos Archipelago off the coast of Ecuador is the renowned laboratory for naturalist Charles Darwin's 19th-century research on the evolution of species. As it also embraces marine protected areas and World Heritage sites, the Ecuadorian government is developing policies to protect seals, manta rays, hammerhead sharks, and other marine life from destructive fishing, tourism pressures, and shipping traffic.

The semiaquatic Galápagos marine iguana is unique to the archipelago.

Earth's Dynamic Ring of Fire

Nowhere on Earth is there more evidence of the powerful processes that reshape the planet than around the rim of the Pacific Ocean. Where continents and oceanic plates collide, there is a necklace of active volcanoes behind deep-sea trenches. Aptly called the Ring of Fire, the region gives rise to many of Earth's most powerful volcanic eruptions and earthquakes.

East of Japan, the great Pacific plate crunches into, and is forced beneath, the Asian continent along the Japan Trench, sending more than a thousand perceptible tremors a year across the nearby islands, including an occasional catastrophic quake.

On the land, nearly every mountain has been climbed, measured, and documented from the bottom to the top, but most of their undersea counterparts have yet to be seen, let alone explored, with many known only from their peaks, the depths below still shrouded in mystery. The tallest mountain on Earth is also the tallest volcano, Mauna Kea, in Hawaii. The height from its fiery peak to sea level is 4,205 meters, but continuing to its deeply submerged base, it extends 6,000 meters more—significantly taller, overall, than Mount Everest's 8,850 meters. Undersea mountains—seamounts—are mostly volcanic formations that rise above the seafloor 1,000 meters or more. Lesser peaks, between 500 and 1,000 meters, are termed knolls, and bumps in the seafloor that are 500 meters or less are referred to as hills. The existence of seamounts has long been known, but only about 1,000 have been surveyed well enough to be named and featured on charts.

Satellite sensors now collect precise measurements of the height, or altimetry, of the sea surface—that is, the sea level—and measurements of the gravitational field across the oceans. Variations in the gravitational pull of rock masses at the bottom of the ocean are reflected on the sea surface—submerged mountains appear as very slight bulges, and trenches on the seafloor show as depressions. Various ways of interpreting these data have led to estimates of 15,000 to as many as 100,000 seamounts worldwide.

Many thousands of small, rocky islands and reefs occur in every major ocean basin. Larger islands such as Madagascar, Sri Lanka, Cuba, and the British Isles rest on pieces of continental crust pulled away from their parent continent through the ponderous movements of seafloor spreading, or separated from their parent landmass through sea-level rise or erosion of the rocks connecting them.

Most oceanic islands are formed by volcanoes in three areas: at seafloor-spreading ridges, above subduction zones in volcanic island arcs, and above stationary hot spots in Earth's mantle. Many islands formed by hot spots can be seen on maps of the ocean, arrayed in long, linear chains created one by one. In the Pacific, the Hawaiian and Galápagos Islands are

"IN THE DEEP AND TURBULENT RECESSES OF THE SEA ARE HIDDEN MYSTERIES FAR GREATER THAN ANY WE HAVE SOLVED."

—RACHEL CARSON,
THE SEA AROUND US

ABOVE: Hawaii's Kīlauea, in Hawai'i Volcanoes National Park, shoots molten lava from an underwater lava tube. OPPOSITE: Eruptions in the park send lava streaming into the sea.

clear examples of this process, especially when the older submerged parts of the chain are considered. Long before hot spots were recognized, James Dana, a geologist working with the United States Exploring Expedition to the Pacific Islands in 1840, concluded that in the Hawaiian Islands there was an increasing degree of erosion, and thus greater age, among the islands to the northwest—Molokai, Kauai, and the distant Midway Islands—than in the volcanically active islands of Hawaii and Maui to the southeast. Mauna Kea, the highest peak, is on the youngest island, Hawaii.

Other islands are formed independently as lone undersea volcanoes break through the water's surface. In the tropics, fringing reefs may evolve into atolls as the above-water peaks erode away, leaving a living circle of coral around a volcanic husk. Whatever the cause of formation, over millennia many volcanoes have grown into islands, and many islands have eroded or sunk to become submerged seamounts, a process that continues.

Across the Pacific, hundreds of undersea volcanoes tap into Earth's molten outer core, some rising thousands of meters to break the surface. Most of these landforms, whether underwater or extending above, are in the Pacific's great middle section, between 30° N and 30° S latitudes. The islands

VOLCANIC ISLANDS

Volcanoes rising from the ocean floor—at seafloor-spreading ridges, above subduction zones, and above stationary hot spots—give birth to new islands below.

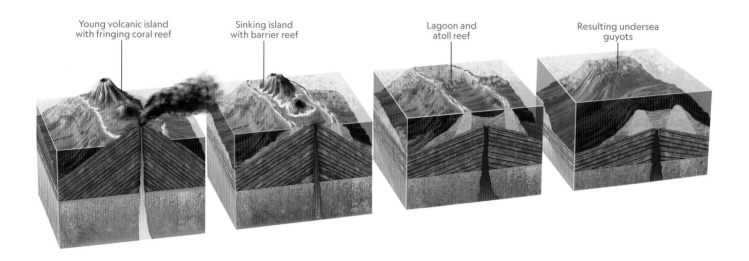

Young volcanic island with fringing coral reef

Sinking island with barrier reef

Lagoon and atoll reef

Resulting undersea guyots

DEEPER DIVE

Quakes and Waves

Deep-sea geology can tell us a lot about seismic activity, including what causes earthquakes and the deadly tsunamis they sometimes trigger. The deepest known places in the world—Sirena Deep (10,809 m) and Challenger Deep (10,984 m), in the Mariana Trench off Guam—are the sites of groundbreaking research that scientists believe will guide them in understanding the process. In November 2018 geologist Patty Fryer, of the University of Hawaii, co-led a team of scientists aboard the research vessel *Falkor*. They sent a battalion of full-ocean depth landers—mechanized, refrigerator-size platforms carrying cameras, sensors, and collection devices—down into the trench. At the deepest spots, rock grabbers snatched samples from the subduction zone, the area where one tectonic plate dives beneath another, an action that can prompt massive seismic movement. Readings have ranged from 4.5 to 8 in magnitude. Here, the angle at which the Pacific plate dives may be especially steep because of a tear in the plate, which may add to a quake's severity. By analyzing the rock samples and reading seismometers placed in the trench to measure tremors, Fryer hopes to determine what triggers the earthquake-tsunami cycle.

Are Volcanic Eruptions Increasing?

. . . .

It's a perennial news flash: Hawaii's Kīlauea; Guatemala's Fuego and nearby Pacayo; Indonesia's Merapi; Bali's Agung. Some 1,500 volcanoes are considered active around the world today, mainly around the Pacific Ring of Fire. At any given time, an average of about 20 are reported to be erupting, with between 63 and 80 eruptions per year. That number has been stable for centuries. But a 2017 study says we may have reason to be concerned about increased eruptions as

the globe warms. Geologists from England's University of Leeds studied the increase in volcanic eruptions in Iceland during the Holocene period, as Earth warmed naturally between 7,000 and 5,000 years ago. Here's what they theorize: As heavy glaciers pressed down on volcanic cones and the magma chambers beneath them, the magma stayed put. As the glaciers melted, pressure eased, opening the door for magma to launch up through Earth's surface in volcanic eruptions. The Holocene warming was likely caused by natural changes in Earth's orbit; today's global warming is a different story— but both lead in the same direction.

. . . .

ABOVE: Indonesia's Mount Bromo erupted from November 12, 2016, to November 12, 2017. TOP LEFT: In the Seychelles, Saint Joseph Atoll is a marine protected area.

here are called Oceania. Australia is typically regarded as part of Oceania, along with three major island groups: Melanesia, Micronesia, and Polynesia.

Culturally more connected to nearby continents are thousands of other islands not regarded as part of Oceania: most of the islands of Indonesia; Japan, including the Ryuku Islands; Taiwan; the Philippines; the islands of the South China Sea; the Galápagos; and the Aleutian Islands, as well as those within the intricately dissected coastal waters of Alaska, western Canada, and southern Chile.

Three oceanic plates converge in the region where the western Pacific meets the Indian Ocean, with dynamic processes giving rise to more than 13,000 islands that make up Indonesia, more than 7,000 Philippine islands, and many other islands in New Guinea, Malaysia, and nearby waters. Borneo, the third largest island in the world, and Taiwan arise from the shelf that borders continental Asia, along with arcs of islands including Java, Sumatra, Bali, and the Lesser Sunda Islands. Curved island chains such as the Marianas Islands along the western margin of the Pacific are volcanic island arcs fringing a trench and subduction zone.

The island country Australia, which lies half in the Pacific Ocean and half in the Indian Ocean, is a large continental piece of the ancient continent dubbed Gondwanaland, once joined to Antarctica. New Zealand, with its two major and numerous small islands, is part of a large, mostly submerged landmass called Zealandia, which meets the definition of a continent according to a 2017 report by Nick Mortimer et al. in *GSA Today*. Clearly, as exploration of the deep ocean progresses, so will understanding of the processes that underpin the geography of the land and sea.

Continental Margins

From the shore seaward, most large landmasses are bordered by a continental shelf that is narrow in some places, hundreds of kilometers wide in others, and has been periodically inundated or dry, depending on whether the planet is going through a warm or a cold phase. Ice ages have come and gone repeatedly, and in the process have tied up water in glaciers or released it during meltdowns that have lowered or raised sea level globally.

Nearly all continental shelves were above sea level during the maximum extent of ice cover in the Pleistocene era 18,000 years ago. At the time, England was connected to mainland Europe; the island countries of Sumatra, Borneo, and Java were joined; several of California's Channel Islands were one; and Florida was about twice the size it is today. Florida's Gulf Coast waters are still so shallow that it is possible in some places to stand on the bottom several kilometers offshore with your head out of the water. The presence of bones of now extinct land mammals such as mastodons at Gray's Reef, a submerged ridge offshore from the coast of Georgia, and in areas far at sea along Florida's west coast provide a glimpse into a time when sea level was much lower than it is now. Much higher sea level is also evident in the occurrence of marine fossils far inland throughout coastal regions globally.

Groundwater from the land seeps into the sea, and freshwater springs well up several kilometers offshore along both coasts of Florida, the Yucatán Peninsula, the Mediterranean Sea, and in the Persian Gulf near Bahrain. Around the Pacific rim in Chile, Hawaii, Guam, American Samoa, Australia, and many other places there are offshore springs, some with subterranean connections to rivers on the land. Ancient rivers carved pathways through now flooded coastal areas, and all rivers have carried masses of minerals, silt, and sediment from far inland to the sea. Submarine canyons are preserved by the flow of turbidity currents—slurries of water and suspended material from the land—that periodically gush through them in underwater landslides, eventually depositing their sediments on the deep seafloor.

At the edge of the continental shelf, there typically is a break at a depth of about 200 meters, with a sharper drop leading to the great depths below. This is the continental slope, a feature that extends into the water to depths of 3,000 to 4,000 meters, then merges near the base with the more gently angled continental rise. Submarine canyons that cut through continental shelves often continue through the slope into deep water beyond. The slope and rise together occupy about 8.5 percent of the ocean, stretching over about 300,000 kilometers. They are the recipients of current-driven sediments that cascade from the shelf, resulting in thick deposits of materials that accumulate over time. The transition from thick continental crust to the thinner ocean crust is likely to occur in this region. Beyond are the broad abyssal plains.

VISIONARIES

Walter Munk
Einstein of the Ocean

Sometimes called "Einstein of the Ocean," an epithet he modestly declined, Walter Munk was nevertheless a diviner of the sea. His exceptional ability to predict when and how ocean waves would break on shore guided U.S. troops to land on Normandy beaches on D-Day. Munk was among the earliest to use scuba, and his background in mathematics and geophysics led him to solve problems about waves, tides, ocean circulation, and even Earth's rotation. His measurements of sound traveling through ocean basins showed that their waters were warming at different rates, a harbinger of climate change. By drilling into the deep seafloor, Munk believed we could learn from life past—archived in middle Earth. From a small research ship that collected deep-sea cores off the coast of Baja California in 1961 grew a global initiative that concerns climate, geology, and life over eons, revealing our origins and helping us predict our future.

OPPOSITE ABOVE: India's continental shelf continues past the Ganges-Brahmaputra Delta into the Indian Ocean. OPPOSITE BELOW: Florida's shelf ranges in width from two kilometers in the Atlantic to 250 kilometers in the Gulf of Mexico.

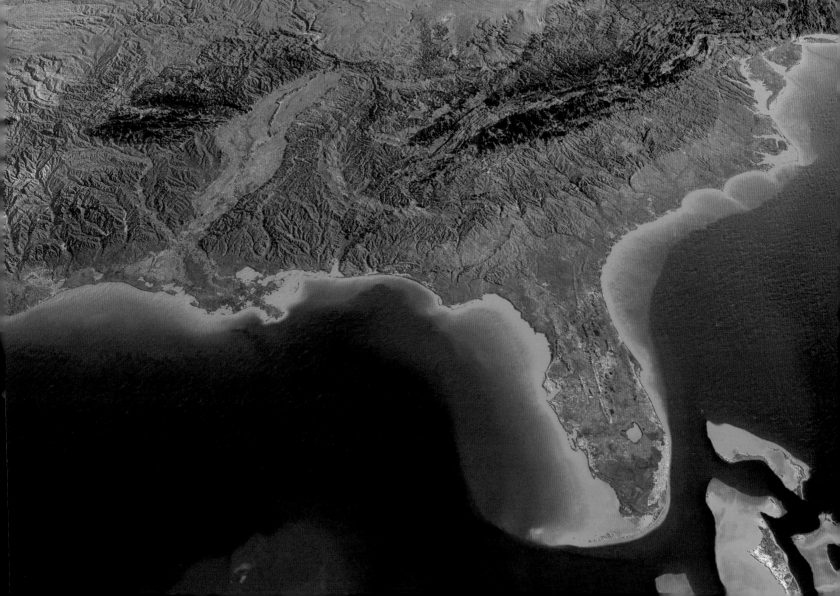

It's One Ocean

Some speak of seven seas; others count five. In truth, there is but one continuous ocean, laced with currents; stirred by winds; rocked by tides; fringed with bays, smooth shores, and intricate, rocky coasts, and, throughout, in every drop, something not known to exist elsewhere in the universe—life.

From the shore, the ocean appears immense, vast, and ever changing but constant in its presence, a formidable barrier separating people and places, but also a liquid bridge connecting everyone and everything on Earth.

There is nothing quite like it in all of the universe. Every living thing exists because of it, from high-flying birds and desert cactuses to microbes living deep within cracks in the ocean floor. Humans tend to believe that the ocean is so vast, so resilient, that nothing could change its basic nature; but evidence gathered largely since the middle of the 20th century clearly shows the magnitude of shifts that have taken place over time through natural

Offshore from South America's Patagonia, dusky dolphins communicate with whistles and clicks.

processes, and how human actions in recent decades have had impacts of geological magnitude.

The ability to look at the ocean as a whole is possible as never before, with growing insight about the significance of the living ocean's dominant role in shaping "ecosystem Earth." Now we know that the Pacific Ocean is the oldest, deepest, widest part of the global ocean, occupying about half of its total area. The Atlantic is about half as large and holds about a quarter of the ocean's water. The third largest and also the youngest is the Indian Ocean, and fourth is the unruliest and least explored, the Southern Ocean. Smallest is the Arctic Ocean, currently the target of intense scrutiny by many nations owing to the Arctic's oversize impact on climate as well as its new opportunities for transportation and exploitation of places heretofore inaccessible.

The most unknown, unexplored part of the ocean is the ocean itself—not the top, not the bottom, but the liquid living ocean between. Millions of ships travel over the surface and millions of divers explore the upper 50 meters or so. Instruments deployed from ships and fleets of new robotic devices are acquiring data about the physical, chemical, and biological nature of that vast ocean space, and are beginning to bring it into focus. But only in recent decades has the microbial realm been acknowledged as an essential basic element in the biogeochemical processes that shape the ocean and the planet overall. Only now is the phenomenon of vertical migration of creatures in the ocean's "twilight zone" being linked to carbon capture, climate impact, and ocean nutrient cycling.

Overall, more has been learned about the nature of the global ocean and the life it holds in the past 50 years than during all preceding history, and fortunately, the pace is picking up.

A 2000 National Academy of Sciences report, *Discovering Earth's Final Frontier,* concluded with the observation "It is impossible to predict when we embark upon the voyage of discovery exactly how that benefit will be manifest. Discovery is the prelude to new paradigms; it jolts us out of the ruts of incremental scientific progress and fuels the great leaps forward."

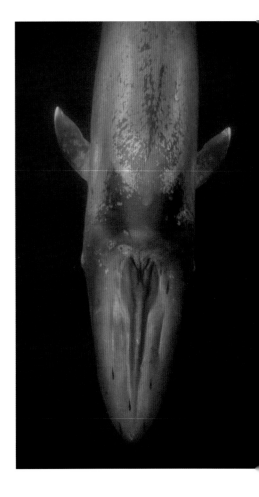

A blue whale *(Balaenoptera musculus)*, Earth's largest inhabitant, glides in Costa Rican waters.

MISSION BLUE

Measuring Migration

Some 200 years ago, when hauling up their nets, fishermen discovered that certain kinds of marine life frequented different depths at different times of day. Later, scientists learned more about this global phenomenon: After an upward migration of hundreds of meters, some animals reach the surface to gorge themselves on plankton; then, in fairy-tale fashion, they disappear with the sun, retreating to 200 to 1,000 meters below the surface. This pattern of ascent and descent has been coined "diel vertical migration" (DVM)—taking place during 24 hours, including a day and the adjoining night. Its purpose may be to avoid surface predators in daylight, but there is likely more. In the 20th century, ship-mounted acoustic systems more accu-

rately recorded evening arrivals and morning departures: The clearer the water at the surface, the deeper the animals would descend to reach the cover of darkness—and the faster they would travel. Today, equipment such as acoustic Doppler current profilers and satellite-mounted, light-detection-and-ranging (LIDAR) instruments measure not only travel times, the extent of the migrations in certain areas, and their global occurrence, but also their greater impact on ocean biogeochemistry. By feeding at the surface at night, then metabolizing the food in deeper realms, marine life efficiently spreads nutrients, carrying carbon from surface waters and sequestering it in the depths below.

At the Top of the World

Polar regions have much in common. From space, both glisten dazzling white, both cap land and sea with snow and ice, and both have major impacts on Earth's climate, weather, and basic temperature regime, the planet's built-in "air-conditioning system." Each pole is centered on an invisible line, the axis around which Earth rotates at an angle that is tilted 23.5 degrees facing the sun. The oblique angle of sunlight reaching the poles causes them to be colder than the rest of the world. As the planet journeys around the sun, each polar region has six months pointed toward the sun and nearly continuous daylight, then six months of darkness. Winter for one is summer for the other. They are the only places on Earth where water occurs year-round in all three forms: frozen, liquid, and in vaporous clouds. Both are extremely cold much of the year and beneath the surface largely unexplored. The name, Arctic, is derived from the Greek *arktos,* meaning "bear," thus the place where bears are; Antarctica is where bears are not.

Although similarities abound, the polar seas are also strikingly different. The Arctic Ocean is largely surrounded by land: northern Canada, Greenland, Russia, Norway, and the United States. The deepest place found so far is 5,669 meters, but because of the unusually broad continental shelf areas surrounding the Arctic, the average depth is only about 1,205 meters, much shallower than any other ocean. In winter months, the ocean is largely blanketed with ice generally two meters thick, but in places 30 meters, with east-to-west-bound currents that cause the ice to slowly rotate. The Arctic Ocean also surges with a rhythm of tides that contribute to the fracturing and movement of sea ice. Global warming, linked to increased emissions of carbon dioxide, is dramatically reducing Arctic sea ice, with ice-free summers predicted by mid-century.

The motion of the Arctic's ice pack was first confirmed when a U.S. naval vessel was crushed in ice north of the Bering Strait in 1881 and the remains arrived three years later on the southwest coast of Greenland. With a crew of 12 men aboard a custom-built, ice-worthy wooden vessel, *Fram,* Norwegian explorer Fridtjof Nansen deliberately sailed into an ice pack, where the ship was frozen in place near New Siberian Island in 1893. *Fram* and its crew drifted in the ice to Spitzbergen, arriving in 1896. Numerous research stations on and under the ice now document critical aspects of the Arctic environment.

Nansen and his crew would likely be astonished if they could see what has now been discovered in the depths below: three huge mountain ranges that cut across the Arctic Ocean basin like a giant extended claw. One, the center of the Arctic's spreading center, is known as the Nansen or Gakkel Ridge, the ocean's deepest and slowest spreading ridge. Volcanic action and hydrothermal vents have also been found, including huge chimneys pumping out extremely hot water as well as fields of low-temperature vents, slightly warmer than the surrounding seawater.

VISIONARIES

Verena Tunnicliffe
Doing Deep Time

....

The force behind Canada's first marine protected area—the Endeavour Hot Vents on Juan de Fuca Ridge off British Columbia—marine biologist Verena Tunnicliffe prefers "deep time" to "down time." A Canada Research Chair in Deep Ocean Research at the University of Victoria, she is growing her extensive log of exploration and research expeditions by scuba and manned and remotely operated submersibles. She has identified 80 new marine species and pursues her passion to learn how they function in extreme undersea environments around seamounts, hydrothermal vents, and volcanoes. "By understanding how life adapts to the edges of existence," she says, "we better understand the capability to adapt." Her digital tool Ocean On-line provides real-time information from the undersea observatory VENUS to onshore scientists helping marine communities survive and thrive with climate change.

....

OPPOSITE ABOVE: Blue ice borders Franz Josef Land, a Pristine Seas research area. OPPOSITE BELOW: Radiant in the Arctic's summer sun, anemones thrive in a kelp forest with encrusting coralline red algae.

From One Pole to the Other

Earth's magnetic field is generated within liquid iron in Earth's core with effects that extend far into space, shielding the planet from incoming solar radiation. The field is constantly changing in strength and direction, most notably with a full reversal a few times in a million years. The term "magnetic north pole" refers to the northern focus of Earth's magnetic field and is the basis for compass readings. Since Earth's axis wobbles slightly, the precise location of the poles shifts over an area of about 9 meters. Owing to motion within Earth's core, the magnetic poles also move 5 to 10 kilometers a year, or, as in recent times, more rapidly, at 50 to 60 kilometers a year.

The geographic poles—the places that mark the ends of the axis around which Earth rotates—are constant. U.S. Navy Adm. Robert Peary and explorer Matthew Henson were the first to reach the geographic North Pole, in 1909. However, the location on the seafloor below was not attained until 2007 when the two Russian submarines, *Mir I* and *Mir II,* transported six explorers to the bottom, where they planted a titanium Russian flag at what they termed "the real North Pole." Whether on the icy surface or on the seafloor more than 5,600 meters below, the North Pole is technically the northernmost place on Earth, the point from which all directions point south and where all lines of longitude meet at latitude 90° N. At the other end of the world, lines of longitude converge at the South Pole, first reached in 1911 by Norwegian explorer Roald Amundsen.

Transit to the geographic (and magnetic) North Pole and across the Arctic Ocean under the ice pack was first accomplished by the U.S. Navy's nuclear submarine *Nautilus,* in 1958. Currently, submarines from several countries travel under the Arctic ice for research, and commercial ship traffic at the surface is growing. The activity will likely increase dramatically with the decrease of ice cover driven by current planetary warming.

A few intrepid scuba divers have plunged into the waters under the ice at the North Pole, including the former chairman of the board of the National Geographic Society, Gilbert M. Grosvenor, and his diving companions, Canadian explorer Joseph MacInnis and photographer Al Giddings. MacInnis recalls swimming where all meridians meet and every direction is south, where "translucent plates of ice glittered . . . a huge rampart of white loomed . . . plunging deep into violet waters and gradually disappearing into purple nothingness far below."

Actually, life abounds in the ocean at the North Pole. In summer months, photosynthetic microorganisms create lush, miniature upside-down forests on the underside of the ocean's ceiling of ice. Other creatures prosper in the water column and on and within the sea bottom below. At the South Pole, large animals have been absent for at least 15 million years, but since 1957, humans have arrived and now occupy the Amundsen-Scott South Pole Station, where researchers work even during the dark winter months of June, July, and August.

Sunset bathes Greenland's Ilulissat
Icefjord, a UNESCO World Heritage site.

The Ocean at the Extreme South

While the Arctic is notable as an ocean largely surrounded by land, the Southern Ocean is distinguished as the ocean that surrounds a continent. At 13.2 million square kilometers, the landmass in the planet's extreme south is the coldest, driest, windiest, highest, and wildest continent on Earth. The existence of the Antarctic continent was predicted long before it was first sighted in 1820 by Russian explorer Fabian Gottlieb von Bellingshausen. No indigenous people have ever lived there, but research stations established in the 1950s now include 70 bases from 30 countries with a summer population greater than 4,000. Thousands more visit or work on fishing vessels from more than 20 countries.

At 3.2 kilometers high, the 225-million-cubic-kilometer ice sheet covering Antarctica is so heavy that it forces much of the continent some 3,400 meters beneath sea level. This permanent ice sheet holds more than 90 percent of Earth's fresh water. Take away the ice, and the archipelago of Antarctica would appear as a single large island fringed by hundreds of smaller ones. Much of what appear to be the continental margins of Antarctica are actually frozen ice shelves—the seaward extension of glaciers and the great continental ice sheet. The immense weight of the three-kilometer-high inland ice continually pushes the coastal fringe outward.

The northern boundary of the Southern Ocean is defined by the powerful Antarctic Circumpolar Current (ACC), where the ocean meets the warmer southern reaches of the Atlantic, Pacific, and Indian Oceans. The 60° S latitude line was chosen to coincide with the northern boundary of influence of the original United Nations Antarctic Treaty of 1959, but the water mass itself often ranges well north of this line.

Between 49 million and 17 million years ago, as moving plates tore apart today's continents of South America and Antarctica, an opening developed and deepened that is now called Drake's Passage. It made possible the west-east flow of the Antarctic Circumpolar Current around the new landmass as it settled at the southernmost place on Earth. During its drift there in the late Cretaceous period, the once temperate Southern Ocean began a cooling process, and temperatures now range from 2°C to minus 1.8°C, from surface to seafloor.

A complex view now exists of vast outer banks, deep channels, submarine canyons, and basins separated by ridges that extend like spokes of a wheel from around the landmass. Ridges underlie the Southern Ocean and surround the Antarctic continent. In the far south of the Atlantic Ocean there is a seaward extension of three continental plates—the Antarctic, South American, and African—abutting at a place near Bouvet Island (at 54° 26' S and 3° 24' E) known as the Bouvet Triple Junction. Three active seafloor-spreading ridges angle away from this dynamic juncture. The beguiling symmetry of the Australian and Antarctic coastlines provides a graphic demonstration of seafloor spreading and continental drift.

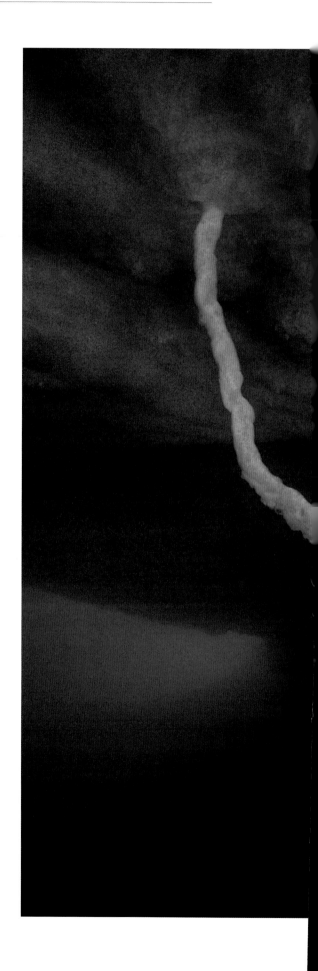

Ice-covered trails of brine, called brinicles, stream into the Southern Ocean from sea ice off the coast of Antarctica.

GREAT EXPLORATIONS

Finding Inner Earth

Inhabiting a core sample from a *JOIDES Resolution* expedition, fossilized diatoms—ancient algae—help scientists determine a seafloor's past.

If the sediments might tell us of past history, what mysteries might the heart of the Earth, deep below the sediment-rich seafloor, share? In 1957 oceanographer Walter Munk at the University of California's Scripps Institution of Oceanography imagined that handling a pristine chunk of mantle rock plucked from below Earth's crust could help us understand Earth's makeup, age, and history.

Munk and his colleagues launched their first drilling expedition, Project Mohole, in 1961 aimed at the Mohorovicic discontinuity, or Moho, the area between Earth's crust and mantle. Off Baja California they sent their drill to a depth of 3,566 meters, then another 183 meters into the crust to extract five cores. The mantle was still thousands of meters away, but the cores revealed for the first time the makeup of Earth's crustal layers: Miocene-age

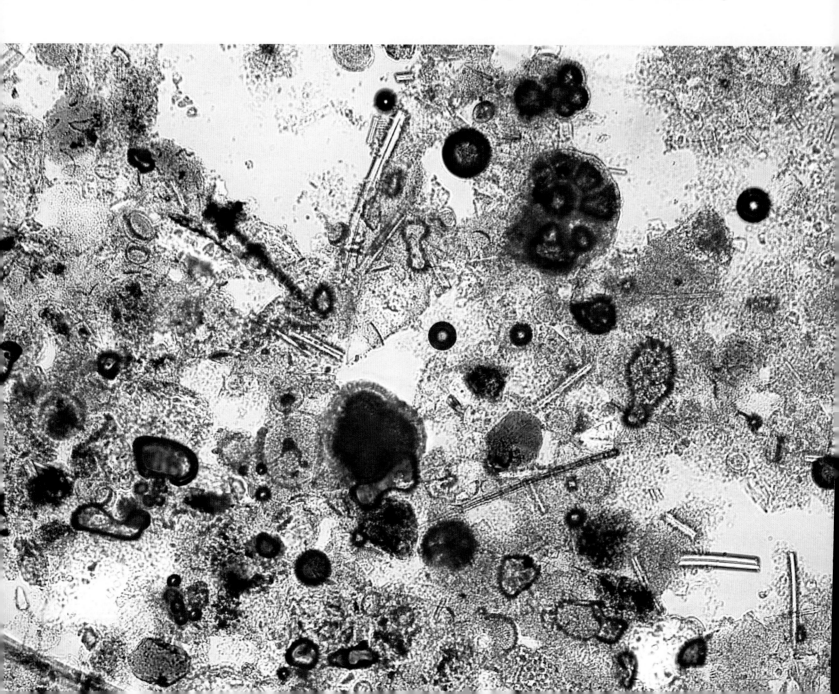

ABOVE: Inside the blue tower on the *Chikyu*, a vessel for Japan's Agency for Marine-Earth Science and Technology, the deep-sea drill awaits duty. RIGHT: Geophysicists study core samples.

sediment in the top layer and a lower layer of basalt, or cooled lava, ejected from a mid-ocean ridge.

Building on Munk's work, in 1968 scientists of the Deep Sea Drilling Project (DSDP), representing six nations, aboard the *Glomar Challenger* obtained core samples from the Gulf of Mexico. It was the same year that the U.S. Apollo 8 mission pioneered exploration of Earth from thousands of kilometers above, paralleling the groundbreaking efforts within the seafloor below.

Early drilling was precarious. In 1973 *National Geographic* magazine writer Sam Matthews described the *Challenger* as "a gangly, improbable seagoing drill tower ... designed ... to lower more than 20,000 feet of pipe in the open ocean, bore into the seafloor, and bring up bottom cores or samples. The technical feat has been likened to drilling a hole in a New York sidewalk with a strand of spaghetti dangled from the top of the Empire State Building."

But over time experts honed the technique of dynamic positioning—balancing propeller and thruster action to keep a ship stable in currents and waves—to drill accurately. Soon the drills of *Glomar Challenger* and other ships were bringing up hundreds of core samples, filling in the puzzle of Earth's past.

Today, under the International Ocean Discovery Program (IODP), scientists use the research drillship *JOIDES* (Joint Oceanographic Institutions for Deep Earth Sampling) *Resolution* to bring up core samples that reveal a thriving biosphere in sediments 2.4 kilometers beneath the seafloor and, even deeper, inside the crust. There, microbes survive on minuscule amounts of energy, a behavior that may unlock the secrets to understanding the origins of life.

The jewel in the crown? In 2015, drillships in the Atlantis Banks, off Madagascar in the Indian Ocean, reached halfway to the mantle, ever closer to the pristine rock sample crystallized from the mantle's magma that Munk dreamed of. Such a treasure would answer countless questions—and likely pose countless more.

TIME LINE

Seeking Inner Earth

1957: Walter Munk, Scripps Institution of Oceanography, plans drilling toward Earth's mantle for core samples.

1961: Drillship *CUSS 1* (named for the consortium of Continental, Union, Superior, and Shell oil companies) develops "dynamic positioning" technique.

1961: Munk launches Project Mohole aboard *CUSS 1;* takes five core samples.

1961–64: *CUSS 1* targets drilling to sample life beneath seafloor.

1966: Project Mohole ends. Successes are dynamic positioning and groundbreaking core samples.

1968–1983: Deep Sea Drilling Project (DSDP) launches aboard *Glomar Challenger;* takes and studies cores from global ocean.

1985–2002: DSDP, renamed Ocean Drilling Program (ODP), aboard *JOIDES Resolution* has 110 global drilling missions revealing plate motion, climate change, seafloor age, microbial secrets.

2003: Twenty-six-nation Integrated Ocean Drilling Program (IODP) uses updated *JOIDES Resolution,* Japan's *Chikyu,* and others to make 52 global expeditions.

2013–2023: IODP becomes International Ocean Discovery Program to explore climate, deep life, geodynamics, geohazards. Sensors in "discovery sites"—cored holes—yield ongoing information.

2016–18: *JOIDES Resolution* drilling at Atlantis Bank approaches mantle.

2019: *JOIDES Resolution* drilling in Amundsen Sea West Antarctic Ice Sheet helps predict shrinking ice sheet and global sea-level rise.

WATER—
GIVER OF
LIFE

······································

Chapter Two

"I BELIEVE IN MAGIC; IT IS IN THE OCEAN."

—MICHAEL AW, ELYSIUM ARTISTS FOR THE ARCTIC

"First find water," says Christopher McKay, a NASA astrophysicist who for decades has been involved with the search for life elsewhere in the universe. He adds, "To find life on other planets, we start by looking for water. It is the single nonnegotiable thing life requires."

Earth is blessed with enough liquid water—1.3 billion cubic kilometers of it—to cover more than two-thirds of the planet's surface with an average depth of 3.7 kilometers, a maximum of almost 11 kilometers. On land, sunlight sparkles from the faces of lakes; rivers curl across continents like silver ribbons; water-laden clouds wreathe the entire planet; frozen water frosts mountaintops and gleams starkly white at the poles. Earth is essentially a marine system. We and all other forms of life—from rainforests to desert scrub—are as dependent on the ocean for our existence as any coral reef or deep-sea fish.

Water makes up more than 98 percent of most jellyfish, 85 percent of a prickly pear cactus, and about 70 percent of a human being. Although we drink fresh water, the blood that courses through our bodies is salty, as are our tears. Life on the land and in lakes, rivers, and streams needs fresh water to survive, but most of life on this planet, both in terms of mass and diversity, is marine. Only about half of the major divisions of animal life that have ever existed occur on the land and in fresh water. Life originated in the sea, and essentially all basic forms of animal life have representation there as living creatures or now extinct ancestors. The vast majority of Earth's microbes are marine, from the sea surface to its greatest depths and down into the water that seeps far below the seabed.

Water bound to minerals also occurs deep within Earth's mantle. Referred to by some as "an ocean beneath the ocean," the amount of water locked in a hydrous mineral, ringwoodite, far from Earth's surface, may exceed the liquid water available for life—and could be linked to the origin of Earth's ocean.

Pure water in nature is rare. Dissolved in seawater is a full library of Earth's naturally occurring elements, held within a vast array of inorganic and organic compounds. And, almost always, where there is water, there is life. While discovering the importance of the ocean to our existence, we are beginning to grasp the shocking magnitude of our ignorance of its nature.

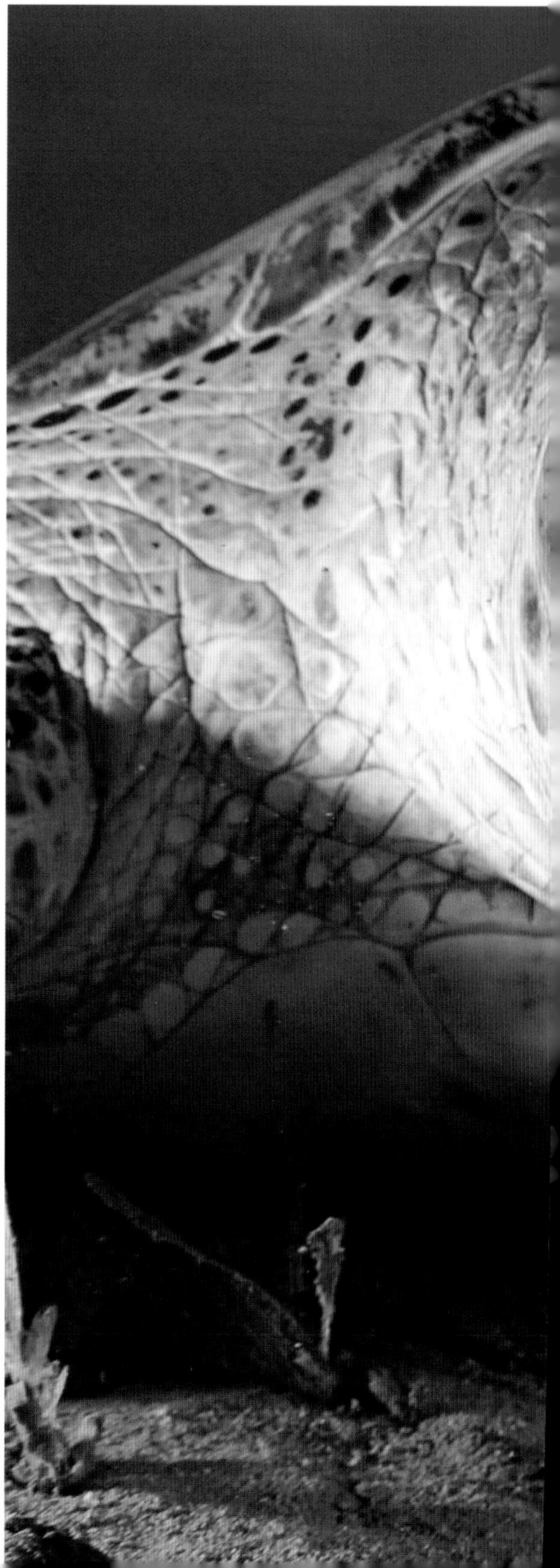

RIGHT: An herbivorous green sea turtle eyes a grassy meal in the Red Sea. PAGES 54-55: Like an indigo eye set in the turquoise waters of the Mesoamerican Reef, Belize's Blue Hole is a steep, deep sinkhole that flooded with rising seas.

Where Did Water Come From?

With technologies that could not have existed until recent decades, astronomers are detecting the presence of plenty of water, mostly frozen or in clouds, far beyond our solar system. The formation of water vapor in space likely appeared in pockets across the universe a billion years after the big bang—much earlier than had previously been thought possible, according to researchers from the Harvard-Smithsonian Center for Astrophysics. Although hydrogen was present in the initial stages of the universe, oxygen is believed to have been produced later, during the formation of the earliest stars. Exactly when and how water's two elements—hydrogen and oxygen—first came together is not known, but evidence now confirms that life's essential ingredient, water, is widely present throughout the cosmos.

Despite the presence of water in our solar system, throughout our galaxy, and in the universe beyond, nowhere else have living organisms been discovered but here on Earth. Water can exist without life, but life as we know it cannot exist without water. Since more than 97 percent of Earth's water is ocean, it should be no surprise that 97 percent of the biosphere—the space on the planet where life occurs—is the ocean.

Although vital to our existence, the nature of water remained mysterious throughout most of human history. Regarded by the ancient Greek philosopher Aristotle as one of the four "elements" that make up the world—along with fire, air, and earth—not until the 1780s was it proved by the British scientist Henry Cavendish that water is a compound made up of two parts hydrogen and one part oxygen, represented by the chemical formula H_2O.

Simple though it appears, H_2O has exceptionally complicated properties that set it apart from any other known substance. It starts with this structure: Two small atoms of negatively charged hydrogen are bonded very strongly to one larger atom of positively charged oxygen in each molecule and arranged not in a straight line but rather at an angle. Hydrogen and oxygen have a strong affinity for each other. Once fused, it takes an enormous amount of energy to break the bonds that hold the atoms together. Those bonds also give water a high degree of surface tension, a feature used by water-striding insects and spiders that are able to "skate" on water without breaking the surface.

In water molecules, each hydrogen atom can be shared between two molecules, together creating a kind of liquid latticework. In a mass of liquid water, the opposite charges attract and lightly attach, then break apart in a moving flow, a game of tag with billions of molecules briefly touching one another, then moving on to engage with others. Science writer Robert Kunzig likens the action to a dance: "When water is heated, the pace of the dance picks up until the bonds break apart, and individual molecules fly

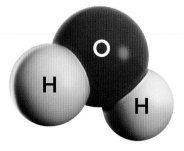

ABOVE: Two molecules of hydrogen with one of oxygen produce water, the foundation for life. OPPOSITE: After its fiery beginning, the formation of Earth's living, liquid ocean set it apart from any other place in the universe.

off as a gas." Water has a high boiling point at 100°C, and a low freezing point of 0°C.

When cooled, the dance slows down, and water, like most substances, contracts, becoming heavier and denser. But near freezing, hydrogen atoms in molecules of water attach to each other in a six-sided ring that creates a solid—ice—that occupies more space and is lighter than the liquid from which it came. Like snowflakes, ice crystals are typically six sided, each one miraculously different from all others. Fortunately for aquatic life, frozen water floats, a phenomenon that shields liquid water in ponds and lakes, in much of the Arctic Ocean, and beneath the large icy shelves surrounding the Antarctic continent.

New peculiarities of water have been discovered by scientists at Stockholm University. A researcher there, Anders Nilsson, notes, "The new remarkable property is that we find that water can exist as two different liquids at low temperature where ice crystallization is slow." Better understanding of water properties may lead to practical applications for human uses, and also to solving ongoing mysteries about the chemistry underlying the origin of life.

The amount of Earth's water has remained fairly constant for hundreds of millions of years. We, like all other forms of life, are part of what is called the hydrological cycle; that is, the continuous movement of water on, above, and within Earth. Though relatively stable in mass, water is in constant flux, shape-shifting from liquid to ice or vapor in processes happening in milliseconds and over millennia. Water in the ocean, lakes, rivers, and streams, and within terrestrial life everywhere, eventually sinks into the ground and is returned to the surface as springs—or released as vapor to the sky, forming clouds that send water back to the land and sea as rain, sleet, snow, and hail. Consider that the water we drink today may have been consumed by dinosaurs or been home for deep-sea fish when the world was younger and very different from what it is now. The ice in a refreshing lemonade today may someday land as snowfall on Mount Everest or erupt from a hydrothermal vent at the Mid-Atlantic Ridge.

On its journey, water has special significance in shaping the nature of Earth as a hospitable place for life: its capacity to absorb and store a large amount of heat from the sun. The ocean regulates and distributes water of different temperatures, moderating what otherwise would be extreme shifts of temperature.

Technologies enabling access to the deep sea and exploration far into the underlying seabed have greatly expanded knowledge of the water cycle. We now know that water sinks into deep cracks in the seabed along subduction zones, where it becomes so hot that rocky material dissolves in it. That material eventually jets back into the ocean through hydrothermal vents and into the atmosphere via volcanic eruptions. An amount that is thought to exceed the volume of water in all of the ocean above occurs in the crystalline structure of minerals within the crust and mantle below. The relationship between deep-Earth water and the cycling of water above is unclear, and highlights how much more there is to explore and learn about the ocean's dynamic processes.

THE WATER CYCLE

RIGHT: In constant motion, groundwater feeds lakes, streams, and sea; evaporation forms clouds; clouds rain back to Earth. ABOVE: A torrential rainstorm over the Fijian Island Taveuni vividly illustrates the water cycle in action.

Not-So-Pure Water

W ater is vital to life. And a pure water cycle is as well. It's a simple equation: The water on Earth today is mostly the same that has been here since life began. It falls to Earth as rain, snow, sleet, or hail, then evaporates back into the atmosphere in a continuous cycle. Some 97 percent of Earth's water available to life is contained in the ocean, and when the ocean is compromised by human contaminants such as plastics, sewage, and toxic chemicals, the water cycle suffers. Interactions among ocean, air, land, and the biosphere no longer work smoothly. A new cycle takes over: one of dysfunction and damage to life.

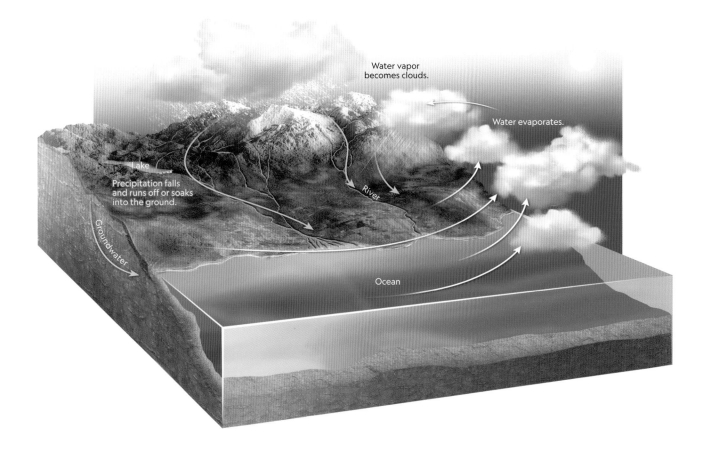

Water vapor becomes clouds.

Water evaporates.

Lake

Precipitation falls and runs off or soaks into the ground.

River

Groundwater

Ocean

Why Is the Sea Salty?

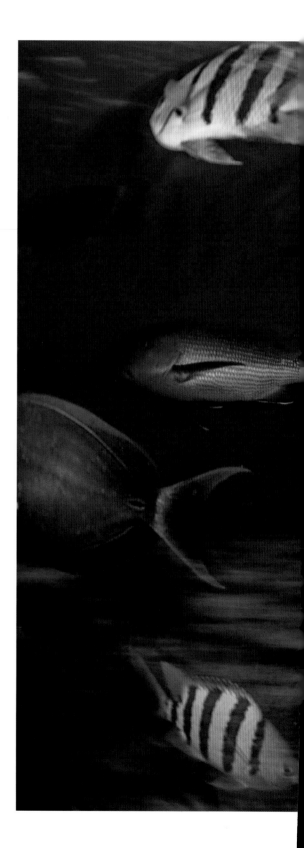

Water as ice and vapor is essentially salt free, but because of water's electrically polarized properties, molecules of water readily attract not just one another but also other substances, a feature that makes water a near-universal solvent. So readily do materials dissolve in it that pure water rarely occurs in nature. Snow, sleet, and rain fall to Earth together with hitchhikers picked up along the way, from soot and dust to extra atoms of oxygen and molecules of carbon dioxide. Rivers gradually dissolve the very rocks over which they flow and carry millions of tons of suspended minerals, sand, sediment, and organic matter. Much of this settles out along the way, but river mouths are notoriously rich with materials that originate upstream and are taken far out into the ocean. This accounts for some of the mineral content in seawater, but most of the saltiness is ancient, originating in the mineral-laden water ejected from volcanoes and hydrothermal vents early in the formation of the planet. Most ocean water is only 96.5 percent water, with the balance consisting of sodium chloride (salt) and varying amounts of other materials.

The density and weight of water varies with temperature and saltiness: Salt water is more dense than fresh water, and cold water is heavier than warm water. Layering occurs where rivers join the sea, with fresh water "floating" over salty ocean water. If the fresh water is cold enough, however, it will sink below the warmer salt water; and, conversely, warm, highly saline water can sink beneath colder, less saline water. These density differences drive currents at all levels in the ocean. In the open sea, the salinity is about 35 parts per thousand parts of water, but that amount can vary owing to rainfall, surface evaporation, and the flow of fresh water from rivers. Where rivers meet the ocean, salinity may be only 20 parts per thousand or less, a favorable environment for marshes, mangroves, seagrass meadows, oysters, and other animals that can thrive within wide-ranging saltiness.

Over the ages, massive deposits of salt have formed when parts of the ocean have been left to dry from shifts in the land and changes in sea level. Masses of salt from ancient oceans underlie places such as the Great Salt Lake, the Mediterranean Sea, and the Great Lakes region of the United States, where salt is mined. Deep within the Gulf of Mexico, ancient salt deposits leak brine several times saltier and therefore heavier than the surrounding seawater. Brine pools form lakes within the ocean that are lethal to most animals, but associated methane seeps power microbial life and are often fringed by dense beds of mussels that shelter crabs, shrimp, and numerous other creatures.

Saint Joseph Atoll lagoon in the Seychelles swirls with scissortail sergeants, twinspot snappers, and blacktip reef sharks.

The Sea Is Alive

Where air and sea meet, down through the water column, and far into the seafloor below, the ocean is alive, an orderly turmoil of eat-and-be-eaten creatures. Even a cupful of what looks like perfectly clear water may be a bustling minestrone of millions of individual cells of bacteria and other microbes. Many live suspended in water, but others occur as symbionts within other organisms, just as we humans have personal microbial biomes in our bodies that aid in digestion and other basic functions. Viruses abound within the cells of bacteria and probably most other marine organisms. The entire life cycle of ocean dwellers—from the smallest microbe to the largest turtle and greatest of whales—occurs partly or entirely while suspended in water, a substance that is 830 times as dense as air and 60 times more viscous. The structure, lifestyles, life histories, even the size and shape of organisms that live in the sea, are influenced by the fundamental characteristics of ocean water. As agile as ballerinas in the ocean, dolphins are helpless on the land, and pulsing veils of jellyfish tissue dissolve into quivering blobs when washed ashore.

Earth's geology and hydrology—its rocks and water—have provided the essential cradle for the existence of life, but over the ages, living organisms have interacted with these basic elements, shaping them into a dynamic, integrated system unlike any other in the universe. Biogeochemistry, with an emphasis on the "bio," sets Earth apart. For example, phytoplankton floating near the ocean's surface emit large molecules of dimethyl sulfide around which water droplets form that in turn evaporate to form clouds that return not only water to the land and sea but also traces of sulfur, one of the vital elements of life. The atmosphere of Earth during the first two and a half billion years or so had only a small amount of oxygen, the rest consisting largely of carbon dioxide and nitrogen. The gradual buildup of oxygen resulted from photosynthesis in tiny organisms with chlorophyll that used sunlight and carbon dioxide to generate sugar and emitted oxygen into the sea and the atmosphere above. There is uncertainty about when the current proportion of gases in the atmosphere—21 percent oxygen, 78 percent nitrogen, with small amounts of other gases including carbon dioxide (CO_2)—was attained, but clearly, the origin of abundant oxygen in the atmosphere is a result of the origin of life.

LEFT: Like many sea creatures, this jellyfish is mostly made of water. ABOVE: Microscopic algae growing on the bottom of Antarctic sea ice feed swarms of shrimp-like krill at South Georgia Island.

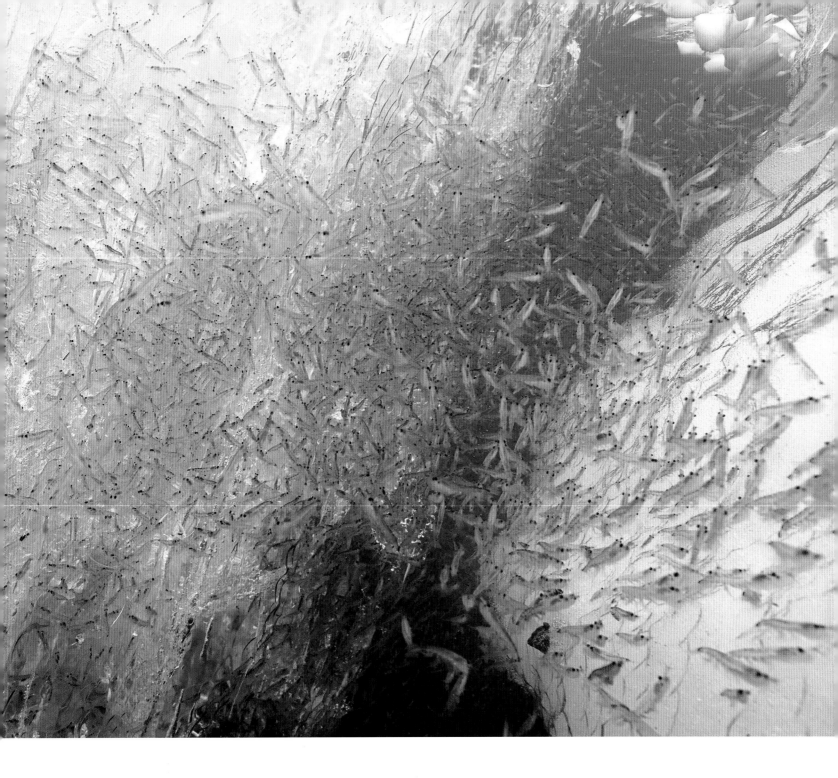

Microbes Rule

Dip a tablespoon into the ocean, and you will likely hold millions of bacteria, some Archaean species and about a hundred million viruses—all microbes of many kinds. The first life on Earth, microbes are vital to the existence of multicellular creatures. In their home of origin, the ocean, they are Earth's most prolific organisms, vital conduits in the flow of energy and nutrients. Scientists collect them for medicines. They influence weather and climate. And their deep-sea populations, living on chemicals from Earth's interior, may provide a guide to ancient life on other planets. Still, relatively little is known about microbes. The International Census of Marine Microbes, an arm of the Census of Marine Life, is delving deeper, with a mission to catalog as many species as possible and to study their connection to our evolution and environmental concerns. In 2018 researchers at the University of Texas, Austin, announced the discovery of 24 new microbes in the Guaymas Basin in the Gulf of California—microbes so different from species already identified that they may represent new branches in the tree of life. Their specialty: eating pollution. Most of the new species are nourished by hydrocarbons such as methane and butane, meaning less of these gases leaks into the atmosphere. Their future as environmental protectors may include mopping up oil spills.

Biogeochemistry in Action

Biogeochemistry in action is illustrated by layered mounds, columns, and sheet-like rocks known as stromatolites. Sticky cells of microbial cyanobacteria bind together grains of minerals, eventually forming durable living rock formations. Stromatolites currently thrive in Western Australia and in a few places in the Indian Ocean and the Bahamas, but they were a dominant feature of shorelines globally for about two billion years. There are several theories for their sharp decline. One is the rise of grazing organisms; another is changing seawater chemistry. Yet another may relate to the growing abundance of foraminifera, single-celled organisms with calcium carbonate shells that produce finger-like pseudopods—"false feet" used to engulf prey.

In a lecture to the Carpenters Union in Norwich, England, in 1857, the distinguished scientist Thomas Huxley described the geological history of the North Atlantic Ocean as "written" in the fossils of organisms in a piece of limestone chalk carved from the White Cliffs of Dover. Limestone rock typically originates from the remains of foraminifera and other organisms embedded in soft lime muds that have become compacted and layered over time, forming beds of sedimentary rock. Coral limestones are derived from the remains of coralline algae and a veritable zoo of other organisms with a calcium carbonate structure, including bryozoans, brachiopods, annelids, echinoderms, and mollusks, as well as stony corals.

In the 1960s, British scientist James Lovelock observed that the organic and inorganic components of the planet have evolved together in what appears to be a self-regulating manner, an integrated ecosystem that tends to hold the planet steady within a range of conditions suitable for life to exist. He was at the time looking at other planets and determined that the habitability of Earth was derived not only from its being at a favorable not-too-hot, not-too-cold distance from the sun, but also from the way that its living systems keep conditions just right. He called this the "Gaia Theory," named after Gaia, the Greek goddess of the Earth. Originally postulated in a way that resulted in skepticism among many scientists, Lovelock's basic concept that "the whole thing, life, the air, the oceans, and the rocks . . . the entire surface of the Earth including life is a self-regulating entity" now enjoys widespread acceptance.

Over the 4.54 billion years of Earth's existence, enormous geological, chemical, and biological changes have shaped the planet in ways that have allowed life to prosper, but the overall ecosystems have not remained constant. Throughout the history of life, waves of extinction and renewal are evident in the geological record, with pulses of relative consistency shattered by drastic change. It is evident that conditions favorable to life as we know it are vulnerable to circumstances both within and beyond our control, but one way or another, as long as there is an ocean, it is likely that Earth will continue to be alive.

VISIONARIES

Samantha "Mandy" Joye
Making Connections

. . . .

After the *Deepwater Horizon* drilling platform explosion spewed oil into the Gulf of Mexico in 2010, Samantha "Mandy" Joye was on the scene. The professor of marine sciences at the University of Georgia dived in the submersible *Alvin* to the Gulf seafloor to determine how the oil and the chemicals used to disperse it were affecting microbial communities. The answer: badly. An expert in methane and hydrocarbon dynamics in the Gulf and in ocean policy, Joye testified before Congress, advocating for Gulf cleanup and restoration. Since then, she has led the consortium Ecosystem Impacts of Oil and Gas Inputs to the Gulf (ECOGIG), which monitors the spill's effects. She also studies the relationship among natural hydrocarbon seeps on the Gulf seafloor, its microbes, and its ecosystem. To fully restore damaged habitats, she says, "We've got to develop an understanding of the biological system and how it's interconnected."

. . . .

OPPOSITE ABOVE: Minerals glued by cyanobacteria formed Earth's oldest fossils, stromatolites, here in cross section. OPPOSITE BELOW: Living stromatolites, shown here, were first discovered at Western Australia's Shark Bay.

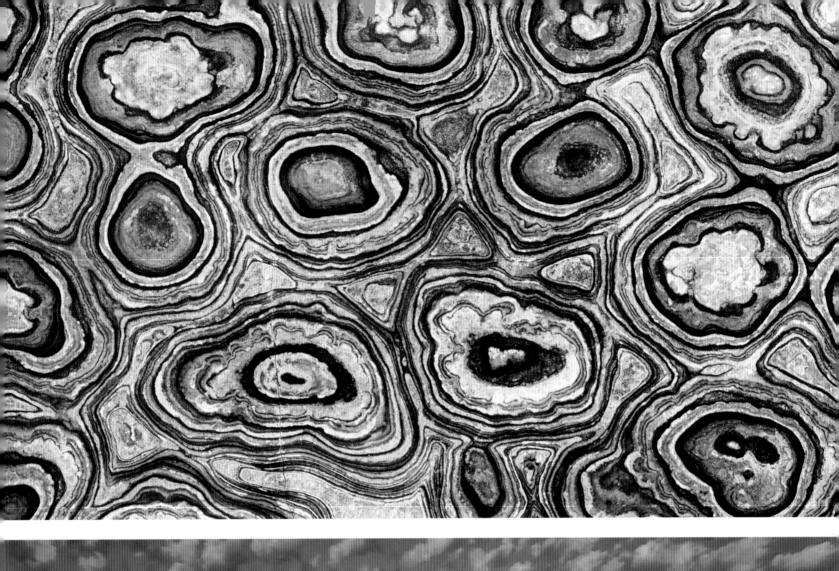

Making Food in the Light

A popular slogan among botanists is "Have you thanked a green plant today?" It is a reminder that we should be mindful of where the oxygen we breathe comes from, where food originates, and what the source is for many products we use and much of the energy that has powered civilization to the present level of prosperity. The basic principles of photosynthesis were figured out by a British scientist, Jan Ingenhousz, in the late 1770s.

New insights continue to be discovered about how photosynthesis drives the great ocean food webs, a process involving chlorophyll-bearing bacteria, protists, and plants that use the sun's energy to convert carbon dioxide and water into simple sugar and oxygen. Chlorophyll is the pigment involved in most photosynthesis, but other pigments help capture light energy, and there are bacteria with a rhodopsin-retinal protein complex that take up and use light energy to break down organic carbon compounds. Estimates vary, but well over half—perhaps as much as 70 percent—of the oxygen in the atmosphere has been generated by photosynthetic organisms that mostly occupy the relatively well-illuminated upper 10 to 30 meters or so of the ocean. Although some green algae *(Chlorophyta)* occur in depths of at least 200 meters, calcareous red algae *(Rhodophyta)* have been documented on the seafloor in depths as great as 268 meters, and cyanobacteria have been observed even deeper in the water column.

It has long been known that seaweeds, seagrass meadows, salt marshes, coastal mangroves, and especially diatoms and other phytoplankton are critical to the basic processes of planetary carbon capture and oxygen production, but one of the most significant photosynthesizers escaped notice until 1986. Minute in size but large in terms of impact, the cyanobacterium *Prochlorococcus* turned up when Massachusetts Institute of Technology scientist Sallie W. "Penny" Chisholm and her colleagues applied a new method of analyzing plankton in clear ocean water near Bermuda. Variations on this mighty microbe are now known to be widespread globally and are the source of as much as 10 percent of the oxygen generated by photosynthetic organisms every year. These microbes are a critically important source of sustenance for the zooplankton that in turn are consumed by legions of larger animals.

Until recently, it seemed obvious that photosynthesis requires sunlight. That assumption changed when a research team led by Thomas Beatty of the University of British Columbia discovered photosynthetic green sulfur bacteria that use the dim radioactive glow emitted by hydrothermal vents nearly 2,400 meters deep in the sea coupled with sulfur to obtain energy. One of the researchers, Robert Blankenship, notes, "These organisms are the champions of low-light photosynthesis"; as he explains, efficiently collected light is transferred to the organism's reaction center, where photosynthesis takes place. The discovery may provide clues about the origin of photosynthesis and its occurrence elsewhere without sunlight.

PHOTOSYNTHESIS

Using the sun's energy, marine plants convert carbon dioxide and water into sugar, while emitting life-giving oxygen into the ocean and atmosphere. Marine animals then dine on energy-rich plants and one another.

OPPOSITE: Kelp and other marine organisms produce more than half of Earth's atmospheric oxygen.

Making Food in the Dark

Serendipity is defined as "the phenomenon of finding valuable or agreeable things not sought for"—a perfect description of what happened in 1977 when a team of geologists unexpectedly encountered thriving ecosystems of previously unknown animals surrounding hydrothermal vents in water more than 2,000 meters deep on the Galápagos Rift in the Pacific Ocean. The source of energy for the richly populated communities turned out to be chemosynthesis, the production of food by microbes that utilize chemicals—in this case hydrogen sulfide— rather than sunlight as a source of energy. This discovery had a dramatic impact on long-held concepts about the nature of life in the deep ocean and highlighted an alternative to photosynthesis for food production.

Previously it was assumed that life in the deep sea was relatively sparse and powered only by the organic material that fell from the productive sunlit waters near the surface of the sea. But here there was another source of sustenance: chemosynthetic bacteria growing within the tissues of giant

A densely packed community of mussels and shrimp thrives on chemicals spewed from the seafloor.

tubeworms, large clams and mussels, and clouds and mats of bacteria being grazed upon by legions of small crustaceans, snails, annelids, and others. Moreover, a new kind of microbe was discovered, one that superficially resembles bacteria but is so different genetically that an entirely new domain of life, the Archaea, was designated. Archaea now are also known to occur in plankton and in the digestive systems of cows and humans, among other varied habitats.

Since the 1977 expedition, more hydrothermal vent areas supporting communities of organisms powered by chemosynthetic microbes have been discovered, mostly along plate boundaries in depths as great as nearly 5,000 meters. The first vent communities in the Indian Ocean were located in 2000 by a team of Japanese scientists, and numerous sites have been found in the Arctic in recent decades. In 2018 an international team came upon a new hydrothermal vent field in relatively shallow water—570 meters—near the Azores, a mid-Atlantic archipelago. Chemosynthetic bacteria also occur in hot springs at Yellowstone National Park, in coastal sediments, in caves, and in the dead bodies of whales.

Heat was thought to be essential for supporting the abundant life associated with hydrothermal vents until similar forms were found around cold seeps of methane gas bubbling up from the seafloor in the Gulf of Mexico. Energy-rich fluids and gas flow into the ocean there and are acted on by microbes that in turn power fields of slender, long-lived tubeworms, masses of mussels and clams, pale white crabs, brilliant red starfish, and numerous other species. Methane-consuming bacteria perform comparable chemosynthetic action in the vast areas of the ocean harboring gas hydrates—masses of ice and methane that form in the sea at low temperature and high pressure.

Discovery of chemosynthetic microbes that thrive in wide-ranging temperature environments has excited those seeking evidence of life beyond this planet. There is speculation that somewhere deep within Europa—a moon of Jupiter that appears to have an ocean of liquid water under a thick layer of ice—there may be a counterpart to Earth's sunless oases in the deep.

Bacteria absorb hydrogen sulfide, feed other life

Bacteria reproduce, feed vent life

CHEMOSYNTHESIS

Spewed from hydrothermal vents on the seafloor, hydrogen sulfide fuels microbial communities that in turn provide sustenance for other vent life.

DEEPER DIVE

Dining in the Deep

What will it be? A light-bathed lunch on the ocean surface? A fizzy spritzer of molten metals on the seafloor? From top to bottom, seawater is a test kitchen, hosting extraordinary feats of food prep for all life-forms. In the top 10 to 30 meters, the sun's energy pours down on chlorophyll-rich phytoplankton—bacteria, protists, and plants, driving their conversion of carbon dioxide and water into simple sugar and oxygen. This in turn provides food for larger life in the marine food web. Some kinds of phytoplankton—the aquatic cyanobacteria *Prochlorococcus* and macroscopic algae with light-absorbing pigments—can synthesize sunlight at depths of 200 meters or more. In the sunless depths,

microbes coating or living inside rocks and lining the tissues of tubeworms, mussels, and clams absorb chemicals such as methane and hydrogen sulfide spewing from hydrothermal vents and turn them into carbohydrates. Across and below the seafloor, the buffet continues. Microbes on the hulls of shipwrecks grow by tapping into the electron flow between different metals. Rocks under the seafloor hold microbes that may pull energy from chemical reactions with the rocky basalt itself. In 2019, Woods Hole scientists announced another food source: bits of ancient proteins from long-dead organisms buried in deep-sea sediments 25 million years old. Basically, some microbes dine on DNA.

Lighting the Deep

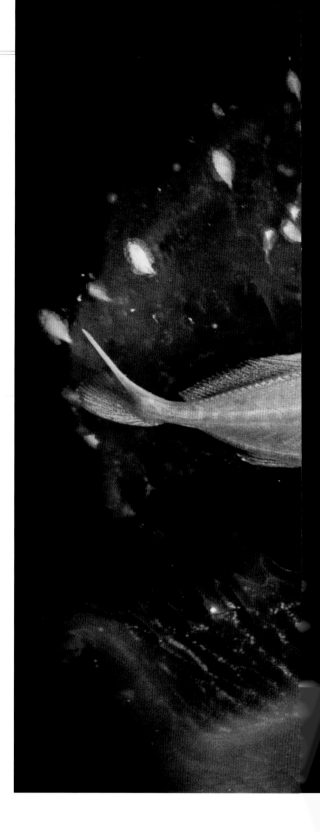

Light from the sun travels 150 million kilometers to reach Earth's atmosphere, where some is reflected back into space by various gases, water vapor, clouds, dust, pollen, soot, and other ingredients that make up air. Whatever actually reaches the sea surface is then so effectively absorbed that barely one percent penetrates to a depth of 100 meters, even in the clearest, calmest ocean water. The photic, or euphotic, zone is defined as the part of the water column that is illuminated by sunlight. At 600 meters, available light is equivalent to faint starlight; at 693 meters, the intensity is approximately one ten-billionth of that at the surface; and below a thousand meters or so, there is darkness—the aphotic zone. Light is so important to humans that it is strange to think that all of life on Earth lives in the dark some of the time. And, since 97 percent of the biosphere is ocean and the average depth of the ocean is 3,700 meters, the vast majority live in the dark all of the time.

Divers who venture below 10 meters or so know that the quality as well as the quantity of light diminishes with increasing depth. In his book with Frédéric Dumas, *The Silent World,* ocean explorer Jacques Cousteau expresses wonder at the way red blood appears green below 20 meters

or so and how at 50 meters, the world is distinctly blue. Zoologist William Beebe described in *Half Mile Down* the changes in light he witnessed in the open sea near Bermuda. Peering through the porthole of the two-person *Bathysphere,* he observed: "As we descend, there vanish in turn the red, orange, yellow, green, and blue, leaving only the faintest tint of violet. Finally . . . looking out into the water, the eye sees only a blackish-blueness, which darkens until . . . every trace of light disappears." The exact place in the ocean where sunlight fades and darkness begins depends on the time of day, the season, the weather, surface roughness, and water clarity.

In the ocean, many bacteria, protists, and animals produce living light—bioluminescence—or harbor symbiotic luminous bacteria in special light-emitting organs. Another light phenomenon, biofluorescence, normally invisible to humans, occurs when organisms absorb light, transform it, and re-emit it in a different color. Human eyes can perceive a range of the electromagnetic spectrum in wavelengths from 380 to 740 nanometers—essentially embracing only the colors of a rainbow. Animals in the sea have a wide range of adaptations that allow them to sense and use light, some with eyes capable of discerning visible, polarized, and ultraviolet light.

LEFT: Northern right whale dolphins swim toward Pacific sunlight.
ABOVE: Traveling partners in the open sea, a juvenile yellowfin trevally shelters amid the tentacles of an inverted jellyfish.

MODIS: Eye in the Sky

Information on the health of the ocean relies in part on data received from an instrument circling the planet: MODIS, for Moderate Resolution Imaging Spectroradiometer. Aloft for two decades through NASA's Earth Observing System (EOS) program, the sensor works tirelessly aboard twin satellites to transmit a continual stream of data on ocean color, temperature, and more. Color sensors can monitor the growth of certain phytoplankton that form toxic blooms called "red tides" that are linked to fouling by sewage runoff or excessive nutrients from agricultural fertilizers. Red tides kill marine life and can affect humans. Temperature sensors track warm and cold currents to predict weather patterns, monitor ice growth and melt, and find healthy phytoplankton. MODIS's 24-hour broadcast is available on NASA's website to anyone with the right equipment to read it, opening the door for experts from Tokyo to Miami to identify and determine next steps for troubled waters.

On a Sour Note

I f given a mandate to disrupt the systems that set Earth apart as a place where life as we know it can prosper, there are three closely connected actions that would have swift and serious impact: change the temperature, alter the chemistry, and eliminate wildlife. Without deliberately trying, and mostly in ignorance of the consequences, our species is successfully doing all these things, profoundly altering the nature of nature, and in so doing, changing the systems that support our existence.

One of the most alarming ways that humans are changing Earth's chemistry is by increasing the acidity of the ocean, the result of carbon dioxide (CO_2), which, in excess, becomes carbonic acid (H_2CO_3). Since 1800, coincident with widespread use of coal and other fossil fuels for energy, acidity of the ocean has increased by 25 to 30 percent, faster than any known change in ocean chemistry in the past 50 million years. Called the "evil twin" of climate change, acidification makes seawater more corrosive, impacts physiological processes, and reduces the amount of carbonate ions needed by organisms with carbonate structures.

Acidity and alkalinity are measured on a pH scale ranging from pH1 (highly acidic, such as the acid in your stomach) to pH14 (the alkalinity of bleach). Distilled water is neutral, at pH7; anything above is alkaline, and below is acidic. Seawater has normally been fairly stable with a pH of 8.2. The current value is about 8.1. Scientists at the Bermuda Institute of Ocean Sciences (BIOS) have measured the acidity of the Sargasso Sea accurately and periodically over the past 37 years—the longest running series of measurements anywhere in the world ocean. They have observed a 17 percent increase in the acidity of Sargasso seawater since 1983, and their frequent measurements show the rate of change is increasing.

This shift toward increased acidity will likely favor some organisms, but among the numerous forms being harmed are oysters, clams, crabs, larval fish, and many of the corals and coralline algae that make up tropical coral reefs. Not only are their existing structures being eaten away but young organisms find it difficult to build carbonate materials in an increasingly acidic environment. Coral reefs may be destroyed faster than they can be built, if current trends continue.

"UNDER BUSINESS AS USUAL, THINGS ARE LOOKING RATHER GRIM."

—KEN CALDEIRA, CLIMATE SCIENTIST, CARNEGIE INSTITUTION FOR SCIENCE

ABOVE: Rising temperature and acidification destroy once healthy staghorn corals on Australia's Great Barrier Reef. RIGHT: Over time, acidic waters dissolve the shell of a pteropod, a "swimming snail."

Fast-Acting Acid

For eons the ocean has absorbed carbon dioxide, turning some into life-sustaining minerals such as calcium carbonate—used by corals, clams, and a host of microscopic animals to build shell—while releasing the rest back into the atmosphere. This natural process has created a balanced ocean pH, allowing life—not just shelled animals but also phytoplankton, clownfish, and the animals that eat them—to thrive in water neither too acidic nor too alkaline. But that is changing, as experts estimate that, mainly by burning fossil fuels, humans are adding an amount of carbon dioxide equivalent to a train-car load of 100 tons of coal to the blue depths every second, too quickly for water and life to process such intake. As a consequence, the pH drops, the waters grow acidic, and life-forms that depend on calcium carbonate for their shells break down. Projections say that by the end of the century ocean acidity will be 150 percent greater than at the start of the 18th-century industrial revolution. This matters because life in the ocean—and on the rest of Earth—cannot sustain such rapid and life-altering change.

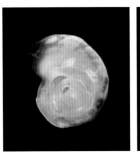

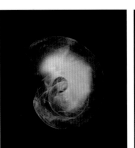

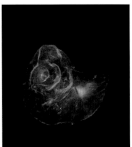

DAY 0   DAY 2   DAY 16   DAY 26   DAY 45

Troubled Waters

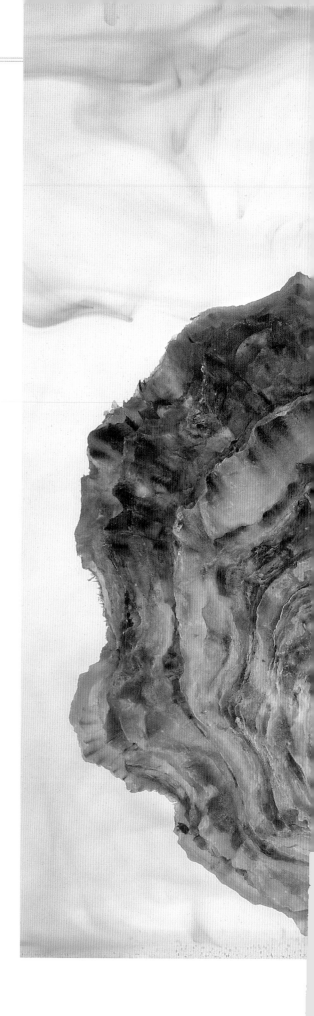

"The current rapid rise in atmospheric carbon dioxide levels caused by the intensive burning of fossil fuels for energy is fundamentally changing the chemistry of the sea, pushing surface waters toward conditions that are more acidic," according to Scott Doney, chemical oceanographer at the Woods Hole Oceanographic Institution. The result is a trend toward acidification of the ocean. Former administrator of the National Oceanic and Atmospheric Administration (NOAA) Jane Lubchenco has other names for it: She calls it "osteoporosis of the sea," and the "equally evil twin" of climate change. Both are driven by the rapid increase of carbon dioxide in the atmosphere and the decrease of natural carbon-capturing and sequestering processes.

Oyster farmers in the Pacific Northwest are struggling with losses of oyster larvae related to ocean acidification, according to researchers at Oregon State University. Dave Garrison, leader of a study on oyster shell formation, noted, "The failure of oyster seed production . . . is one of the most graphic examples of ocean acidification effects on important commercial shellfish." Less obvious, but with more far-reaching consequences, are the changes in shell formation in coccolithophorids, planktonic organisms that have a major role in the regulation of climate, according to Patrizia Ziveri, a researcher at the Universitat Autònoma de Barcelona. She notes that these tiny organisms "form the basis of the aquatic trophic chain, and through calcification and photosynthesis . . . regulate atmospheric and oceanic CO_2 levels." Their individual cells are cloaked with layers of lacy disks made of calcium carbonate that en masse have accumulated in such quantities that they represent a major means of sequestering carbon deep within the ocean.

Paradoxically, increased carbon dioxide appears to be linked to exceptional blooms of coccolithophorids in parts of the North Atlantic. A team from Johns Hopkins University analyzed data over large ocean areas and noted a tenfold increase in their abundance between 1965 and 2010. Foraminifera, another kind of plankton with a calcium carbonate shell, is vulnerable to the ocean's increased acidity. Some forms are predicted to become extinct in the next few decades if increases in temperature and acidity continue.

Pteropods, a group of planktonic snails known as "sea butterflies," are

LEFT: The calcium carbonate shell of a sea butterfly, a mollusk with "wings," has little defense in an acid ocean. ABOVE: A Pacific Oyster releases its cloudy sperm.

already in trouble in the North Pacific and in the waters around Antarctica. With increasing acidity, their spiraled shells are becoming more fragile. While individually small, usually less than one centimeter long, collectively they have a large impact on ocean food chains. They are critically important "middlemen," small enough to dine on microscopic algae and zooplankton, but large enough to be consumed by fish, birds, and other animals that cannot feed directly on microplankton.

Another threat to the stability of ocean ecosystems is deoxygenation—the depletion of oxygen over broad areas of the ocean. Dissolved oxygen tends to be highest near the ocean's surface and decreases with depth. In some areas, particularly along the western coasts of continents, at depths between 200 and 1,500 meters, current flow and biological activity combine to create regions known as oxygen minimum zones (OMZs)—where oxygen falls from a normal range of 4–6 mg/liter to 2 mg/liter.

The actual concentration of oxygen in the sea varies widely, depending on temperature, current flow, wind mixing the water, and the amount of oxygen-consuming activity that is taking place. Since warm water holds less oxygen than cold water, a warming planet translates to a reduced capacity of the ocean to retain free oxygen. Warming also increases stratification that in turn prevents mixing between surface waters and the deeper ocean.

Many more measurements of oxygen levels in the deep sea and the requirements of animals that live there are needed to better understand and explain the abundance and diversity of organisms that live where oxygen levels are relatively low. Numerous kinds of microbes thrive in the absence of oxygen, but invertebrates such as starfish, sea cucumbers, polychaete worms, anemones, and others require at least some oxygen. The oxygen requirements of deep-sea squid, octopus, and fish and most of the rest of life in the sea are simply unknown. What is known, however, is that the levels of oxygen in the ocean are decreasing overall due to human activity, and the cause is not just driven by rising temperature.

Since the 1950s, increasing amounts of fertilizers and synthetic pesticides applied to large-scale farms as well as domestic lawns, fields, and golf courses have been carried by rivers and groundwater into the sea, driving the creation of hundreds of coastal "dead zones." Enhanced levels of nitrates and phosphates favor blooms of a few species that rapidly deplete the available oxygen, causing other organisms, from small crustaceans to large fish, literally to suffocate. In a few decades, the number of human-induced dead zones has increased from none to more than 500 globally.

Overall, there has been a reduction in the amount of phytoplankton in recent decades. Researchers at Dalhousie University in Canada have reported a 40 percent decline since 1950, based on a review of thousands of samples taken globally. Geological changes move at a stately pace, and over time, species come and species go. But the rapid change in basic planetary systems is cause for concern. As ocean temperature and chemistry change, so will the number and kinds of organisms that can live in the sea. There will be winners and losers depending on the ability of organisms to adapt—or not. Dinosaurs had no control over the circumstances that made Earth uninhabitable for them, but we do.

BELOW: A satellite view of the Mississippi Delta at the Gulf of Mexico reveals green "dead zones," where harmful algae rob water of oxygen. RIGHT: Like most marine life, the Pacific Northwest sunflower star needs oxygen to thrive.

"SURVIVAL IS THE
ABILITY TO SWIM IN
STRANGE WATER."

—FRANK HERBERT, *DUNE*

Ocean Depth & Pressure

At sea level, we live at the bottom of an ocean of air that extends more than 97 kilometers high and weighs 14.7 pounds per square inch. At the highest point on land—the top of Mount Everest, 8,850 meters high—the weight diminishes to 4.5 pounds per square inch. At 11,000 meters in the sky—the cruising altitude of many passenger aircraft—the weight of the atmosphere is less than a pound per square inch. At 2,000 meters under the sea, the pressure is 200 times the atmospheric pressure at sea level, and at 11,000 meters, the pressure is 16,000 pounds per square inch.

When H.M.S. *Challenger* embarked from England in 1872 to explore the global ocean, there was widespread belief that beyond a certain depth there was, as C. Wyville Thomson wrote in *The Depths of the Sea,* "a waste of utter darkness, subjected to such stupendous pressure as to make any kind of life impossible." As nets, dredges, and baited hooks were deployed and returned brimming with exotic-looking fish, starfish, sponges, octopus, jellyfish, and much more, the concept of "wasteland" gave way to what is now known to be home to an abundance and diversity of life. In 1960 Swiss explorer Jacques Piccard peered out of the porthole of his bathyscaphe, *Trieste,* and reported, "I saw a wonderful thing. Lying on the bottom just beneath us was some type of flatfish, resembling a sole, about 1 foot long and 6 inches across . . . Here, in an instant, was the answer [to the question] that biologists had asked for decades. Could life exist in the greatest depths of the ocean? It could!"

But the next question was, and still is: "How?" Some answers have emerged.

Air, a combination of gases, is compressible, but water and oil, far less so. Sharks and rays have large, oily livers that help them maintain buoyancy underwater, but fish with gas-filled swim bladders are vulnerable to rapid changes in pressure. In his classic manual *Deep Diving and Submarine Operations,* Sir Robert Davis explains: "If such a thing as a plum or grape be lowered into deep water and pulled up again, its delicate skin will be found to have sustained no damage because the watery contents are incompressible . . . but if an airtight tobacco tin be lowered, it will be crushed by the pressure, because the air it contains is compressible and yields when the pressure is transmitted to it through the thin tin walls."

Marine mammals, sea turtles, marine iguanas, diving sea birds, and humans are constrained by the need to return to the surface for air—and by having compressible air spaces in their lungs and sinuses. Nonetheless, with adaptations honed over millions of years, leatherback sea turtles regularly dive to more than 1,000 meters and elephant seals are known to dive to more than 2,000 meters. Sperm whales go even deeper to dine on deep-sea squid, but the deepest diving air breather is the mysterious, little-known Cuvier's beaked whale, a champion diver able to descend to more than 3,000 meters in the open sea.

VISIONARIES

David Gruber
Descend to Discover

....

National Geographic 2014 Emerging Explorer, 2016 and 2020 grantee, marine biologist, coral reef and photosynthesis expert, underwater photographer, submersible designer, and professor at City University of New York David Gruber discovered with collaborators that proteins that allow coral reefs to mysteriously glow may also keep the coral alive and may help humans with cancer or brain injuries. Called fluorescent proteins, they absorb light of one color and emit light of a different color. Attached to proteins inside a cancer or injured brain cell, they can reveal how the cell behaves, and potentially lead to treatment. In addition, Gruber and Harvard roboticist Rob Wood developed "Squishy Robot Fingers." Attached to a sub or ROV (remotely operated vehicle), the robot uses flexible fingers to gather study samples from delicate animals like jellyfish, while keeping them and their environment safe.

....

Using a special filter, biologist David Gruber captures the biofluorescent hues of creatures with proteins that take in blue light and emit it in vivid green, orange, and red colors.

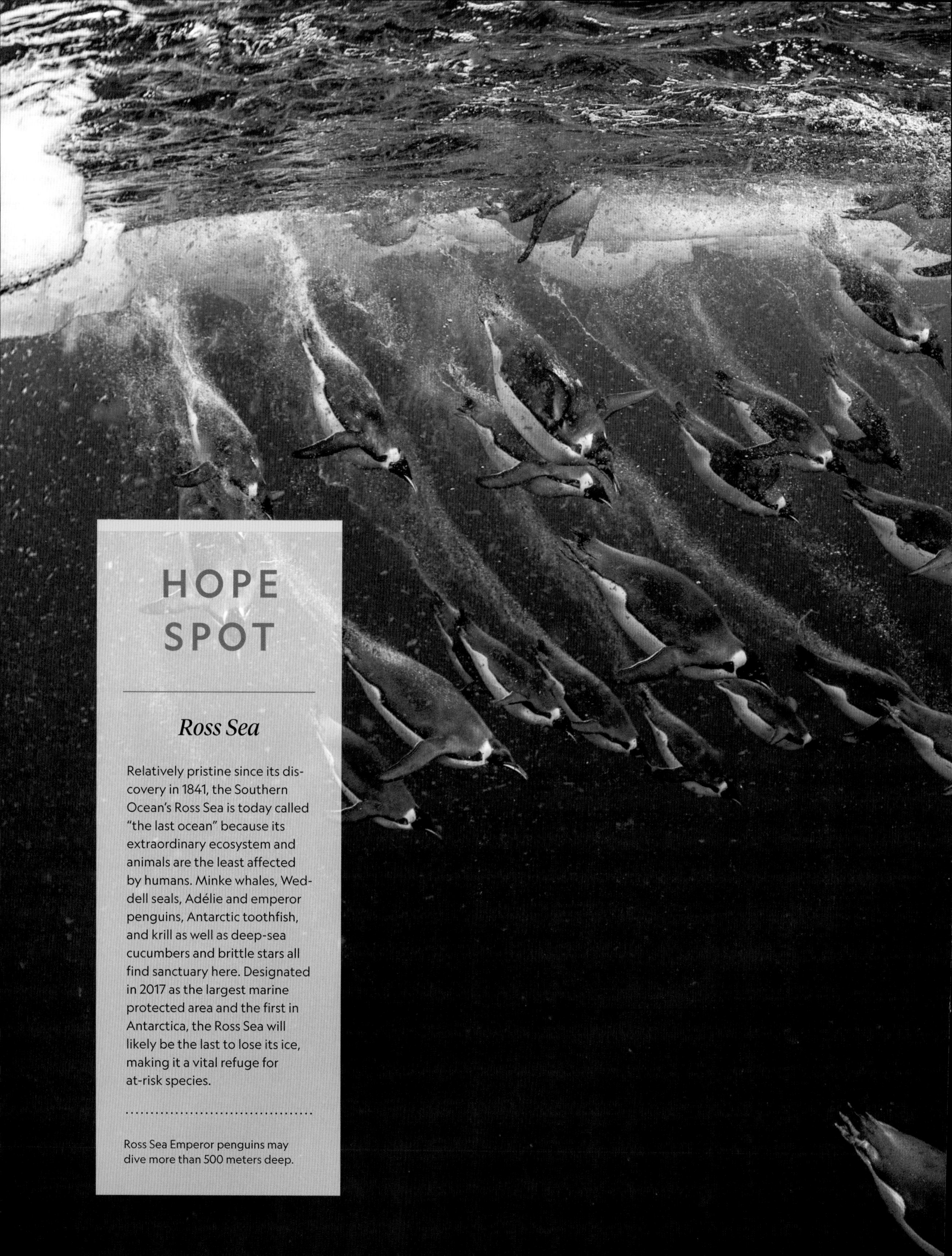

HOPE SPOT

Ross Sea

Relatively pristine since its discovery in 1841, the Southern Ocean's Ross Sea is today called "the last ocean" because its extraordinary ecosystem and animals are the least affected by humans. Minke whales, Weddell seals, Adélie and emperor penguins, Antarctic toothfish, and krill as well as deep-sea cucumbers and brittle stars all find sanctuary here. Designated in 2017 as the largest marine protected area and the first in Antarctica, the Ross Sea will likely be the last to lose its ice, making it a vital refuge for at-risk species.

....

Ross Sea Emperor penguins may dive more than 500 meters deep.

Seeing Through Water

I f you open your eyes underwater without a face mask, even in a swimming pool, what you see is blurred, the images distorted. How do fish and other animals see in the sea, especially in the darkness of the deep ocean? Can they see clearly when we cannot? The answers lie in the properties both of water and of the eye.

For terrestrial animals, sunlight refracts as it passes from air through the cornea and lens to project a clear image at the back of the retina. But light passing through water into the eye bends differently, making the image fuzzy unless corrected with a mask. Seals, otters, and penguins have relatively flat corneas like a human's but spherical lenses similar to those of fish. Seals actually see better underwater than above. Their lenses are accompanied by a large iris that fully opens when submerged, aided by a lining that reflects and amplifies available light.

While constructed much like a human eye, the eyes of fish are water specific. The lens, rounder than a human's, helps the cornea refract the light

One set of sophisticated eyes under the sea belongs to the octopus, whose retina shifts back and forth, sharpening its focus.

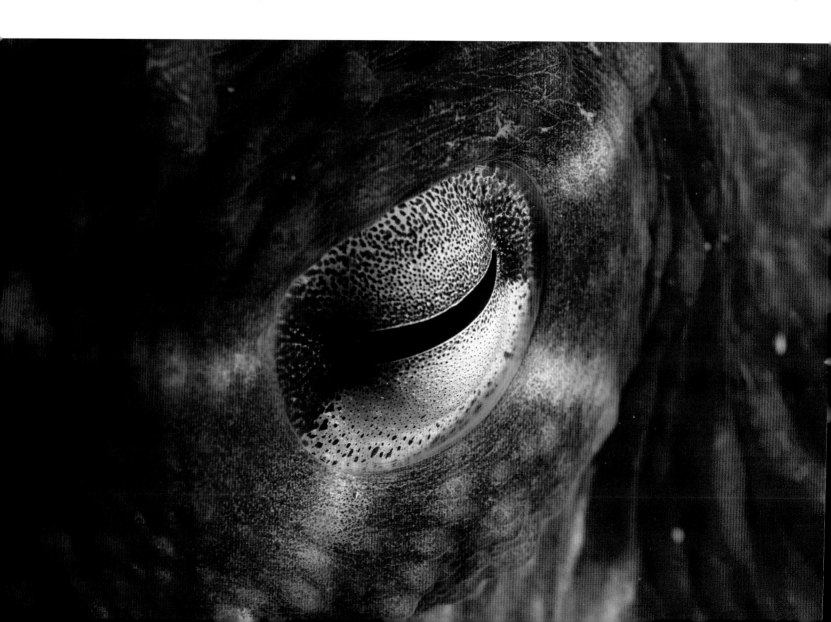

coming through water, and special muscles move the lens closer or farther from the retina to fine-tune the focus.

The 33,000 or so species of fish have wide-ranging adaptations of eye anatomy, but the spookfish or barreleye *(Macropinna microstoma)* is especially noteworthy. Using a remotely operated vehicle, Bruce Robison and Kim Reisenbichler, researchers at the Monterey Bay Aquarium Research Institute, spent hours observing the black fist-size fish at depths of 600 to 800 meters. Suspended motionless in the water, the fish can rotate its two tubular eyes from a vertical, upward-looking position—suitable for finding prey and avoiding predators—to straight ahead, making it possible to see what it eats. A clear canopy filled with transparent fluid covers the fish's head, allowing more light to reach the eyes and providing protection from the stinging cells of its gelatinous prey.

Among the most sophisticated eyes in the sea are those of cephalopods—octopuses and squid. Similar in structure to the eyes of vertebrates, the eyes of an octopus achieve clear focus by adjusting the lens through muscular action. The giant squid *(Architeuthis dux)* has eyes the size of a dinner plate with a lens as big as an orange, capable of capturing whatever light—often bioluminescent—that may exist at 500 to 1,000 meters below the surface, where it susses out prey and detects predators.

The sea for many animals is a veritable show of visible, polarized, and even ultraviolet light. Where human eyes carry just three color receptors, the mantis shrimp, with 12 receptors, navigates a kaleidoscopic world. Like its trilobite ancestors, its compound eyes hold thousands of individual cells working in harmony to capture both polarized and ultraviolet light. The deep sea is home to most of life on Earth, a realm that to humans appears eternally dark. But we might think otherwise if endowed with the visual superpowers of a mantis shrimp, giant squid, or spookfish.

Beware the barreleye, or brown-snout spookfish. Inside its domed head, tubular eyes with sharp vision rotate to quickly spot prey.

DEEPER DIVE

Photographing Through Water

When experts like National Geographic photographers David Doubilet, Brian Skerry, and Paul Nicklen photograph subjects underwater, the result is beauty, motion, personality, and magnificent color. Their secrets of success are many after years of experience working in the unpredictable ocean—cold water and swift currents, color change with depth, curious or elusive sea life, and equipment glitches. But all agree that the basic challenge is lighting. "It was and still is a challenge to make images in a world where you can see a hundred feet on a good day," Doubilet says. Water absorbs light, and the deeper a photographer descends, the more the view diminishes as light is gradually filtered out. To compensate, an on-camera flash might work near the surface but is too weak to overcome the dull blue hue that colors images at depths. An underwater strobe helps restore brilliant hues. Even with light, however, there can be problems. Plankton and silt can cause backscatter; photographing at a distance requires more powerful strobes; and, to eliminate harsh shadows and capture breathtaking dimension and color, more than one strobe may be needed, requiring an assistant. The price is small given the importance of the mission. Says Doubilet, "I keep swimming and taking pictures because images have the power to educate, celebrate, and honor."

Sound Waves Through the Water

Leonardo da Vinci wrote in 1490, "If you cause your ship to stop and place the head of a long tube in the water and place the outer extremity to your ear, you will hear ships at great distance from you." Sounds in the sea travel four times faster than in air: 1,500 meters per second in the realm of fish compared with 330 meters per second in the world of birds and humans. But radio waves—a tried-and-true method of communicating in air—do not function well in the ocean. It is possible to pick up a phone and speak to an astronaut on the International Space Station or on the far side of the world, but that same phone cannot connect with an individual who is three meters underwater.

What limits transmission of radio waves in seawater is conduction and absorption. In the sea, long waves travel, or conduct, better than short waves. The U.S. Navy has used extremely low frequencies (ELFs) to communicate underwater, imitating something fin and blue whales have been doing for millions of years—using long-wave, low-frequency calls to connect with whales hundreds of kilometers away.

In his classic book with Frédéric Dumas, *The Silent World,* Jacques Cousteau suggests that the sea is a quiet place, but it is hard to escape sounds under the ocean. Bubbles, spray, waves, rain, wind, cracking sea ice, undersea volcanoes, and other natural phenomena, combined with many human-made sources of noise, now infuse the ocean. In addition, all marine mammals vocalize, and it is possible that all fish do, too. Fish known as croakers croak, grunts grunt, and groupers produce powerful low-frequency thrumming that can be felt as well as heard. Fish have a trio of ear bones (otoliths) that function for balance and hearing but also serve as a timekeeper. Successive rings are laid down on the bones as a fish grows, much as rings on a tree record its annual growth. Many invertebrates are notoriously noisy, notably the well-named snapping and pistol shrimps. Spiny lobsters purr and, like many other crustaceans, bristle with sensitive hairs that may detect sound vibrations. Mantis shrimps are able to close their front claws with such force that the sound created can stun prey and even break the glass of aquariums.

Sound waves in the ocean were used in the 1990s to develop a method for tracking large-scale changes in ocean temperature associated with global warming and events such as El Niño. Conceived by Walter Munk at the Scripps Institution of Oceanography and Carl Wunsch of the Massachusetts Institute of Technology, the project, called Acoustic Thermometry of Ocean Climate (ATOC), involved sending low-frequency sound signals from underwater speakers and tracking how long it took for them to reach receivers moored to the seafloor thousands of kilometers away. Because sound travels faster in warmer water than cool water, a long-term series of tests that recorded increasingly faster travel time could indicate that the ocean is warming. Although brilliant in concept, the frequency and intensity of the sound signals raised concerns about the impact on marine life.

VISIONARIES

Shane Gero
Whale Hero

Long-term studies of marine animals require exceptional ingenuity, patience, and willingness to stay wet much of the time. For nearly two decades marine biologist Shane Gero, founder of the Caribbean-focused Dominica Sperm Whale Project, has done just that, obtaining remarkable new insights into the behavior of these highly social animals. "I have come to know whales as individuals," he says, "as brothers and sisters, as mothers and babysitters, and as a community of families, each with their own way of doing things." Gero has spent thousands of hours tracking over 30 whale families and recording their behaviors and ways of communicating. Focusing on a family he named "The Group of Seven," he has watched as they hold their breath for more than an hour and dive three times as deep as a nuclear submarine. He has also recorded the dialect unique to them. In ongoing fieldwork Gero plays back to whales the sounds he has recorded, to determine how they recognize one another, their families, and their larger clans. His bigger mission: to reframe conservation policy to recognize whales' cultural diversity as biodiversity.

OPPOSITE: Southern right whales off the Argentina coast communicate among family pods through belches, growls, upcalls, downcalls, and down-upcalls.

SOUND WAVES

This diagram shows the natural and human-made sounds that disrupt communications between whales and among other kinds of marine life. Noise above 120 decibels can disturb behavior.

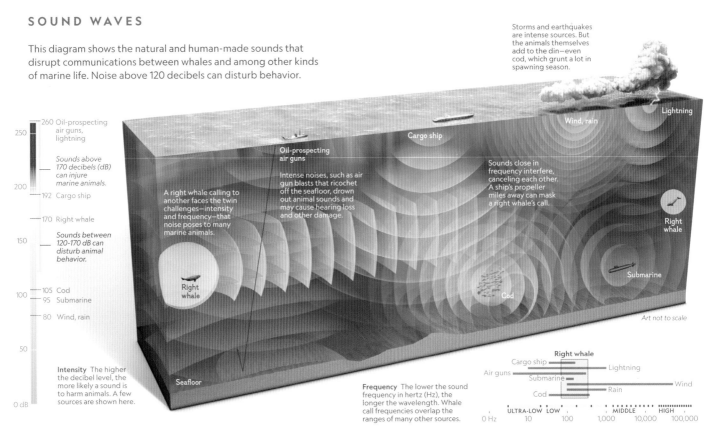

Storms and earthquakes are intense sources. But the animals themselves add to the din—even cod, which grunt a lot in spawning season.

Lightning

Wind, rain

Cargo ship

Oil-prospecting air guns

Oil-prospecting air guns

Intense noises, such as air gun blasts that ricochet off the seafloor, drown out animal sounds and may cause hearing loss and other damage.

A right whale calling to another faces the twin challenges—intensity and frequency—that noise poses to many marine animals.

Sounds close in frequency interfere, canceling each other. A ship's propeller miles away can mask a right whale's call.

Right whale

Right whale

Submarine

Cod

Art not to scale

Seafloor

Intensity The higher the decibel level, the more likely a sound is to harm animals. A few sources are shown here.

- 260 Oil-prospecting air guns, lightning
- 250
- *Sounds above 170 decibels (dB) can injure marine animals.*
- 200
- 192 Cargo ship
- 170 Right whale
- *Sounds between 120-170 dB can disturb animal behavior.*
- 150
- 105 Cod
- 100
- 95 Submarine
- 80 Wind, rain
- 50
- 0 dB

Frequency The lower the sound frequency in hertz (Hz), the longer the wavelength. Whale call frequencies overlap the ranges of many other sources.

Right whale

Cargo ship — Lightning
Air guns —
Submarine — Wind
Cod — Rain

0 Hz	ULTRA-LOW	LOW		MIDDLE	HIGH
	10	100	1,000	10,000	100,000

Songs of the Sea

n the 1960s zoologists Roger Payne, Katherine Payne, and Scott McVay correctly identified humpback whales as the source of mysterious and melodious sounds recorded by the U.S. Navy at a submarine tracking station in Bermuda. Later analysis showed that these haunting, rippling notes, ranging from deep rumbles to canary-like chirps, followed patterns that can be likened to stanzas of music. Now known as "the singing whales," humpbacks produce songs that are regionally specific and change slightly from year to year. A knowledgeable humpback whale song specialist can distinguish general geography and year class from a recorded whale song. To enhance an article Roger Payne wrote about humpbacks in the January 1979 issue of *National Geographic* magazine, a record of humpback songs was tucked inside the back cover. At 10.5 million copies, it exceeded platinum status.

Baleen whales, from bowheads and Bryde's to blues and grays, produce a wide range of sounds both lower and higher than those audible to humans. Toothed whales, dolphins, belugas, orcas, sperm whales, and others in that group emit rapid-fire staccato sonar beeps, enabling them to "see" with sound—sensing size, shape, and distance of underwater objects through analysis of sound that bounces or echoes back.

Every bottlenose dolphin has a personal whistle, or "name"—a distinctive high-pitched call that is used when greeting another dolphin. There has been little success deciphering what the dolphins' extensive repertoire of whistles and rapid sonar notes mean from a human standpoint, but clearly they mean something to the dolphins. The largest toothed whales, sperm whales, also have distinctive calls, known as "coda." Since 2005, whale researcher Shane Gero has observed and gotten to know individual sperm whales that tend to make the Caribbean island of Dominica their home, and he can recognize their distinctive calls.

A musician, Bernard Krause, was inspired during the 1970s to begin documenting the collective sounds produced by entire ecosystems, above and below the sea. The sound signatures change over time, as species flourish or fade. The soundscapes that he and others have recorded over the past four decades provide a measure of species diversity and ecosystem health. Coral reefs that sizzled, popped, and snapped with an almost deafening cacophony at night 25 years ago are quieter now, an audible measure of decline that is also visible to the eye.

Meanwhile, intrusive waves of noise began scaling up as seismic surveys for oil exploration and research, as well as commercial ship traffic across the world, greatly increased. At present, about 90 percent of the goods marketed globally are transported by massive cargo ships, significantly altering the soundscape of the sea. Leonardo da Vinci would not need a listening tube to hear them coming from far, far away.

Called singing whales, male humpbacks grunt, whoop, and whistle—their melodies, hauntingly beautiful, differ by region and change over time.

"THE POD IN CHORUS, BLOWS, AND IN A
MOMENT'S GRACE—CREATION SINGS."

—MARGARET WENTWORTH OWINGS, *VOICE FROM THE SEA*

GREAT EXPLORATIONS

Salt: A Sea Story

W hy is the sea salty? Aristotle in fourth-century B.C. Greece, Pliny the Elder in first-century A.D. Rome, inventor Leonardo da Vinci in 16th-century Italy, Robert Boyle in 17th-century England—all had their own theories.

In 1715 astronomer Edmund Halley proposed that rivers leached salt and other minerals from rock as they ran to the sea, where they concentrated over time, giving seawater its salinity. Not until 1951 did William W. Rubey and his contemporaries have a more integrated view: Halley was half right, but in addition, sodium had been leached from the seafloor as the ocean formed, joined by hydrochloric acid spewed from undersea volcanoes and hydrothermal vents, which added chloride, magnesium, calcium, potassium, and sulfates. Scientists today believe that seawater's makeup has been fairly stable for billions of years.

Together, these dissolved salts make up 3.5 percent of the weight of seawater. If they could be removed and spread across the continents, they

The ribs of a sunken ship mark its grave on the bottom of the Black Sea.

RIGHT TOP: As Borneo's Baram River water feeds the sea, some collects in a catchment.
RIGHT BOTTOM: The H.M.S. *Challenger* global research mission (1872–76) confirmed the consistency of the ocean's saltiness.

would form a layer some 152 meters high, as tall as a 40-story building.

We do not drink seawater, but we do need salt. Salt in our bodies balances fluid and electrolytes and helps nerves and muscles function. Creatures in the sea provide almost half the oxygen on the planet, and seawater supports most of life on Earth. Some marine animals have bodies adapted to take in a continuous flow of water and excrete excess salt through gills, kidneys, skin, and special glands. The marine iguana of the Galápagos Islands drinks seawater and snorts salt out of its nose.

Deep ocean currents are driven by water's salinity and temperature. As polar sea ice forms, it leaves behind salt, making the surrounding water saltier and denser. The salty water sinks, initiating deep ocean currents that drive the global conveyor belt, moderating weather and climate. The ocean also absorbs excess heat and carbon dioxide from the atmosphere.

The denser salt water becomes, the less oxygen it holds. As it sinks and drives currents, it may, in some areas, also form anoxic—non-oxygenated—layers. These layers support little to no life, but they create ideal environments for preserving human history. Over the course of two decades, archaeologists George Bass and Fredrik Hiebert, along with geophysicist Robert Ballard and international teams, have used remote imaging to discover more than 40 preserved wooden shipwrecks—one 2,400 years old and more than 1,600 kilometers deep—settled in the anoxic layer of the Black Sea. Norwegian archaeologists are finding similarly preserved wrecks in the deep brackish waters of the Baltic. Unlike wrecks in oxygen-rich salt water, where wood succumbs to the gnawing of worms, bacteria, and other organisms, these remain near-pristine crucibles of life past.

Informing our future is the key to NASA's Salinity Processes in the Upper-ocean Regional Study (SPURS). Since 2012 this collaboration with Woods Hole Oceanographic Institution and experts in Argentina, France, Spain, and other nations (some aboard the International Space Station) has tracked changes in the upper ocean's salinity and its effect on the water cycle and climate using satellites, glider robots, and other technology. Findings help scientists monitor, understand, and model the water cycle across the global ocean, how it affects us, and what we can do about it.

TIME LINE

A Time Line of Brine

4.54 billion years ago: Earth is born.

3.8–3.1 billion years ago: Ocean forms; photosynthetic organisms oxygenate atmosphere; continents form.

580 million years ago: Multicellular life evolves.

4th century B.C.: Aristotle says seawater has salty and bitter elements; distills it to gain "sweet" water.

1st century A.D.: Pliny the Elder says salt water sinks, fresh water floats.

ca 1500: Leonardo da Vinci theorizes the "saltness of the sea" comes from fresh water passing through "mines of salt" to the ocean.

1715: Edmund Halley says rivers deposit salt in the sea.

1865: Danish chemist Johan Georg Forchhammer determines seawater includes chloride and sodium.

1872–76: British H.M.S. *Challenger* expedition and German scientist William Dittmar confirm seawater elements.

1951: William Rubey confirms and adds to Halley's theory.

1976: Deep Sea Drilling Project (DSDP) scientists propose anoxic ocean conditions prompted early extinctions.

1976: Willard Bascom of Scripps Institution of Oceanography proposes Black Sea anoxic waters could preserve ancient shipwrecks.

1999–present: George Bass, Fredrik Hiebert, and Robert Ballard identify shipwrecks in Black Sea's anoxic layer.

2003: Peter Agre and Roderick MacKinnon win Nobel Prize for determining how salt and water move in and out of cells.

2010: British scientist Hugh Jenkyns links anoxic conditions to global warming.

2012–present: NASA SPURS project monitors water cycle, currents, and climate.

2018: Salt deposits from 2.3-billion-year-old seas two kilometers beneath northwest Russia show seawater carried life-giving oxygen.

MOTION
IN THE
OCEAN

Chapter Three

"THIS STRANGE WORLD . . . ITS COSMIC CHILL . . .
THE ETERNAL DARKNESS, THE INDESCRIBABLE
BEAUTY OF ITS INHABITANTS."

—WILLIAM BEEBE, *HALF MILE DOWN*

An ocean of air rests on the sea, merging at the interface—a continuum of water and gases defined by differences in density, lighter above, heavier below but joined as one tightly coupled system wrapped across the face of the planet. Where the sea touches land, the atmosphere extends beyond, taking the essence of the sea far inland. The atmosphere, the water, the land, and all living things together make up the biogeochemical miracle called Earth. These elements are fully engaged, inextricably linked to one another and to energy from the sun in ways that have yielded the only place in the universe hospitable to life as we know it.

There are times when the ocean is glassy calm, mirroring clear skies that seem at the horizon to merge as one with the sea. But the ocean is never truly quiet. Forces are imposed by the sun, moon, and neighboring planets, coupled with wind, Earth's tilt and rotation, the undersea terrain, earthquakes, and volcanoes. There are differences in temperature, salinity, and even the motions of the small, medium, and large forms of life living within. Add to that the disturbances created by human activities, especially in the modern motorized world of shipping, and the result is an enormously complex interacting system of flowing water, constantly on the move.

Storms rile the ocean's surface and propel energy in the form of waves that strike surprisingly distant shores. Melting ice, collapsing rocks and glaciers, and shifting shorelines can introduce significant variables. Tides are predictable except when they are not—when a collision of subtle or sudden shifts above or below the ocean alter what is anticipated in unexpected ways. Great waves—tsunamis—are occasionally triggered by underwater earthquakes, overwhelming regular tidal rhythms.

Polynesian sailors developed uncanny skills of wayfinding across the uncharted Pacific Ocean, navigating by the stars, currents, and waves, inferring location and direction with knowledge acquired long before compasses, marine chronometers, sextants, and satellites existed. Technologies today make possible a global overview from high above the sea surface and, increasingly, from deep within, are at last dispelling much of the mystery about what drives waves, tides, and currents, and how to predict their behavior with increasing accuracy.

RIGHT: While devastating the eastern seaboard in 2012, Hurricane Sandy also dug a new inlet at the east end of New York's Fire Island. PAGES 92-93: Majestic and powerful, ocean waves off U.S. coasts generated energy equal to more than half the nation's electricity in 2019.

Oceans of Air & Water

The ocean absorbs and holds a thousand times more heat than the atmosphere, and is the principal regulator of planetary temperature, distributing warm and cold water through a global system of interconnected surface and deep-sea currents known as the "conveyor belt." Currents that are obvious at the surface are linked to deep circulation below. There are currents in the atmosphere as well, and both realms are in constant motion, driven and shaped by winds. Ocean currents are also moved by Earth's rotation and tilt, and by differences in temperature, physical geography, gravity, pressure, and salinity.

Solar radiation striking the Earth is greatest at the Equator, with lesser levels toward polar regions. When the molecules of gases in the air are warmed, they move faster, spread apart, and rise upward, leaving a low-pressure area at the Earth's surface. The molecules in colder air slow down, move closer together, and sink, creating a high-pressure area. Atmospheric wind is caused largely by the flow of air from high-pressure regions to areas of lower pressure. The greater the difference between the pressures, the faster the movement, and therefore the stronger the force of the wind. Earth's spin influences the direction of most wind, deflecting it to the right in the Northern Hemisphere and to the left in the Southern Hemisphere, a phenomenon known as the Coriolis force.

Early 20th-century British scientist Sir Gilbert Walker observed that when air pressure in the western Pacific was high, it was low in the eastern Pacific, and vice versa. In the 1960s, Norwegian meteorologist Jacob Bjerknes connected this oscillating pattern and the appearance of warm water along the western coast of South America known as El Niño, which creates massive kills of marine life and disrupts fisheries. Later, global observations of the ocean and atmosphere from satellites and instrument arrays in the sea rapidly increased insights into the inextricably linked ocean-atmosphere systems. Researchers discovered that when surface pressures are lower than usual over the warm waters of the western equatorial Pacific, surface pressures are higher in the colder eastern equatorial Pacific. Then, El Niño's counterpart, La Niña, drives trade winds from east to west, pushing cold surface water westward. As this occurs, deep water wells up, bringing nutrients close to the surface, powering plankton blooms that yield an abundance of anchovies and numerous other species.

Such air-sea interactions occur according to the laws of physics in ways that are increasingly predictable as exploration of the deep sea continues and refined instrumentation comes into use. However, the many variables and unknowns in major areas of the ocean lead to meteorological and oceanographic uncertainties. What is known is that the ocean is heating, and the pace of warming has accelerated sharply since the 1980s. A 2019 issue of *Advances in Atmospheric Sciences* reported that the previous 10 years were the warmest on record. The consequences are stronger winds, more intense storms, rising sea levels, and serious climate disruption.

VISIONARIES

Wallace Broecker
Earth's Conveyor Belt

....

A geologist at Columbia University's Lamont-Doherty Earth Observatory, Wallace Broecker termed the world ocean's connected system of currents the "global conveyor"—a vast river of water that transfers heat as it loops around the planet from the warm surface to the cool depths and back to the warm surface. Broecker pioneered the use of carbon isotopes and other trace compounds to map the circulation. A disruption in its flow, he projected, could catalyze an ice age resembling Europe's "little ice age" some 12,000 years ago—a concept Hollywood turned into the film *The Day After Tomorrow*. Broecker observed the planet and its climate as an integrated system including ocean, atmosphere, and ice. He was also one of the first scientists, in 1975, to sound the alarm about climate change, popularizing the term "global warming." He warned: "The climate system is an angry beast, and we are poking it with sticks."

....

High winds and battering surf tear through an erosion-control fence and pummel oceanfront homes on the Outer Banks of North Carolina.

Going With the Flow

I f the ocean is Earth's blue heart, ocean currents are its veins and arteries—a global circulatory system that transports heat and cold, nutrients and energy, while serving as liquid highways for organisms large and small, shallow and deep. There are many types of currents in the world ocean, driven by different forces. As tides rise and fall with the position of the sun, Earth, and moon, tidal currents rush into shore and out again. Wind, the sun's energy, and the Earth's rotation help to form powerful surface currents in the open ocean, like the Gulf Stream. Coastal currents, as they sweep along the edges of continents, are shaped by variations in landforms and local winds, and may push aside surface waters, allowing deeper, nutrient-rich waters to rise in a phenomenon called upwelling. Cold temperatures and high salt concentration can cause surface waters to sink, or downwell, and create deep-water currents. Storm winds can drive water toward the shore, causing an abnormal rise of water, "a storm surge," that is sometimes responsible for rapid and extreme flooding.

Mapping and measuring ocean currents is critically important for those who travel on and under the sea, for all who seek to understand the ocean's role in shaping climate and weather, and for all people, everywhere, as a fundamental part of understanding the nature of the planet we call home.

Knowledge of ocean currents has come a long way from 1769, when Benjamin Franklin and his sailing captain cousin, Timothy Folger, jointly published the first sketch of the Gulf Stream current, derived from knowledge gained by merchant and whaling captains who took advantage of riding the stream from New England to British ports and avoiding the current on the return.

A century later, during a 110,844-kilometer voyage around the world, scientists aboard the British research vessel H.M.S. *Challenger* recorded the first global assessment of ocean currents, temperature, and much more. Seafarers eventually became familiar with tides and currents regionally, but even the venturesome Polynesian wayfarers who navigated for millennia over thousands of kilometers of the Pacific Ocean could not know about the three-dimensional nature of Earth's system of currents and the impact they have on shaping global climate, weather, and interactions among physical, chemical, and biological processes.

Satellite technologies are greatly improving knowledge of the basic structure of ocean currents by measuring and transmitting data from high above the ocean and receiving information from instrumented buoys, drifters, and other ocean sensors. Since 2000, a global array of instrumented seagoing floats called Argo (for the ship of Greek mythological hero Jason) has been deployed in collaboration with some 30 nations to measure changes in temperature and salinity with depth. By 2020, 3,943 floats were at work,

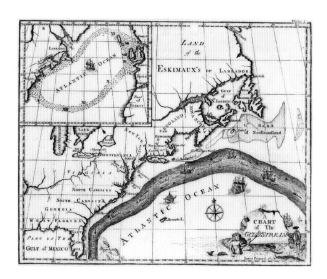

In this 1769 sketch, Benjamin Franklin and Timothy Folger first defined the Gulf Stream along the U.S. Atlantic coast.

The Great Pacific Garbage Patch

During a 1997 Pacific yacht race from California to Hawaii, Captain Charles Moore suddenly found his boat surrounded by millions of pieces of plastic. He had entered a new ocean zone: the Great Pacific Garbage Patch. Ranging across 1.6 million square kilometers of ocean between Hawaii and California, this mass of plastic bottles, packing materials, carryout containers, shoes, and fishing nets lost or discarded by humans is trapped in place by the circular currents of the North Pacific Subtropical Gyre (NPSG). It's one of five major plastic garbage patches in the world ocean, coinciding with five major subtropical gyres, and smaller

patches are being discovered elsewhere—one recently in the Caribbean Sea. They may appear contained, but these sleeping behemoths tamper with every facet of the web of life. Marine animals choke on plastic bits, ingest pollutants the plastics let off, or are strangled in nets; pollutants inhibit oxygen-producing photosynthesis; and plastic pieces now being found in humans will wreak havoc on our own systems.

LEFT: Bull kelp sways off British Columbia, where river runoff joins ocean currents and upwelling that support rich marine life. ABOVE: Plastics dominate the Great Pacific Garbage Patch.

gathering more than 100,000 data points on temperature, salinity, and sound from locations in the upper 2,000 meters of the global ocean spaced three degrees apart.

The data from the combination of Argo floats and the Jason satellite altimeter system are providing, for the first time, a fuller picture of the physical state of the upper ocean in near-real time, made publicly available for use in ocean data models. Benjamin Franklin would surely appreciate the advances in mapping ocean currents in the past 300 years. He also likely would wonder why, with such technological capabilities, the deep ocean currents below 2,000 meters—and their impact on the world—still remain largely unstudied. It is a serious gap in knowledge, sure to be given increased attention during the new global mapping initiatives and the UN Decade of Ocean Science for Sustainable Development commencing in 2021. Ocean exploration and research are fundamental to being able to achieve the ambitious goals set for action concerning climate change and ocean care by 2030.

The Ups & Downs of Tides

Why the ocean predictably advances and retreats along the shore in a rhythmic dance with the sun and the moon has challenged great minds throughout the ages. If Earth did not rotate, had no atmosphere, no gravity or magnetic forces, and the temperature and salinity remained forever constant, the ocean might have a stable, serene surface. But Earth is a dynamic system with complex interactions that continue to baffle even the most sophisticated technologies and the world's greatest minds. Galileo Galilei did not take the moon into account when he tried to explain the tides in 1595, but he correctly suggested that the sun had a notable influence. Others, from Nicolaus Copernicus in the early 1500s to modern scientists with satellites, networks of sensors, and number-crunching computers at their disposal, still cannot fully anticipate the behavior of tidal phenomena.

Earth's gravity keeps water on the surface of the planet, but the gravitational force of the sun and the moon inexorably draw the ocean in their direction. Given the size of the sun—27 million times larger than the moon—and the sun's gravitational pull on the Earth—177 times greater than that of the moon, it might seem that the sun would dominate the pull on the ocean's water. But the sun is 390 times farther away from Earth than the moon is, diminishing its impact by 59 million times. The sun's impact on tides is therefore about half that of the moon's.

When Earth and the moon in their orbits are in alignment with the sun (a phenomenon called syzygy), lunar and solar influences combine and produce the largest tidal ranges. These occur during the full moon, when Earth is between the moon and the sun, and during the new moon, when the moon is between Earth and the sun. These are called the "spring" tides, when the range of tidal highs and lows is significantly greater than at other moon phases. When the sun and moon are at right angles to each other (called quadrature), the moon is in its first or third quarter, and the gravitational pulls from the sun and moon tend to cancel each other out. That's when a reduced, or "neap," tide occurs. In any lunar month—from one full moon to the next—there are two distinctive spring and neap tides. As the tide rises and falls, currents form—"flooding" when water moves toward the shore and "ebbing" as it retreats. In between, there is typically a slack period that may last for seconds or minutes before the flow shifts direction.

The largest tides and most notable tidal currents occur in the Bay of Fundy in Nova Scotia. There, at high tide, water levels can rise as high as a five-story building—16 meters—and the current can speed along at more than 18 kilometers per hour, sometimes catching wading tourists off guard who find themselves racing against the tide for safety.

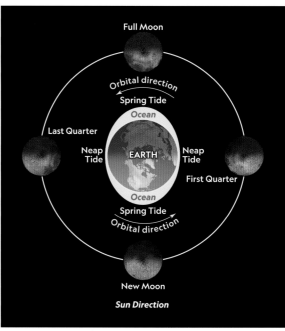

TIDES AND THE MOON

ABOVE: As the moon orbits Earth, its gravitation sends ocean tides to and from shorelines. OPPOSITE: When sun and moon align—at new and full moons—their combined pull brings the largest tidal ranges, as with high tide on coastal Maine.

Wind: Rivers of Air in the Sky

You can feel but not see the wind, but like currents in the sea, atmospheric wind has weight and substance, and both are powered largely by Earth's motion and differences in temperature and pressure. Both move water over vast distances, and in the sky as in the sea, various forms of life, from pollen and spiders to birds and bees, hitchhike along as aerial plankton.

Meteorologists who study the atmosphere to understand and predict weather and climate take into account the influence of the sea on the air above. Similarly, oceanographers consider the bonds between the movement of water, gases, and life, as well as the configuration of land—above and below the ocean's surface. Those connections are most obvious where prevailing winds shape the direction of water's flow in depths as deep as 400 meters, directly influencing climate and weather as currents distribute heat and cold around the planet.

Global wind patterns blow consistently in the same direction, driven by the unequal heating of Earth by the sun, Earth's rotation, and the Coriolis force caused by Earth's spin. Trade winds move east to west around the Equator in a band stretching from 30° N to 30° S, and westerlies move from west to east in the middle latitudes. In both polar regions, winds move east to west in a band from the poles to 50° and 60° north and south of the Equator. These winds and Earth's spin power the great circular gyres that swirl over thousands of kilometers across entire ocean basins, in the North and South Atlantic, the North and South Pacific, and the Indian Ocean, turning clockwise in the Northern Hemisphere and counterclockwise in the Southern Hemisphere. Each of these gyres includes a well-defined western boundary current where the Earth's eastward rotation piles water up against major landmasses, and a weaker, broader eastern boundary current. The western boundary currents—the Gulf Stream in the North Atlantic, the Kuroshio in the North Pacific, the Agulhas off southeast Africa, the Brazil Current in the South Atlantic, and the East Australia Current in the South Pacific—are the world's strongest major currents, moving at 40 to 120 kilometers a day and conveying as much as 100 times the total amount of water discharged by all the world's rivers.

In the waters around Antarctica, cold polar water flows west to east in a continuous circle, unimpeded by any landmass, but it rubs against the warmer waters of the great ocean systems to the north. This merging of water masses forms the Antarctic Convergence, a moving ring of water averaging 40 kilometers wide that can migrate seasonally between about 48° S and 61° S latitude, with rich biological activity and distinctive air and water temperature ranges. These broad patterns of air and water movement remain unchanged even during large storms because they are dependent on consistent physical forces. Superimposed on this orderly base are curling

Invisible but influential, sea-surface winds power sailboats off Grenada and drive the currents that carry them.

eddies that spin off and sometimes last for weeks or months adjacent to major currents. Coastal currents are more variable, shaped by the land and winds that in turn impact local weather. Recurrent patterns of rain, storms, and dry periods are largely influenced by the effect of ocean currents on temperature and humidity.

Wind also drives hurricanes, typhoons, and cyclones—powerful tropical storms by whatever name—that form over warm ocean waters within 30° latitude north and south of the Equator. In the north, such storms spin counterclockwise, and in the south, clockwise. As air just above hot ocean water warms and rises, a low-pressure area forms at the surface. Fresh air moves in that also warms and rises in a cycle that spins and grows. The system captures water evaporated from the sea surface and creates a storm with bands of clouds around a low-pressure region called the eye. The system picks up speed and strength as long as rising warm air continues to create a low-pressure area at the center. When winds reach 62 kilometers per hour (km/h), the system is called a tropical storm. At 119 km/h, it becomes a hurricane, cyclone, or typhoon. Above that are categories of increasing power, ranked from one to five, with top wind speeds reaching well over 290 km/h.

The intensity of a storm is measured in two ways—maximum wind speed and barometric, or atmospheric, pressure. At sea level, when the weather is calm, the weight of the atmosphere above is 101 kilopascals (14.7 pounds per square inch), giving a barometric reading of 1,013 millibars.

"THE WINDS AND WAVES ARE ALWAYS ON THE SIDE OF THE BEST NAVIGATORS."

—SAMUEL SMILES, SCOTTISH AUTHOR AND GOVERNMENT REFORMER

With winds as high as 314 kilometers per hour, Super Typhoon Haiyan decimated the central Philippines and its Tacloban City in November 2013.

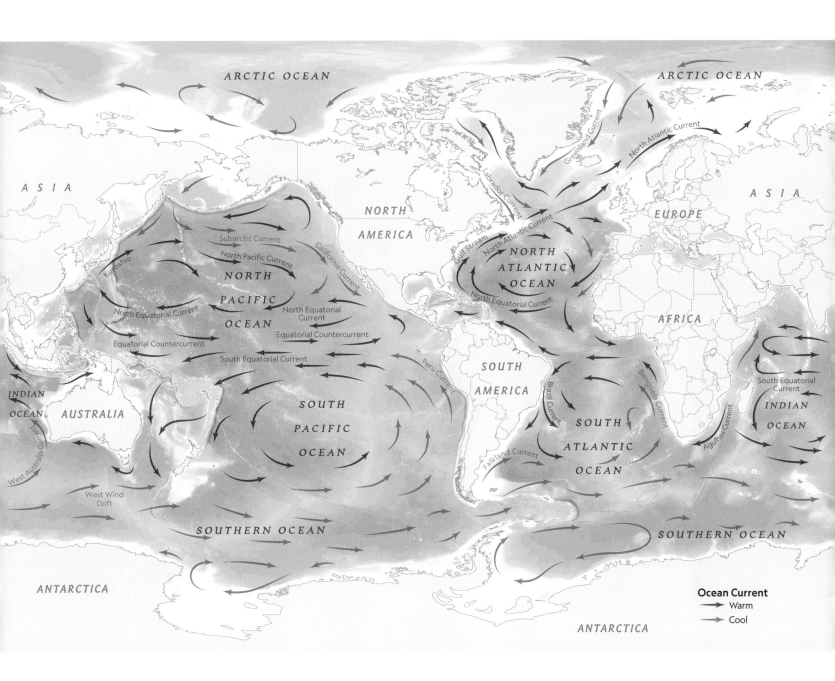

When a low-pressure system forms, a result of rising warm air, barometric pressure decreases. The lower the pressure, the more powerful the resulting storm.

The largest storms grow in the broad reaches of the Pacific. The most massive storm ever recorded was Cyclone Tip, a system that reached a diameter of 2,220 kilometers and swept across the Philippines in 1979; the barometric pressure was 870 millibars and winds reached at least 305 km/h. The highest recorded winds—346 km/h—were measured in an intense hurricane named Patricia that rocked the eastern Pacific off the coast of Guatemala in 2015 with an unusually low barometric pressure of 882 millibars. There is evidence that increasing global temperatures are giving rise to more frequent storms of greater intensity. In the Atlantic, 2020 marked the most active season on record, with 30 named storms and 13 hurricanes.

OCEAN CURRENTS

Swirled by the spin of the globe, shaped by landmasses, and driven by winds, surface currents in the ocean's upper 400 meters make up 10 percent of all ocean water.

Welling Waters

Ask a seabird, "Where are the best places on the planet to go to feast on fish?" and it would likely say, "Look for upwellings." The phenomenon of upwelling occurs where coastal winds push aside surface waters, allowing deep, cold, nutrient-rich waters to rise to the sunlit surface. The nutrients power photosynthetic plankton, which in turn feed legions of zooplankton and small fish, which then provide sustenance for seabirds, larger fish, and numerous other marine animals. Upwelling areas account for less than five percent of the surface area of the ocean, but they have a disproportionately large role in the dynamics of coastal chemistry.

In the Pacific, permanent upwelling areas are linked to cold eastern boundary currents—the Humboldt Current off South America and the California Current in North America. Counterparts in the Atlantic are the Benguela and Canary Currents off the coast of western Africa. In the northwestern Indian Ocean, upwelling occurs as a seasonal phenomenon in the Somali-Arabian Sea system, driven by Asian monsoons.

Cold water that originated around Antarctica travels northward along the coast of South America and rises to the surface off the coast of Peru. It fuels plankton blooms that support extraordinary numbers of small fish and huge numbers of cormorants, boobies, and pelicans that breed and roost nearby. Until late in the 20th century, the birds were so numerous that their guano, accumulated for centuries, was gathered and sold as a particularly potent fertilizer, rich in nitrates and phosphates. Periodically, during El Niño years, warm water comes close to Peru's shore, preventing the upwelling of cold water and causing gloomy weather, lots of rain, and fewer fish. These cycles are normal, but in combination with overfishing in the early and mid-20th century, the robust ecosystem that had developed over many millennia collapsed in the 1970s. Although upwelling continues and modest populations of seabirds still thrive there, the system now fosters a fraction of its former level of sea life. Similarly, in central California, overfishing of sardines and other animals in the early to mid-1900s contributed to the collapse in the 1960s of a system of legendary richness. Wildlife in the waters around Monterey, California—the location of major fishing and canning operations—has been gradually recovering, thanks to various protective policies, with increasing numbers of sea otters, whales, herring, and sardines.

Spanning the Equator about 965 kilometers west of mainland Ecuador, around the Galápagos Islands, a complex collision of winds and currents results in areas of upwelling, especially along the western islands, Isabela and Fernandina. Normally, immense populations of small fish provide sustenance for large populations of boobies and other seabirds, but as in coastal Peru, during El Niño years, warmer water yields less phytoplankton, fewer fish, and lean times for the larger animals that rely on them for sustenance.

In the cool, nutrient-rich upwellings of Alaska's Frederick Sound a busy pod of humpback whales ensnare krill by blowing bubbles to form a net.

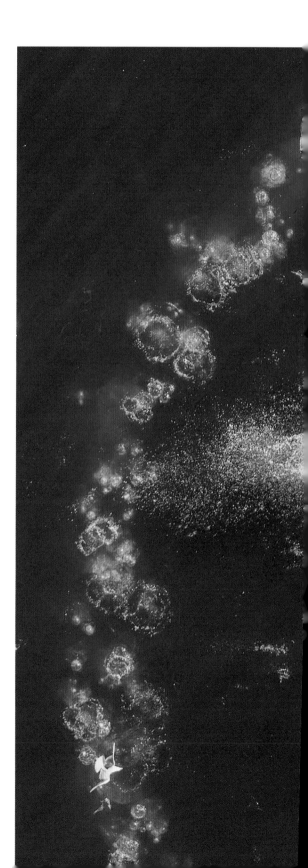

UPWELLING

Winds (near right) push surface water offshore, drawing nutrient-rich cold water from the deep to support marine life. In El Niño years (far right) the warm-water layer is thicker, and upwelling draws from less productive seas.

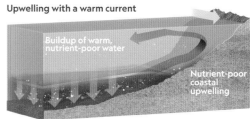

Surface winds

Normal upwelling

Warm, nutrient-poor water

Cool, nutrient-rich water

Coastal upwelling

Upwelling with a warm current

Buildup of warm, nutrient-poor water

Nutrient-poor coastal upwelling

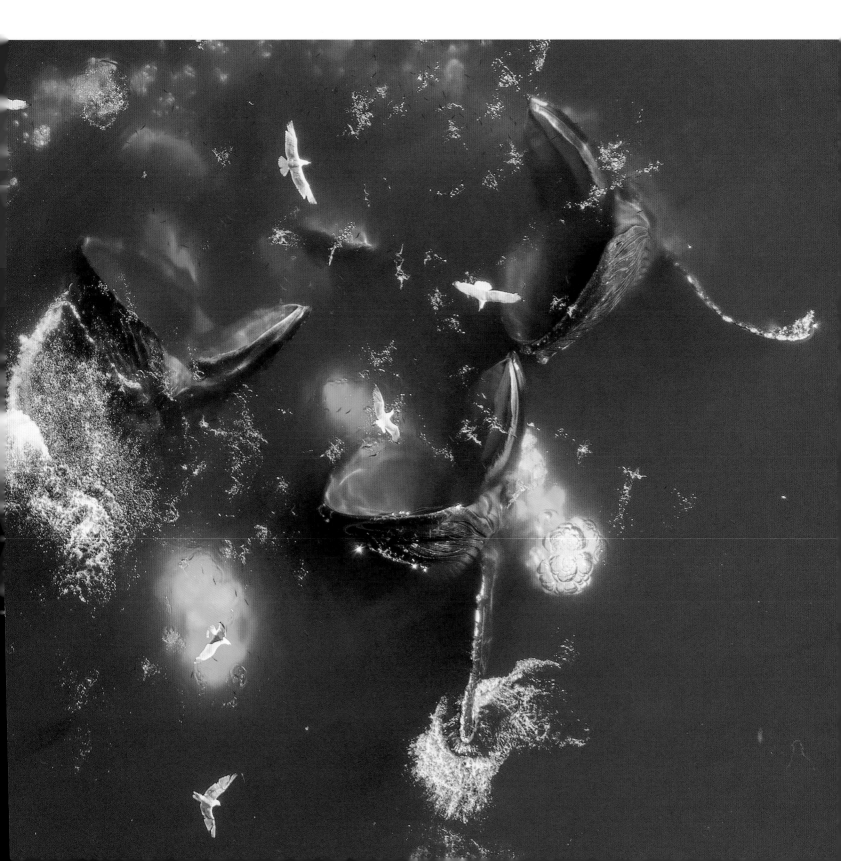

Deep Ocean Motion

At the surface, ocean currents are largely influenced by winds, but deep ocean currents are driven mostly by differences in density caused by temperature and salinity. The higher the salt content, the greater the density, and thus the weight of the water.

But water density also varies with temperature. As temperature increases, water molecules move apart, decreasing density. When cooled, the molecules huddle together, increasing density. Over broad areas, the ocean is layered with cold, dense water below and lighter, warm water above. However, in the Arctic and Antarctic regions, as seasonal sea ice forms, salt is forced out, increasing the salinity of the seawater below. The heavy, more saline, cold, dense water sinks to the bottom then slowly moves, following the contours of the deep seafloor. Thermohaline ("thermo" = temperature, "haline" = salt) circulation initiated in polar regions drives the movement of water globally, visualized as a great looping conveyor belt of dense cold and lighter warm currents of varying salinity.

Layering of water masses occurs where waters of different densities converge. Low-salinity water floating above saltier water below sometimes forms a sharp delineation called a halocline. Where warm- and cold-water masses come together, a distinct density-driven thermocline can form. These vertical layers can serve as liquid boundaries as effective as walls, separating some organisms based on their tolerance for temperature and salinity differences.

Gaining knowledge about the forces underlying the movement of water globally has not been easy. By sampling more than 200 locations around the world, scientists aboard the H.M.S. *Challenger* in the 1870s were the first to confirm the relative stability of the open ocean's salinity—about 35 parts of salt (mostly sodium chloride) per thousand parts of water, or 3.5 percent. Instruments now record salinity by measuring conductivity—the capacity of salt water to pass electrical current—which varies with the concentration of salt ions in the water. Millions of measurements have been taken in recent decades by ocean scientists using instruments called CTDs because they simultaneously determine conductivity, temperature, and depth from the surface to great depths.

Imagine the reaction of *Challenger* scientists to the Acoustic Doppler Current Profiler (ADCP), a technology developed a century after their laborious sampling efforts. ADCP measures ocean current velocity using sonar "pings" that echo back from materials suspended in the moving water. Since 2011, satellites have been sweeping the ocean's surface with sensors measuring salinity and its variations globally. In addition, Remote Ocean Current Imaging Systems (ROCIS) record ocean surface current velocities from an airplane in near-real time over broad areas, feats that would astonish pioneering oceanographers even a few decades ago.

OPPOSITE ABOVE: In *Deep Rover,* biologist Samantha Joye collects samples from a brine pool in the Gulf of Mexico. OPPOSITE BELOW: Supersaline water flowing from under the seafloor forms a concentrated brine lake and channel.

The Global Conveyor Belt

Ocean currents that transport water and heat around the world were described by Columbia University geologist Wallace Broecker as a continuous looping system that he compared to a moving conveyor belt. A parcel of water that sinks to the seafloor in the Arctic may return there in about a thousand years after making a journey to the other end of the planet, cycling around the globe, and eventually becoming part of the warm Gulf Stream water leading back to the Arctic. Although this general pattern has been confirmed, much remains to be defined about the nature of water flow in the ocean's vast, deep, three-dimensional space.

At the surface and for many meters below, where wind and waves create significant mixing, stratification occurs based on rainfall, blooms of plankton, and the effect of sunlight or clouds on temperature. But below the surface in most lakes, coastal waters, and the open ocean, a sharp temperature difference—called the thermocline—forms between the warm surface waters and the cold water below. The thermocline can be such a marked change in water temperature over a small depth interval that it acts as a physical barrier separating different species of ocean life, stopping the penetration of sonar waves, and locally blocking the vertical movement of ocean currents. Depth of the thermocline typically ranges from about 200 meters to as deep as 1,000 meters but varies in some areas seasonally. It is usually stable in the tropics, variable in temperate areas, and rare in polar areas, where the temperature is cold from top to bottom.

Starting in 2000, instruments known as Argo floats began recording temperature, salinity, and depth data in vertical transects over large areas of the ocean, refining knowledge of ocean circulation. Recently, a team of scientists discovered a way to use the data to also measure the speed of deep currents. Hiroshi Yoshinari, Nikolai Maximenko, and Peter Hacker at the Asia-Pacific Data-Research Center calculated the time and distance traveled when floats moved with currents in deep water and then rose to the surface and transmitted information via satellites in real time to land-based data centers. The data sets, known as YoMaHa'05, are yielding new insights into the location and movement of deep and intermediate-depth ocean currents.

Knowledge about the movement of water masses throughout full ocean depth is increasingly important for understanding and predicting weather and climate. Ships on the surface, as well as underwater construction, laying of undersea cables and pipelines, and deployment of submarines and other equipment, benefit from enhanced knowledge of undersea currents and waves. Most of all, as the dynamic nature of the ocean becomes increasingly well known, so does appreciation and accounting for the ocean's conveyor belt movement of water masses in maintaining a planet suitable for the existence of humankind. There is evidence that the current rapid melting of polar ice could slow or shut down the creation and sinking of the highly saline water that initiates the conveyance of water around the globe.

A specialist in Southern Ocean currents and their influence on biology, oceanographer Steve Rintoul prepares to launch a data-gathering Argo float.

GLOBAL OCEAN CONVEYOR BELT

Circling Earth in constant, three-dimensional motion, the global ocean "conveyor belt" of currents distributes heat and cold around the planet, governing planetary temperature.

Shallow warm water
Middle cool water
Deep cold water

DEEPER DIVE

Sub Protectors

Submarine warfare is the topic of classic films from World War I to the Cold War to today, with the hunt the biggest part of the drama. Since World War I, sub-tracking ships have used sonar devices to locate submerged enemies. Working much like bat echolocation, a device on the surface ship directs sound waves to the depths, where they strike the enemy submarine and reflect back, giving information on location, size, and dimension. In early submarines, reflective steel coatings efficiently returned the sound to the surface. Later, coatings of sound-absorbing tiles stifled incoming and outgoing signals. Savvy captains have learned to rely on natural resources, too. Some hide their subs on the seafloor among rocky ledges

U.S.S. *Nebraska* nuclear submarine

that reflect sound. Others use ocean forces. In the upper part of the ocean, winds, tides, and currents distribute warm temperatures through a mixed surface layer. Below that point—and above the colder ocean below—the change from mixed to consistently colder temperatures prompts the water to vary in density. This in turn forms an acoustic barrier between the upper and lower ocean, the thermocline. The thermocline deflects sound waves from above, so that submarines can shelter below it, hiding from sonic "view." Because thermoclines form at different levels based on how winds, tides, and currents mix surface waters, it's key for mariners to know those forces in different areas and seasons.

HOPE SPOT

False Bay, South Africa

Dense kelp forests, abalone, rainbow-colored reef fish, and small sharks range through South Africa's False Bay, at the convergence of the Atlantic and Indian Oceans, running from Cape Point to Cape Hangklip, near Cape Town. Once a global destination for great white shark sightings, its waters shelter few sharks today, due to fishing pressures and pollution in unprotected parts of the bay. The designation of parts of its 1,091-square-kilometer waters as a marine protected area and no-take zone, however, has created a sanctuary for life to recover and thrive.

....

The star of the documentary *My Octopus Teacher* shimmies up a kelp stalk in South Africa's False Bay.

Polar Ocean Circulation

The dynamics of ocean circulation are in many ways very different at the two frozen ends of the Earth. The waters of the Arctic are largely surrounded by land; Antarctica is a large landmass surrounded by water. But compared with the rest of the planet, both polar oceans have a disproportionately strong and outsized influence on global currents, biology, temperature, and chemistry—and therefore the climate, weather, and habitability of Earth.

In the Atlantic Ocean, as the Gulf Stream reaches the northern end of its journey from the tropics and cools, the water sinks and bends southward and continues for thousands of kilometers as a deep, cold, highly saline current, a key segment of the global ocean conveyor belt. When it ultimately reaches the Antarctic Circumpolar Current, the water masses merge. Enriched with nutrients derived from massive blooms of phytoplankton, krill, and other life, water travels northward into the deep Pacific and Indian Ocean basins.

Since its formation across the Arctic Ocean at least five to 13 million years ago, the Arctic ice has always been remarkably thin. There are pressure ridges in places as much as 30 meters thick, but the average thickness of ice cover across the entire Arctic Ocean is only about two meters. Now the ice is melting, and quickly. Global warming is causing rapid loss of ice that is expected to affect ocean circulation with profound consequences for planetary climate and weather. As Greenland's ice melts and flows into the sea, salinity is reduced. Water that normally would freeze in winter months, causing the water below to sink as it becomes saltier and heavier, now remains at the surface, potentially cutting off a critical part of the planet's thermohaline circulation. Similarly, in Antarctica, increased melting of glaciers is altering temperature and salinity patterns that drive ocean currents.

In 2018, NASA began measuring the extent of polar ice using laser tech-

POLAR CIRCULATION

As ice freezes over Antarctica's continental shelves, the salt it leaves behind turns the cold ocean water so dense that it sinks, becoming Antarctic bottom water, which travels north. Above it, the less dense Antarctic Circumpolar Current continues its flow around the continent.

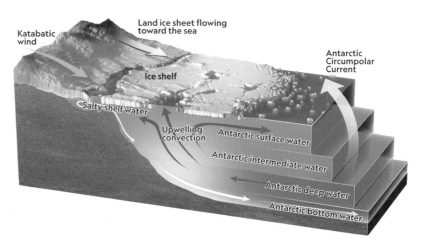

nology that is yielding vitally important data with increased accuracy. Until recently, most of the sinking of the Arctic's cold, dense water was thought to be along the western coast of Greenland; but according to a study led by Susan Lozier, a physical oceanographer at Duke University, the bulk of the action is taking place along the eastern coast of Greenland and across to the Scottish Shelf. Deep-water circulation in the Arctic Basin is not well understood because widespread ice cover inhibits taking measurements, but all of the densest, deepest waters appear to enter from the Atlantic. Some Arctic Ocean water moves through channels among the Queen Elizabeth Islands into Baffin Bay, as well as into the relatively shallow Davis Strait west of Greenland. There is modest flow through the narrow and shallow Bering

Off the Arctic's Svalbard archipelago, glacial ice is sculpted by ocean motion that includes the warm West Spitsbergen and the icy East Greenland surface currents.

Strait between Alaska and Russia. But about 80 percent of the movement of water into and out of the Arctic Basin streams through the channel between Greenland and Spitsbergen. This is also the only deep-water connection between the Arctic and other oceans.

Circulation of water on the Russian side of the Arctic Ocean flows from the Chukchi Sea to the Greenland Sea in the long, east-to-west sweeping arc of the Transpolar Drift. On the Canadian side of the great Arctic Basin, water moves in a wide clockwise swirl larger than the Gulf of Mexico, known as the Beaufort Gyre. Fresh water from ice melt and from northern rivers accumulates in the gyre—sometimes more, sometimes less, depending on the temperature of the air above and the water below.

Called the "flywheel of the Arctic climate engine" by Andrey Proshutinsky, an oceanographer from the Woods Hole Oceanographic Institution, the Beaufort Gyre is also referred to by some as the "great white black hole," owing to the vast unknowns concerning its nature. Joining an international team of researchers from Canada, Japan, and the United States, Proshutinsky has been leading recent research efforts to explore above, below, and within this vast system. Instrument-bearing moorings that measure temperature, chemistry, and currents are deployed under the ice in the Beaufort Gyre, anchored to the seafloor at depths of 3,800 meters, and extend up to within 50 meters of the ocean's ice cap. Other 800-meter-long lines have been deployed through holes in the ice with instruments that travel up and down the line daily, sending data via satellite back to Woods Hole. Data from these instruments and information from other water samples indicate that the waters in the Beaufort Gyre make one grand, clockwise revolution around the basin roughly every four years.

Increased flow of fresh water from the Beaufort Gyre spilling out of the Arctic over the more saline, warmer waters of the North Atlantic can block the release of heat from the warmer waters to the atmosphere, thus inhibiting the tempering effect of the ocean on winters in Europe and North America. According to an analysis by the U.S. National Center for Atmospheric Research, if the global ocean conveyor belt were to shut down completely, the average temperature of Northern Europe, currently warmed by the Gulf Stream, would cool 5° to 10°C.

There is evidence that changes in ocean circulation owing to a freshening of North Atlantic waters have happened before and that those changes have been reflected in shifts in climate. Unlike present trends, however, previous shifts have not been linked to human activities. Two centuries of technological achievements have led to unprecedented impacts on the natural systems that underpin our existence but at the same time have yielded unprecedented knowledge about what can be done to keep them intact.

"I DREAMED OF THE ARCTIC, AN OCEAN WITH A CRYSTAL CEILING."

—DAVID DOUBILET, ELYSIUM ARTISTS FOR THE ARCTIC

OCEAN GYRES

At the top of the world, the Beaufort Gyre—a massive, wind-driven current—captures and pushes fresh water and ice around the Arctic Ocean.

OPPOSITE: The Arctic's largest glacier, Hubbard, calves near Yakutat, Alaska.

The Nature of Waves

and dwellers are accustomed to having a solid, steady surface beneath their feet, a condition that does not exist across more than two-thirds of the planet. To venture into the ocean means coping with everlasting motion, sometimes gentle, sometimes rocky, and occasionally violently rambunctious. From wrinkles on the face of a calm sea to rogue waves towering 10 stories tall, waves share a common anatomy. Each has height measured from its crest (highest point) to its trough (lowest point) and length defined by the distance from the top of one crest to the next. Wave frequency is determined by the number of wave crests that pass a fixed point in a given time. Winds blowing across the ocean are the primary source of energy powering waves; the stronger the winds, the higher the waves. Another factor is the fetch, or the distance the winds and waves travel unimpeded by land. A single wave can travel thousands of kilometers, with individual particles of water moving in a gentle, circular pattern at each site as the wave of energy passes through. Waves at sea have an average height of three meters, but 10-meter-high waves are common during storms. The leading edge of a wave approaching a shallow shoreline is slowed by contact with the bottom, and the wave grows taller until it peaks and breaks, then washes back under oncoming waves.

Waves beloved by surfers typically form when a long fetch is coupled with a steep shoreline that abruptly powers mountainous monsters that inspire most observers to race for the shore. However, millions of wave riders in at least 162 countries race to engage the waves on surfboards. "Banzai Pipeline" on the north shore of Oahu, Hawaii; "Jaws" on Maui, Hawaii's north coast; "Mavericks" on the northern California coast; and Cortes Bank, 1,600 kilometers offshore from Los Angeles, are places where waves form that are the stuff of dreams or nightmares, depending on your perspective.

Few surfers cope with waves above 15 meters high, but Laird Hamilton set a new standard at Teahupoo, Tahiti. On August 17, 2000, he gracefully and successfully slid along the face of what is widely regarded as the "heaviest wave"—a surfer's term for a wave both tall and thick—ever surfed, reverently called "The Millennium Wave." Since then, records have been set for surfing and surviving the tallest known waves near Nazaré, Portugal, including a 23.7-meter record in 2011 by Hawaii's Garrett McNamara and a 24.4-meter-high megawave by Brazilian Rodrigo Koxa in 2017. At the same location on December 14, 2018, British surfer Tom Butler rode a moving mountain of water that appeared to be at least 27.4 meters high. And on February 7, 2019, German surfer Sebastian Steudtner sliced across the face of a liquid Goliath estimated to be 28.9 meters tall.

Giant waves in the open ocean were once thought to be exaggerated "sea stories," but extremely high, steep, "rogue" or freak waves are now well documented. The existence of numerous offshore oil platforms has made

Wind, a long fetch—the distance a wave travels unimpeded—and a steep shoreline sculpt Tahiti's ideal surfing waves.

Internal Waves

A rogue wave—also called a freak, monster, killer, and extreme wave—was likely the subject of Japanese artist Katsushika Hokusai's "The Great Wave off Kanagawa" woodblock print in the 1800s. Those who live to describe these waves echo explorer Ernest Shackleton's 1916 description of "a mighty upheaval of the ocean." Internal waves are another large and globally ubiquitous way the ocean moves—and scientists are gaining new understanding about their nature and occurrence. Triggered when energy is applied to an interface between different-density water layers, internal waves sometimes rise much higher than surface waves—perhaps 60 meters high. Tidal surges—or contact with undersea mountains or ridges—can set an internal wave in motion. When a temperature/density phe-

"The Great Wave off Kanagawa"

nomenon, like a thermocline, is widespread, internal waves can move thousands of kilometers, much as surface waves do. The roles of internal waves in transferring heat, energy, and nutrients are likely of great significance. And they may benefit ocean life. In 2019 a quantitative analysis by scientists from five institutions, led by Alex Wyatt at the Hong Kong University of Science and Technology, showed how the cooling action of internal waves may protect coral reefs from damage in warming waters. But there are challenges in identifying, mapping, and predicting internal waves. While current research includes generating the waves in test tanks and modeling theoretical behavior, the laboratory exercises are not yet matched by knowledge of what happens in the complex reality of the ocean.

possible the accurate documentation of thousands of extreme waves that surge without warning. In her book *The Wave,* Susan Casey reports on observations made from the Statoil drilling platform *Draupner,* 161 kilometers offshore from the tip of Norway. On New Year's Day 1995, a laser wave recorder documented 11.5-meter waves powered by the convergence of two storms. They were impressive, but, Casey noted, "At three o'clock in the afternoon . . . a 25-meter wave came careening over the horizon and walloped the rig at 72 kilometers per hour . . . It was the first confirmed measurement of a freak wave, more than twice as tall and steep as its neighbors, a teetering maniac ripping across the North Sea."

The rig was damaged, but all aboard came through unscathed. The crew aboard the *Ocean Ranger,* a 121-meter-long, 31-meter-high oil platform located 274 kilometers off the coast of Newfoundland in 1982, were not as fortunate. Though the rig was built to withstand 34-meter seas and 185-kilometer-per-hour winds, when struck by a massive rogue wave it collapsed and sank with no survivors. According to Casey, every year, on average, more than two dozen ships sink or otherwise go missing, taking their crews with them. Freak waves are thought to be the cause of at least some of these disappearances.

A very tall open-ocean wave smashed into the British research vessel R.R.S. *Discovery* in 2000 during an expedition aimed at taking rare winter hydrographic data. According to co-chief scientist Penny Holliday, during a 12-hour period on February 8–9, the ship endured a total of 23 waves that exceeded 20 meters, peak to trough, diminutive compared with one that was more than 29 meters tall. Though the ship was damaged and some injuries were suffered, Discovery returned to port with all hands safely on board. An even taller rogue wave—at 34 meters high—reportedly hit the Navy-chartered tanker *Romapo* in the Pacific in the 1920s. Although unpredictable, rogue waves seem to occur where strong winds move against strong ocean currents. A team of scientists from Australia, Belgium, Italy, and the U.K. are using a test tank and computer simulations to understand how ocean winds can generate spontaneous freaks, to better forecast them.

Data collected between 1985 and 2018 show that the wind intensity around the globe has increased between one and two centimeters per second per year, according to a report in *Science* magazine by Colin Barras. That doesn't sound like much, but windier seas are reflected in an increase in average wave height, a trend that will result in stormier seas and rougher waves. A 2019 report by Ian Young and Agustinus Ribal of the University of Melbourne shows that the magnitude of extreme wind and wave events (those in the 90th percentile) has increased significantly in the Southern Ocean. Extreme wind speeds in the ocean around Antarctica have increased by 1.5 meters per second or 5.3 kilometers per hour, an 8 percent jump, and wave height has increased by about 5 percent. Planetary warming underpins these trends, changes that are based not on hearsay or guesses but on the evidence of four billion observations of wind speed and wave height collected by 31 satellites and cross-checked with 80 ocean buoys that collect similar data.

VISIONARIES

Susan Casey
Rogue Seeker

. . . .

"We can split the atom, decode our own genome, go to Mars, but we didn't even know that you can have 100-foot waves in 30-foot seas and that 850-foot ships were being taken out by these waves until 1995," ocean journalist and explorer Susan Casey has said about her extraordinary story in *The Wave: In Pursuit of the Rogues, Freaks, and Giants of the Ocean.* A lifelong competitive swimmer, she was compelled by curiosity to explore and write about the largely unknown ocean depths. Her trademark: Take readers on the ride of a lifetime via in-depth scientific and environmental research about ocean extremes. She ventured into 21-meter storm waves with extreme surfer Laird Hamilton to write *The Wave;* lived in a neighborhood of great white sharks for *The Devil's Teeth;* swam in open water with a pod of dolphins for *Voices in the Ocean;* and investigated the effects of toxins in massive plastic gyres for the *Best Life* magazine article "Our Oceans Are Turning to Plastic, Are You?" All to inspire her readers with the ocean's extraordinary power and influence over us, including the millions who will never set foot in it.

. . . .

OPPOSITE: A freighter churning through heavy seas is on the alert for rogue, or monster, waves—random walls of water in the open ocean, sometimes rising 30 meters.

Really Big Waves

T sunami, a Japanese term for harbor wave, sounds rather peaceful, but the towering walls of water evoked by that name are anything but. Earthquakes, volcanic eruptions, landslides, and asteroid impacts in the ocean can trigger these long but low waves that travel across the ocean at great speed, largely undetected until they reach shallow water, where they may grow into powerful giants.

On the morning of December 26, 2004, the Indo-Australian tectonic plate thrust under the Eurasian plate, releasing stress accumulated over hundreds of years and generating one of the largest and longest earthquakes ever recorded, with a magnitude of 9.1. Devastation from the earthquake was widespread throughout the island of Sumatra, and reports of ground shaking came from places as far away as India to the west and Thailand to the north. The Sumatran earthquake caused an area approximately 1,200 kilometers long by 100 kilometers wide and lying 2,000 meters under the waters of the Indian Ocean to rise by an estimated 20 meters, lifting the ocean surface by a similar amount. A few hours later, tsunami waves traveling at speeds up to 800 kilometers per hour rose as much as 30 meters high along the coast of Sumatra. Impacts resounded around the Indian Ocean and even affected nesting seabirds on Midway Island in the Pacific.

A volcanic island, Krakatau, which had emerged from the sea eons ago in the channel between Java and Sumatra, literally blew apart in August 1883 with a series of explosions estimated to exceed 200 megatons. The blasts were heard 3,200 kilometers away, and only a third of the volcano was left above sea level. The eruptions created nine enormous tsunamis, and walls of dark water as tall as 40 meters engulfed the northwestern tip of Java. The cloud of ash ejected into the sky rose 80 kilometers and encircled the globe, blocking the sun's rays and cooling the world for five years. Other volcanoes around the world, now quiet, could erupt in the future with comparable or greater impact.

On July 9, 1958, Alaska's Fairweather fault slipped, causing 30.58 million cubic meters of rock to slide into a narrow inlet, Lituya Bay. The resulting landslide-induced wave, a thundering rush of water 517 meters high, swept down the length of the bay.

Wave propagation models are used in combination with real-time data reported by a network of 39 tsunami sensors called DART (Deep-Ocean Assessment and Reporting of Tsunamis), strategically located in the major ocean basins to monitor the sea surface every 15 seconds for potential tsunami-inducing changes. In the span of a human lifetime, or even over several generations, tectonic movements and other grand geological processes that have, are, and will continue to shape the planet may go unnoticed as humans complacently build great cities in vulnerable coastal and seismically active regions. Nothing humans can do can revoke the laws of nature, but armed with knowledge, there are ways to anticipate earthquakes, tsunamis, and volcanic eruptions and to prepare for the consequences.

Tsunami Detection Systems

When a magnitude 7 earthquake rocked Anchorage, Alaska, on November 30, 2018, a NASA/NOAA tsunami detection prototype alerted tsunami warning centers to plan for an incoming wall of water. None came, but local communities had time to prepare. Underwater earthquakes can spawn massive tsunamis, which travel up to 800 km/h, undetected on the open ocean. The waves grow when they hit shallow water and can pulverize coastlines hundreds or thousands of kilometers away. Until the 21st cen-

tury, the only warning was the suction from the approaching tsunami, which lays bare a harbor or shoreline about five minutes before the water strikes. The NASA/NOAA system is one among many global efforts from Rome to Melbourne to Berkeley, California, to develop fast responses to earthquakes and their tsunamis. GPS real-time networks along with other instruments measure quakes; determine tsunami direction, speed, and size; and send immediate alerts—giving 20 or more extra precious minutes for life-saving measures. Along with community evacuation plans, tsunami detection systems can help save thousands of lives worldwide each year.

ABOVE: On Chile's Pacific coast, a sign points to the tsunami evacuation route. LEFT TOP: In 2011 a tsunami triggered by a sub-Pacific earthquake leveled Japan's Rikuzentakata. LEFT BOTTOM: Only a mosque stood in an Indonesian village decimated by an Indian Ocean tsunami in 2004.

Dynamic Duo:
Water & Life

I n the sea as on the land, a great many organisms spend most of their lifetime anchored in place. Kelp forests, coral reefs, oysters, and barnacles, as well as trees, moss, and lichens, come to mind. But in the sea, as well as on the land, currents—whether wind or water—serve as highways, facilitating travel for plants and animals, as well as transporting warmth and cold, water, oxygen, and nutrients. Great migrations of life are shaped by temperature, sunlight, abundance of food or lack of it, and the need to secure suitable places to mate and provide for the survival of future generations.

Jellyfish and other planktonic organisms tend to go with the flow, drifting to wherever currents and tides take them. But they, along with small fish and even tiny copepods, have built-in modes of propulsion that enable them to move within the space they occupy, stirring things up on a small but locally significant scale. These miniature migrations distribute nutrients and influence the nature of the water where they occur. The chemical and physical properties of plankton-filled water are different from those of lifeless salt water. Measurable drag can occur on ships motoring through gelatinous masses of jellyfish.

Notwithstanding the impressive long-distance flights of some seabirds, the largest and longest migrations on Earth actually take place in the open sea.

Some, such as the multiyear pathways of sea turtles and annual journeys of some species of whales, sharks, and seals, cover thousands of kilometers each year. Stanford University scientist Barbara Block and her colleagues have tracked individual great white sharks from California to Japan and back, as well as to a gathering place for these animals, dubbed "the White Shark Café," more than a thousand kilometers from any island or continent.

LEFT: On the Seychelles' Bird Island, mature sooty terns come to breed after traveling continuously for up to five years.

ABOVE: Legions of krill migrate vertically in the water column while Pacific sardines (below) migrate horizontally from Southern California to Canada and back.

MISSION BLUE

Protecting Life in a Moving Ocean

Ocean life moves—drifting with currents or swimming against them—to migrate, feed, and breed. But those natural processes are increasingly disrupted by shipping and fishing. Since the mid-1990s, Woody Turner has been building the biological diversity and ecological forecasting programs at NASA. Through satellite data combined with records from animals with electronic tags, his team can make estimates of ocean currents to track animals along certain highways. This allows NASA to alert ships to adjust their speeds and routes and fisherman to take care casting their nets, to avoid the migrating wildlife. Thanks to the efforts of Turner's group and others, populations of endangered species are on the rise, including blue whales along California, North Atlantic right whales off the northeastern United States, and sturgeon along the Delaware coast.

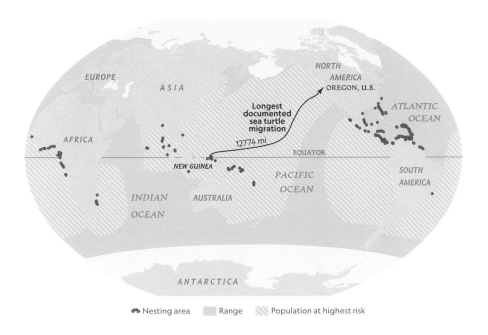

TURTLE MIGRATION

From Indonesia to Oregon in the United States, a leatherback sea turtle migrated 12,774 miles across the Pacific in 2008, a record among vertebrates.

Zoologist Archie Carr spent years puzzling out the migration route of green sea turtles in the Atlantic Ocean from the time hatchlings left nesting beaches, following ancient ocean pathways governed by currents, temperature, and sources of food, until years later, when they returned to the same beaches as adults, ready to perpetuate the next generation of turtles. Cod highways, now mostly gone, once formed visible columns through the waters of the North Atlantic as large, old fish led migrations to favored places for food and spawning. Similarly, once robust migrations of bluefin tuna, visible from ships and aircraft, took advantage of ocean currents, changing water chemistry as they moved, consuming smaller fish, and leaving behind trails of nutrients that in turn powered blooms of phytoplankton.

But far and away the greatest migrations take place over distances measured in meters, not kilometers. Vertical migrations from 100- to 1,000-meter depths are achieved daily by the most abundant vertebrate on Earth, a finger-length fish called *Cyclothone* (or, more commonly, bristlemouth), together with legions of other small fish, crustaceans, jellies, squid, and other invertebrates. First noted by scientists aboard H.M.S. *Challenger* during their global assessment of the ocean in the 1870s, the massive layer of small creatures that thrive out of sight of surface dwellers was not more clearly defined until the 1950s when the U.S. Navy found with sonar echosounders what appeared to be a "false bottom" of the sea at 100 to 1,000 meters, in places that were known to be much deeper.

The chemical and physical character of the ocean provides the framework for what kinds of life can live and move where—but, in turn, life moving through the ocean shapes the ocean's chemical and physical nature.

VISIONARIES

Kakani Katija
Bioinspired

For this National Geographic Emerging Explorer and former member of the U.S. International Figure Skating Team, the ocean holds the future. A background in aeronautics and astronautics and a Ph.D. in bioengineering have led Kakani Katija to work as principal engineer at California's Monterey Bay Aquarium, where she heads the Bioinspiration Lab. The techniques and tools her team develops are inspired by life and their movements in the deep sea. "There is a need for better observational tools and techniques," she says, "so now my efforts have shifted to trying to provide those for the community." Her projects include developing an unmanned vehicle called the *Mesobot,* whose stereo cameras noninvasively track small organisms for more than 24 hours. For larger, soft-bodied invertebrates, an electronic tagging package called ITAG measures both animal behavior and the surrounding environment, giving researchers a unique, integrated picture. Together, these underwater technologies can be used to understand such ocean phenomena as nutrient cycling, ocean mixing prompted by moving marine life, and the worlds of deep-sea jellies.

OPPOSITE: A green turtle swims amid a school of bigeye jacks near its nesting grounds on Malaysia's Sipadan Island.

GREAT EXPLORATIONS

Perils of the Poles

Antarctic explorer Ernest Shackleton's *Endurance* was crushed in pack ice. The crew survived.

Prizes sought by countless explorers, the poles beckon from the center of the Arctic's icy ocean and Antarctica's glacial continent. Places of promise and peril, these frigid waters—the Arctic and Southern Oceans—and surface winds generate the major currents that fuel ocean motion, energy, climate, and, in turn, life. But the same forces also form and move glaciers, icebergs, and pack ice that can crush ships and leave crews stranded in Earth's harshest conditions.

Native peoples who have inhabited the Arctic for millennia know its ways, but early outside explorers—including Greek merchant Pytheas of Massalia in 330 B.C. and Viking settlers a thousand years later—were awed by its lush land and then driven out by ice.

By the 1500s navigators focused not on settling but on finding a northwest

"The lure of the Arctic is tugging at my heart," wrote Matthew Henson (above center), with Inuit guides Ootah, Ooqueah, Seeglo, and Egigingwah. With Robert E. Peary (right) they claimed the North Pole in 1909.

passage for trade through the Arctic sea ice from Europe to Asia. Dutch explorer Willem Barents, and later British navigator Henry Hudson, died trying, their ships crushed by shifting ice. A similar fate met the U.S.S. *Jeannette* in 1879 and inspired Norwegian scientist-explorer Fridtjof Nansen in 1893 to build the *Fram*. Its egg-shaped hull could be pushed to the surface—not crushed—by encroaching pack ice. It worked: Ice carried the ship for three years, an example for future floating Arctic research stations.

At last the Northwest Passage was mapped by Norwegian Roald Amundsen from 1903 through 1905. Then, in 1909, Americans Robert Peary and Matthew Henson, led by native Inuit guides, reported reaching the North Pole by land. A sea route remained unconquered until 1958 when the nuclear sub U.S.S. *Skate* motored under the ice and surfaced at the top of the Earth.

Meanwhile, to the south, British scientist Edmund Halley in 1700 had identified the major Antarctic Circumpolar Current (ACC), as did British navigator Captain James Cook in his 1772–75 global expedition. Not until 1911 did Roald Amundsen sail his *Fram* into the Southern Ocean's Bay of Whales and then trek overland for 1,287 kilometers to the South Pole. In 1912 Ernest Shackleton tried to follow. But after his *Endurance* was crushed by ice, he spent two years bringing his crew safely home.

Beyond adventure, these expeditions allowed scientists to explore why the poles matter to the planet. Since 1937, scientists have studied circulation, wave propagation, ice formation, and underwater acoustics, using, in *Fram* fashion, drifting research stations like the British glacier-built Halley Research Station in Antarctica. Today autonomous buoys, aircraft, and satellites, including NASA's laser-altimeter-equipped ICESat-2, monitor both poles. We now know that the Arctic's complex circulation impacts the global food web and climate; that the Southern Ocean's ACC is Earth's largest wind-driven current, connecting every ocean basin; and that, together, they help maintain planetary balance. Today, the peril of the poles has changed—from reaching them to protecting them.

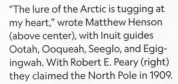

TIME LINE

To the Ends of the Earth

ca 330 B.C.: Pytheas of Massalia describes polar ice north of Britain.

1400s–1500s: Vikings settle Greenland; cold, sea ice later cause retreat.

1594–1610: Willem Barents, then Henry Hudson, seeks Northwest Passage.

1699: Edmund Halley identifies Antarctic Circumpolar Current (ACC).

1725–1742: Mikhail Lomonosov maps North Pole in Arctic Ocean.

1839–1843: James Ross improves description of ACC.

1878: Baron von Nordenskiöld draws Northeast Passage map across Arctic seas of Russia.

1879–1881: U.S.S. *Jeannette* crushed by Arctic sea ice.

1882–83: First International Polar Year (IPY); 22 nations participate.

1892–96: Fridtjof Nansen aboard the *Fram* tests Arctic sea ice motion.

1906: Roald Amundsen is first to navigate Northwest Passage.

1909: Robert Peary, Matthew Henson, and Inuit guides are first to reach North Pole.

1910: Russian Arctic Ocean Hydrographic Expedition uses icebreaker research vessels in Arctic.

1911: Amundsen is first to reach South Pole.

1912–15: Ernest Shackleton's *Endurance* is crushed by Antarctic ice.

1932–33: Second International Polar Year (IPY); 40 nations participate.

1937: Russia sets up first Arctic scientific drifting station, *North Pole-1*.

1957: International Geophysical Year (IGY) uses war technology for ocean research; 67 nations participate.

1958: U.S. nuclear sub *Skate* is the first to surface at North Pole.

1961: Antarctic Treaty signed by 44 nations, agreeing to scientific cooperation and no exploitation.

1970–2010: International drifting research stations monitor Arctic.

2007–2009: Third International Polar Year (IPY); 60 nations participate.

Present: High-tech instruments on autonomous research buoys, aircraft, and satellites monitor polar change.

I LIFE

PART TWO

IN THE OCEAN

2

SWIMMING
IN THE
SEA OF LIFE

....

Chapter Four

"WE ARE ALL MADE OF STAR STUFF . . . IF YOU WISH TO MAKE AN APPLE PIE FROM SCRATCH, YOU MUST FIRST INVENT THE UNIVERSE."

—CARL SAGAN, *COSMOS*

Throughout most of the history of humankind, much of the ocean was thought to be too dark, too cold, and too deep to support life at all. Surprise, surprise! Life exists everywhere in the sea, from the surface to several kilometers beneath the ocean floor. Ocean life follows the water into cracks leading to utter darkness and temperatures hot enough to kill all but microbes and tiny wormlike animals. Nearly all of the major divisions of life are in the sea; only about half are terrestrial. When you swim in the ocean, you are immersing yourself in the history of life on Earth.

Every creature has a comfort zone—a range of environmental circumstances that makes it prosper, or that it finds suitable or at least tolerable. Some like it hot; some cold. Some need a place to anchor; others never stay put. "Home range" defines the area where a species lives. For some it's a few square kilometers, while others range from pole to pole or across entire ocean basins. Animals need food, water, and oxygen. Plants require water, oxygen, carbon dioxide, sunlight, and trace elements. And all forms of life depend on connections to one another.

Globally, about two million individual species have been identified and dignified with proper scientific names. But it is estimated that there may be more than twice that number yet to be discovered on land and in freshwater areas—and even more in the ocean. It is counterintuitive to think that the ocean—97 percent of the biosphere—contains fewer species than are known to occur in the land-based 3 percent. Few bacteria, viruses, and fungi were thought to live in the sea before ways were developed to identify their enormous diversity and widespread occurrence from the ocean's surface to deep within the seabed. New methods of identifying species by analyzing their DNA and by assessing DNA in water samples have resulted in finding a great number of organisms that have not even been given names. Undescribed fish, corals, and even large squid have been discovered recently in the deep sea.

Biologist Jesse Ausubel, whose vision helped create the Census of Marine Life, astutely sums up the current state of knowledge about ocean life with this observation: "The deeper we go, the less we know, but the more new discoveries we find."

RIGHT: Standing on the tips of its tentacles, an octopus nestles amid horse mussels and hundreds of bristling brittle stars. PAGES 130-131: Tiger anemones sway in the Philippines' Bohol Sea. PAGES 132-133: Hovering vertically, sperm whales take 10- to 15-minute naps over 7 percent of their day.

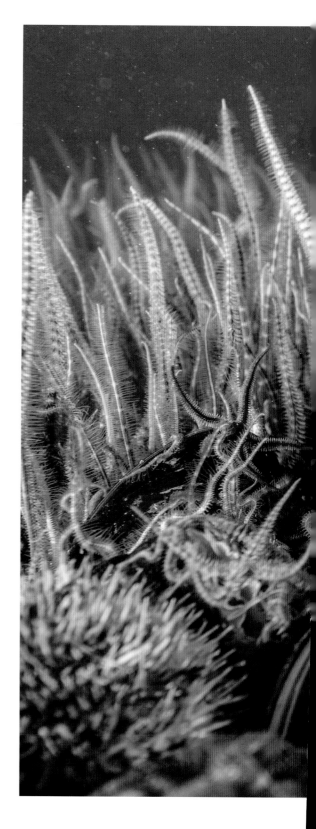

In the Beginning

Although the origin of life is shrouded in mystery and fraught with unknowns, it appears that it all began in the depths of the sea. Deep within the ocean, microbes have succeeded in deriving sustenance and establishing wide-ranging use of the 54 or so elements making up the chemicals in the surrounding seawater. Chemosynthesis is the term for the way many of these small creatures make a living, tapping into energy from hydrogen sulfide or methane in a manner that is likely to have prevailed hundreds of millions of years before life on land.

The most wonderful discovery about life in the sea is that, in effect, all of life on Earth exists because of it—and continues to depend on it. Water is essential, but over the four billion or so years that life has existed in the sea, unicellular organisms morphed water plus carbon, nitrogen, sulfur, and phosphorus into the 30 or so basic molecules that are characteristic of living things.

Prior to about 3.4 billion years ago, when photosynthetic bacteria first appeared, chemosynthetic microbes had the planet all to themselves. These organisms derive sustenance from chemicals in rocks or the surrounding seawater and respire anaerobically, or without oxygen. Even after the appearance of cyanobacteria (often called blue-green algae), for more than three billion years life apparently consisted entirely of bacteria and archaea—prokaryotes that do not have an organized nucleus but do have the characteristics that define life compared with, say, a rock. They move, consume energy, grow, change, maintain functions, respond to the environment, reproduce, and pass traits to their offspring.

For more than two billion years, Earth's warm, shallow seas were dominated by lumpy rock formations, stromatolites, consisting of sticky tangles of threadlike, single-celled cyanobacteria, glued together with bits of sand and sediment into solid, firm mounds. These layered, living rocks eventually gave way to different formations called thrombolites, a term meaning "clotted stones." Microbial ecologist Virginia Edgcomb and geobiologist Joan Bernhard from Woods Hole Oceanographic Institution have found evidence that the appearance and proliferation of single-celled protists, foraminifera, displaced filamentous cyanobacteria as the living glue in these rocky formations.

Photosynthesizers within stromatolites and plankton in the open sea steadily, gradually, released oxygen into the atmosphere and the ocean. Eventually, a level was reached that enabled larger forms of life to develop. The first evidence of multicellular life occurred less than a billion years ago, likely coincident with the displacement of atmospheric carbon dioxide by sufficient oxygen to support larger forms of life. The energy that enables creatures to grow large is derived from the combustion of glucose as fuel in combination with oxygen. All living things burn energy through respiration,

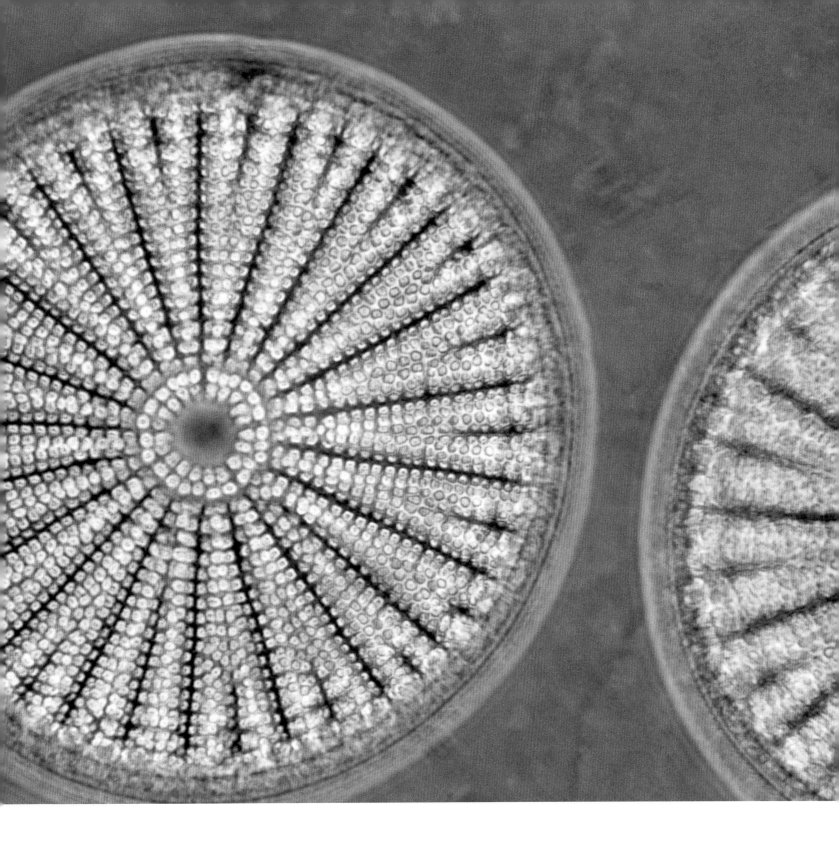

some utilizing oxygen and others, such as sulfur bacteria and archaea, doing so anaerobically.

Oxygen respiration—the result of consuming oxygen and glucose and releasing carbon dioxide and water—is essentially the reverse process of photosynthesis. Compare the chemical formulas:

Photosynthesis (making energy): $6H_2O + 6CO_2 = C_6H_{12}O_6 + 6O_2$
Oxygen Respiration (the burning of energy): $C_6H_{12}O_6 + 6O_2 = 6CO_2 + 6H_2O$

Photosynthesis and oxygen-driven respiration are processes that have made all the difference between the planet that once existed—populated

Microscopic but mighty, the lacey-looking diatoms shown here are among the thousands of species that are so abundant in sunlit seas that they influence Earth's biogeochemical processes.

with an enormous diversity and abundance of life but with no individual larger than the punctuation on this page—and the planet that we know as a place with giant redwood trees, 30-meter-long whales, and more than seven billion humans. The ancestors of all that we think of as "life on Earth" are microbes. And they in turn are deeply rooted in stardust.

The basic recipe for life is found in DNA (deoxyribonucleic acid), the genetic material that gives instructions to organisms from archaea to elephants. DNA is needed to develop, live, and reproduce. The ingredients of DNA consist of nucleotides—molecules of cytosine, guanine, adenine, and thymine, each containing nitrogen, phosphorus, and sugar consisting of carbon, hydrogen, and oxygen. Helical strands of DNA are contained in genes that in turn are arranged within chromosomes. Bacteria and archaea typically have a single chromosome, while yeast has 32. Sea otters have 38, dogs have 78, and humans have 46; our closest relatives, chimpanzees, have 48. A tiny, shrimplike Hawaiian amphipod crustacean has the same number of chromosomes as humans, but the genetic material, the genome, is clearly very different. Genomes determine the distinctive characteristics that make every human, every cat, every tree, tuna, or individual shrimp unlike any that has ever been or ever will be. The remarkable consistency inherent in the chemistry of life is countered by its infinite variety.

Throughout time, marine microbes and the glorious diversity of multicellular creatures in the sea have developed distinctive ways to utilize the many materials in the mineral-rich soup that surrounds them. Vertebrates, including humans, use iron-based compounds in their blood to transport oxygen; but echinoderms utilize a vanadium compound. Horseshoe crabs rely on a copper compound to transport oxygen, resulting in blood that is a delicate shade of blue. Diatoms build their glassy shells using silica extracted from the surrounding sea. Many creatures, from single-celled coccolithophorids, foraminifera, and reef-building corals to giant clams,

> "THE GREAT CARTOGRAPHERS OF TODAY . . . ARE MAPPING THE GENETIC CODE."
>
> —JUAN ENRIQUEZ,
> *AS THE FUTURE CATCHES YOU: HOW GENOMICS & OTHER FORCES ARE CHANGING YOUR LIFE, WORK, HEALTH & WEALTH*

DEEPER DIVE

Ancient Dinos of the Deep

When in 2019 paleontologists finally dislodged the complete skeleton of an *Elasmosaurus*—an ancient aquatic reptile—from an island in Antarctica's Southern Ocean, they had added a near-13-metric-ton piece to the puzzle of life in the ancient sea. As dinosaurs walked the land during the Cretaceous period, the *Elasmosaurus*, of the plesiosaur reptile family, swam the sea below, alongside—and sometimes in pursuit of—fellow reptiles mosasaurs and ichthyosaurs, and other ancient marine life. Imagine a plump manatee's body with a giraffe-like neck carrying a snake-like head and propelled by four webbed flippers. And long, prey-impaling teeth. Now imagine it the length of a school bus. Among the biggest sea creatures of the Cretaceous, *Elasmosaurus* would have con-

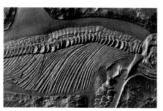

Fossil of ichthyosaur *Stenopterygius*

sumed massive amounts of other marine life to survive. These details help paleontologists draw a picture of the Cretaceous sea as a vibrant, animal-rich ecosystem until mass extinction, likely from an errant asteroid, ended that period. Like its fellow plesiosaurs, including crocodile-headed *Nothosaurus* and 12-meter-long *Styxosaurus* (six meters were neck!), *Elasmosaurus* may have hunted dolphin-like *Ichthyosaurus;* three-meter-long *Protostega,* among the largest sea turtles ever recorded; saber-toothed herring *Enchodus;* massive shark *Squalicorax;* and the sail-finned, mollusk-crushing *Bananogmius.* In the end, a fierce *Mosasaurus* may have laid the Antarctic *Elasmosaurus* to rest—or perhaps it fell victim to a fiery asteroid that fell from the ancient sky.

extract carbonate ions to build their characteristic shells. Most shell-bearing mollusks rely on calcium carbonate for this purpose; but a deep-dwelling snail was recently discovered around hydrothermal vents in the Indian Ocean that, in addition to calcium carbonate, armors itself with overlapping scales made of iron. Known as the scaly-footed snail, or sea pangolin, the little creatures have oversize hearts, likely to accommodate the oxygen needs of the symbiotic bacteria that live in their bodies and provide food in exchange for a safe place to live. Such interconnectedness among species is now known to be the usual way of life on Earth, each individual making its mark and each influenced by the existence of others. Microbiologist Lynn Margulis noted, "Life did not take over the world by combat, but by networking."

Cocooned in transparent spheres, larval horseshoe crabs continue a 400-million-year lineage now represented by just four species that face imminent extinction.

Living in Layered Liquid

From the surface, the ocean appears much the same all over the world. But underwater, life is neither randomly nor uniformly distributed; it is based on the needs and tolerance of individuals, species, and ecosystems. There are zones of life in the sea, but the edges are fuzzy and the layers move dynamically within the ocean's superhighways and corridors. The ocean has liquid islands bounded by light, temperature, salinity, pressure, oxygen, pH, substrate, sounds, nutrients, contaminants, and other conditions as formidable and confining as walls. Only humans, by harnessing various technologies, are able to go from pole to pole and from the surface to the deepest sea, crossing boundaries that inhibit or prohibit other species. Without special arrangements, however, we are as vulnerable under the sea as a fish out of water.

Some animals are notorious for their ability to cross boundaries of distance, depth, light, pressure, and temperature. Whales, sea turtles, some sharks, and tuna range widely, covering many degrees of longitude and latitude in months, diving daily thousands of meters deep to dine on squids, jellies, small fish, and crustaceans. Stowaways cling to the shells of sea turtles, to the feet and feathers of birds, the skin of fish and whales. Others move with rafts of vegetation, mats of volcanic pumice, ship ballast, and, in recent decades, flotillas of plastics and other human-generated debris carried by currents.

OPPOSITE ABOVE: Caves and overhangs in the Indo-Pacific are home for golden sweepers. OPPOSITE BELOW: Young tripodfish such as this one may swim near surface waters at night but as an adult will live in the eternal darkness of the deep sea more than 4,000 meters below.

ZONES IN THE OCEAN

Beneath its surface skin, the ocean has three main zones: photic or epipelagic, bathyal or bathypelagic (includes mesopelagic and abyssopelagic), and hadal.

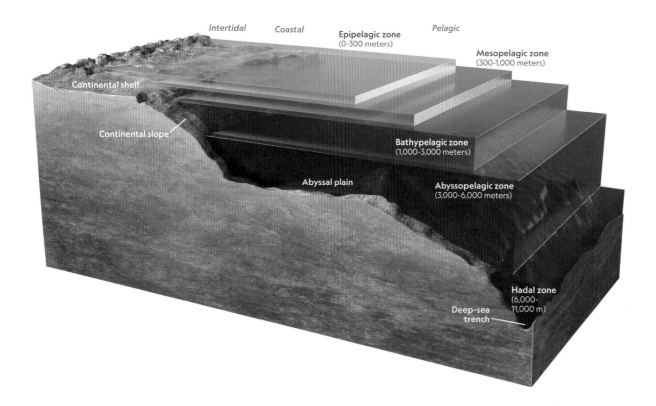

Intertidal   Coastal   **Epipelagic zone** (0-300 meters)   Pelagic

Mesopelagic zone (300-1,000 meters)

Continental shelf

Continental slope

Bathypelagic zone (1,000-3,000 meters)

Abyssal plain   **Abyssopelagic zone** (3,000-6,000 meters)

Hadal zone (6,000-11,000 m)

Deep-sea trench

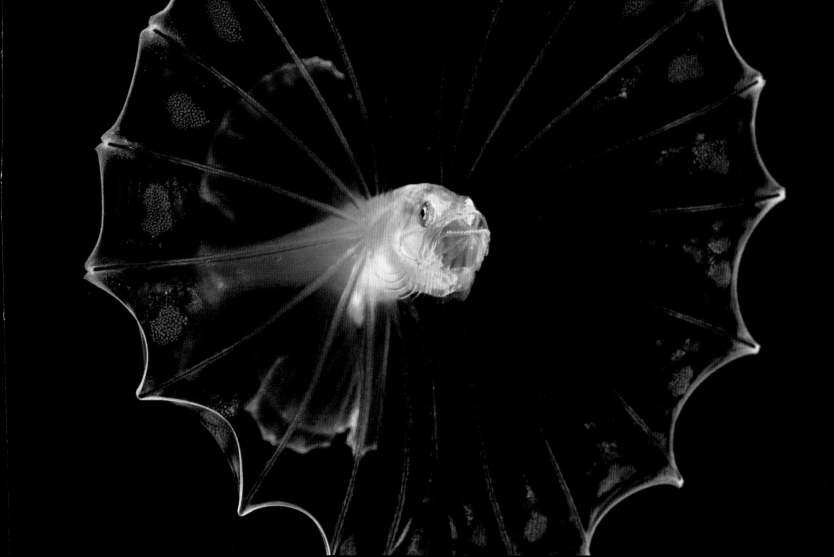

There are zones within zones dictated by requirements important to the creatures who live there but not obvious to human observers. Sponges, corals, barnacles, burrowing animals, rooted plants, and anchored seaweeds typically begin life as drifters, eventually settling down in favored places where they remain for the rest of their lives. Hard surfaces are vital for oysters, deep sand is just right for garden eels, rocky crevices are required for many small animals, and soft mud may be home for a metropolis of life.

Of overarching importance to the distribution of life in the sea are zones of temperature and light extending north and south from the Equator, as well as vertically in the sea. These zones are governed largely by the position of the sun relative to the revolving, rotating, inclined Earth. Temperature, in particular, shapes the fundamental nature of whether an organism is at home in the tropics, warm temperate, cold temperate, or polar regions. Vertically, the temperature of the ocean typically grows colder with depth. A thermocline occurs about 200 meters down, with distinctly colder water below. From there, the temperature is fairly constant to 1,000 meters, and cools to near freezing in the deep sea. Some far-ranging species move through seasons or depths to stay within favorable temperature conditions, or to reach breeding or feeding areas.

At the surface, the ocean has a "skin"—the air-sea interface—about the thickness of a human hair—where organic compounds in the water come in contact with the atmosphere. In late 2016, the Schmidt Ocean Institute focused a research expedition in the Indian and Pacific Oceans on this slim but vital zone, correlating phytoplankton distribution with the development of the microlayer and gas exchange in calm versus rough seas at 17 locations. According to the expedition's chief scientist, Oliver Wurl from Germany's University of Oldenburg, "The sea-surface microlayer plays a vital role in the uptake and release of greenhouse gases, including methane and carbon dioxide, via the ocean."

Often, layers of the ocean are defined vertically as photic (the upper 300 meters or so), bathyal (300 to 6,000 meters), and hadal (below 6,000 meters), but the bathyal zone is often subdivided, giving a total of five unique depth zones. Zones are also defined horizontally, according to the proximity to land: intertidal—the region between highest and lowest tides; coastal—the region bordering landmasses; and pelagic—the open sea.

However, zonation must reflect not only the way the world was centuries, or even decades, ago, but the reality now includes physical, chemical, and biological changes to the nature of the sea, mostly imposed by human actions. Although indifferent to political zones and boundaries, life in the sea is challenged by changes in temperature and chemistry as well as shipping, dumping, industrial fishing, mining, drilling, dead zones, noise, military operations, wind farms, offshore aquaculture, accumulation of plastics, and other impacts. The "new normal" for ocean zonation is "no normal."

A school of razorfish, or shrimpfish, known for their flat, transparent, long-snouted bodies, dart for cover in shallow Indo-Pacific waters.

Dining in the Deep

The ocean is a dynamic, complex, eat-and-be-eaten, give-and-take realm, where the ingredients of life are passed from one organism to another in continuous, often rapid-fire exchanges, every individual making its mark on the chemistry of its surroundings. Food chains in the sea tend to be long, twisted webs that include a wide range of small animals that spend some or all of their time adrift in the open sea, dining on one another and powering phytoplankton production with their nutrient-packed excrement.

From the smallest shrimp to the largest whale, all eat and excrete matter back into the sea that serves as food for some and "fertilizer" for phytoplankton. Ultimately, when animals are eaten or otherwise die, their elements stay in the system, powering new generations. Bacteria, viruses, fungi, and small protists with a big name—Labyrinthulomycetes—break down organic matter, living and dead, and cycle vital nutrients back into the water. In nature, there is no excess and nothing is wasted in the continuous flow of life and death giving rise to new life.

Photosynthesis is the sunlight-driven process that generates sustenance for most of life on Earth. In the absence of light, chemosynthesis by microbes powers life in the deep sea, in deep soil, in marshes, in the water column, and within the tissues of other organisms. Known to be widespread but difficult to measure, the contribution of chemosynthesis to carbon capture and storage is likely to be underestimated.

Seagrass meadows, marshes, mangroves, and benthic algae, large and small, are locally important sources of food for marine animals. But far and away the largest producers of sustenance for sea life originate in the living soup of small organisms collectively referred to as phytoplankton. Numerous species are involved, mostly cyanobacteria—major players in ocean productivity for billions of years—as well as four kinds of microscopic protists. A large portion of the action is driven by diatoms, organisms that are individually very small but together have had a major role in primary productivity in the sea for at least 200 million years. Some kinds of dinoflagellates, known for producing toxic blooms and eerie displays of bioluminescence, are also key photosynthesizers. Among the best known for capturing and storing carbon are coccolithophorids, literally "rounded stone-bearers": minuscule organisms with ornate limestone shells that en masse make up vast, carbon-rich deposits in the deep sea. Another kind of protist, foraminifera, also has complex limestone shells, but many photosynthesize with adopted symbiotic algae as well as capturing other organisms by using sticky extensions that protrude through holes in their shells.

Dramatic changes to ocean nutrient cycles have come about in recent decades owing to climate change, pollution, and the unprecedented extraction of ocean wildlife by industrial fishing. Humans have become major players in ocean food webs and nutrient cycles with consequences to ocean chemistry and to the existence of life as we know it.

VISIONARIES

Gabrielle Corradino
Team Plankton

As marine scientist Gabrielle Corradino prepared for a 2017 expedition to the Gulf of Mexico as part of Team Plankton, friends were curious: Why would you spend 35 days on a ship just to filter seawater? The Global Change Fellow from North Carolina State University was undeterred. Some 236,000 milliliters of seawater and 550 filters later, she had collected millions of the microscopic zooplankton and phytoplankton that thrive in the photic, or sunlit, top layer of the ocean. Corradino's filters, each holding tens of thousands of these organisms, are used by Team Plankton to determine the range, diversity, and activities of these vital microbes and how they benefit both the ocean and humans. A 2019 National Geographic Early Career Fellow, Corradino continues the quest to understand how plankton support the marine food web and the ocean's adaptation to rising carbon levels and temperatures caused by climate change.

OPPOSITE ABOVE: Long-beaked common dolphins lead high-speed lives fueled by sardines and other tiny fish. OPPOSITE BELOW: Diatoms and other microscopic phytoplankton power ocean food webs.

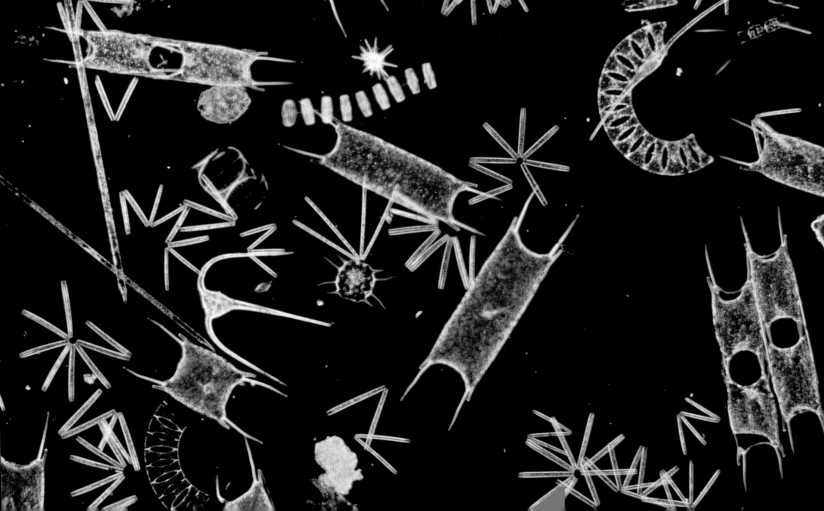

HOPE SPOT

California Seamounts

Along a hauntingly beautiful California coastal seascape of towering underwater mountains and volcanoes, seamounts shelter lush marine life. These biodiversity hot spots are fueling stations for migrating seabirds and endangered blue and gray whales, and home to rare corals that take hundreds of years to grow. Once threatened by trawling, offshore drilling, and ocean acidification, their biological treasure trove has been protected as a Hope Spot since 2019, with an eye to a thriving future.

....

A giant ray wings its way over seagrasses crowning a California seamount.

Longevity in the Ocean

How old is "old" in the ocean? For a bit of phytoplankton, a week might be a long time to exist as an individual among companions that may live for hours before becoming a snack for a small, filter-feeding copepod. Curiously, the short-lived cyanobacteria in phytoplankton are related to microbes that may exist in a dormant state for centuries or even millennia until circumstances favor their awakening. Thousands of kinds of endolithic, or stone-dwelling, organisms—bacteria, archaea, and fungi—that dwell in water-laced rocks many kilometers beneath the seafloor give new meaning to the phrase "growing old," with reproduction measured in centuries and age in millions of years. Such longevity was confirmed in 2020 by a team led by the Japan Agency for Marine-Earth Science and Technology, which reported revival of microbes dormant in deep-sea sediments for 100 million years.

A lucky copepod might survive for more than a year, but few have a chance to get old, since these tiny crustaceans are voraciously consumed by other planktonic animals as well as by baleen whales and other animals equipped to make a meal of these microscopic beings. Another group of crustaceans—85 or so species collectively known as krill—also graze on phytoplankton and in turn are sustenance for a wide range of animals. One kind that abounds in the waters around Antarctica, *Euphausia superba*, takes about three years to mature, and individuals may live for more than 10 years. Millions of tons of krill are consumed annually by marine mammals, fish, squid, birds—and, since the 1980s, by humans equipped with enormous nets that capture krill for commercial markets. Longevity for krill is now a matter of sheer luck. A lucky Atlantic lobster, *Homarus americanus*, might live to be as old as the humans dining on it.

Maturing and reproducing swiftly and abundantly are the keys to survival for species that are a primary food source for animals in the sea. Like Antarctic krill, small fish such as herring and menhaden that dine on phytoplankton take two to four years to mature and may live more than 10 years. But to reach maximum age, these so-called forage fish or bait fish must get through the hungry hordes that need them as middlemen to consume and extract energy from microscopic plankton. They serve themselves up as food in a size and form that the larger animals can find and capture. People, too, take enormous quantities of these small fish, mostly for use as fertilizer or as food for farmed animals—whose natural diet does not include fish.

The deep-dwelling Patagonian toothfish *(Dissostichus eleginoides)*—sold in restaurants as Chilean sea bass—and various species of California rockfish *(Sebastes)* have rapidly been depleted for high-end markets, and few reach their potential age of 50-plus years. A group of deep-sea fish commonly called oreos can live up to 170 years, with calculations based on radiometric dating and examination of ear bones, or otoliths. *Hoplostethus*

Umbellula sea pens anchor in deep-sea sediments in the eastern Pacific, where they dine on passing plankton.

AGE OF MATURITY

Among long-lived ocean predators such as sharks, tuna, and whales, larger, older fish produce more offspring than smaller members of the same species. Ironically, the big, old fish with the greatest potential to reproduce are also the most prized for market.

Atlantic bluefin tuna, maturity 5-10 years, life span 20+ years

Horseshoe crab, maturity 9-12 years, life span 20+ years

American lobster, maturity 5-8 years, life span 50+ years

Patagonian toothfish (Chilean sea bass), maturity 6-9 years, life span 50+ years

Australian orange roughy, maturity 27-32 years, life span 150+ years

Bowhead whale, maturity 20 years, life span 200+ years

atlanticus, known to scientists as the slimehead since 1889 and to gourmet diners as the orange roughy since the 1980s, was determined to mature late (27-plus years) and live a long time (150-plus years) only after its numbers had declined by more than half because, to humans, it tastes good.

Sharks and rays also grow and mature slowly. They have a small number of ready-to-go offspring and tend to live a long time. The whale shark *(Rhincodon typus)* is the largest and among the most wide-ranging of all fish and may live for a century or more. As many as six spiny dogfish pups *(Squalus acanthias)* grow inside their mother for two years and when born into the sea have a chance to live for a hundred years. The all-time champion for longevity among sharks—in fact, among all vertebrate animals—is a large (up to six meters), deep-dwelling (down to 3,000 meters) species related to dogfish, the Greenland shark *(Somniosus microcephalus).* A 2016 article in *Science* reported on a female thought to be 400 years old, with a potential for attaining another 100 years or so. If true, she would first have seen the ocean at about the same time that Galileo Galilei spotted the moons of Jupiter and would have reached sexual maturity 150 years later, at about the time that Captain James Cook arrived in Australia and claimed the entire country for England.

Among mammals, the oldest known is a male bowhead whale *(Balaena mysticetus),* taken by Inuit whalers in 2007. While the whale was being processed for oil, a deeply embedded stone spearhead was found in its tissues that indicated it was truly an elder whale at 211 years. Other long-lived species include the burrowing clams, geoducks, that can be 150-plus years, and red sea urchins *(Strongylocentrotus franciscanus),* at least 200 years. One Icelandic clam was named Ming after it was discovered that its life began 507 years earlier during the Chinese Ming dynasty.

Black coral *(Antipatharia)* in the Gulf of Mexico can be more than 2,000 years old, and a specimen of Hawaiian gold coral *(Kulamanamana haumeaae)* began its deep-sea existence in the year 720 B.C. An Antarctic sponge *(Cinachyra antarctica)* can live 10 times as long as an orange roughy, but it is a youngster compared with glass sponges in the frigid depths of the Southern Ocean that began growing at the time of the Neolithic Revolution, 10,000 years ago. Among the oldest of the old are the so-called manganese nodules formed by bacteria that deposit various minerals around a shell fragment or shark's tooth, forming "living rocks." More than a million years of metabolism may be invested in a still growing, potato-size nodule.

Invasive Species

....

Beneath North America's coastal waters lurk alien animal and plant species: Mediterranean green algae, zebra mussels from the Caspian Sea, lionfish from the South Pacific and Indian Oceans. They may have hitched a ride on the hull of a ship heading across the world, or were released from an exotic-animals aquarium, or were used as bait that escaped in foreign waters. Once enough members of an outside, or exotic, species establish a new territory and reproduce, the species becomes invasive. Lionfish, for instance, may release two million eggs a year. An invasive animal species can eat native plants and animals to a point of dismantling food webs and habitats, threatening biodiversity and ocean health. Global organizations are working to curb invasive species introductions and diminish populations that have already changed address.

....

ABOVE: A South Pacific native, lionfish inhabit parts of the Atlantic Ocean as an invasive species. OPPOSITE: Among Earth's longest lived mammals, the Arctic's bowhead whale can live at least 200 years.

Age of reproductive maturity | Life span

A Century of Loss & Learning

Technological advances in the 20th century accelerated the ability of people to find, take, market, and consume greater amounts of sea life than during any other era in history. But, at the same time, new technologies made possible unprecedented powers of exploration, computation, and communication. As a result, children today have access to knowledge about the ocean that even the wisest people *could not know* a thousand, a hundred, or even 50 years ago. Sonar systems developed to track submarines in World War I were later turned into effective "fish finders." Improved navigation, communication, weather prediction, increased vessel size and speed, enhanced refrigeration, and substantial government subsidies combined to efficiently industrialize the killing of ocean wildlife. The same technologies also made possible knowledge that there are limits to what can be taken out of or disposed into the sea without disrupting basic planetary functions.

More whales were killed for food and products early in the 20th century than during all preceding time, but when extinction threatened, new awareness came into focus about not just *what* whales are, but also *who* they are as individuals, as families, and as cultures. Zoologist Roger Payne's long-term research on southern right whales confirmed that each has a distinctive face and behavior. Work by various scientists confirmed that whales have a complex social structure and sophisticated communications. By the 1980s, the value of whale watching exceeded the value of whale killing, and widespread public support for protecting whales helped bring about a global moratorium on commercial whaling in 1986.

Over time, New Yorkers took so many billions of oysters from nearby waters that fewer than one percent remained by the beginning of the 21st century. In 2008 concerned citizens initiated a "Billion Oyster" restoration project, now involving thousands of oyster-loving volunteers. In the mid-1800s, San Franciscans' appetite for seabird eggs nearly caused the extermination of nesting colonies on the nearby Farallon Islands, but today the birds and their nests are fully protected there, and in many countries globally. There is a long history of humans consuming sea turtles and their eggs, but since the 1970s, protection has grown globally. Although numbers overall are far fewer than 500 years ago, there are more sea turtles now than there were 50 years ago. The Pacific island country Palau regards the tourism attraction of a live shark to be worth a thousand times more than the one-time sale of a dead one. Technology's two-edged sword continues to cut both ways, enabling rapid exploration and exploitation. However, new insight about the value of the ocean as a living system that supports all life on Earth, coupled with awareness of the decline of vital ocean systems, is leading to changes in the relationship between people and the living ocean.

VISIONARIES

J. Frederick Grassle
Census of Marine Life

. . . .

Though he claimed to be "comfortable with ambiguity," Frederick (Fred) Grassle ensured that his work was crystal clear in mission and results. Starting his career as a marine biologist at Woods Hole Oceanographic Institution, where he focused on deep-sea hydrothermal vent ecosystems and hosted the 1977 National Geographic television documentary *Dive to the Edge of Creation,* revealing the vent-rich Galápagos Rift, Grassle went on to establish the Institute of Marine and Coastal Sciences at Rutgers University. There he and colleague Jesse Ausubel brainstormed a global "Census of Fishes," launched as the Census of Marine Life in 2000. A decade, $650 million, 540 expeditions, 2,700 scientists, and some 6,000 newly discovered species later, the Census was named the largest program in marine biology history and included the first global database for marine life. Though formally closed in 2010, the Census continues to inform ocean research and inspire new discovery.

. . . .

The second largest cuttlefish species, the broadclub, hypnotizes prey by flashing and rippling its skin and waving clubbed arms.

HOPE SPOT

Palau

Its 340 islands extending 113 kilometers across the Pacific's blanket of blue, Palau is a diver's paradise, rich with coral reefs, blue holes, and submerged tunnels. Named a Mission Blue Hope Spot in 2009, Palau became the first nation to fully protect sharks, and in 2015, 80 percent of its entire exclusive economic zone (EEZ) was designated a no-take zone, banning fishing and mining. In just a decade, sharks and other populations have increased as much as fivefold.

In scarlet splendor a Gorgonian sea fan graces a Palau reef.

WONDERS: THE DIVERSITY

To bring order to classifying Earth's many forms of life, thousands of scientists contributed to the Catalogue of Life, published in 2015 in the journal *PLOS ONE*, with ongoing digital updates. The online Encyclopedia of Life, managed by the Smithsonian Institution, provides data and images of all known species. Data concerning life in the sea is assembled in the World Register of Marine Species (WoRMS), a digital database hosted by the Flanders Marine Institute in Belgium. Below is a glimpse of some of the large and growing inventory of creatures that together make Earth habitable for humans and the rest of life in the 21st century. Here, three great domains of life are noted: Archaea, Bacteria, and Eukaryota. Within the Domain Eukaryota, five kingdoms are recognized with a selection of phyla, the next level in the classification hierarchy. Phyla are divided into classes, classes are broken into families, families into orders, and orders into genera. Every genus has at least one species, a unit of life defined as a group of organisms that share similar genetic makeup, can mate with their own kind, and produce fertile offspring.

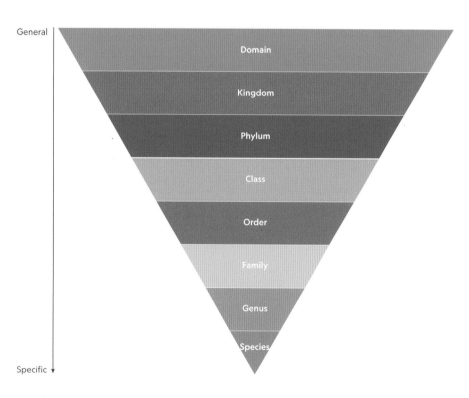

General

| Domain |
| Kingdom |
| Phylum |
| Class |
| Order |
| Family |
| Genus |
| Species |

Specific

HOW ANIMALS ARE CLASSIFIED

Taxonomy classifies each living thing using several levels. Consider the bowhead whale: in the domain Eukaryote; kingdom Animalia; phylum Vertebrata; class Mammalia; order Cetacea; family Balaenidae; genus *Balaena*; species *mysticetus*. OPPOSITE: California's intertidal zone shelters living treasures from crabs to starfish.

OF OCEAN LIFE

VISIONARIES

Edward O. Wilson
Preserving Biodiversity

Often called "a Darwin for the modern day," Edward Osborne Wilson began his career studying animal behaviors, specifically of ants. His groundbreaking discoveries have since shed light on all members of the animal kingdom, from bacteria to blue whales. As a young professor at Harvard in the 1950s he identified chemical communication between animals via pheromones. By the 1970s he had coined the word "biophilia"—the human need to connect with nature—and had become a pioneering conservationist and Pulitzer Prize–winning author. In the 1990s he promoted another term—"biodiversity," or the variety of life on Earth—as well as the idea that all life is interconnected. Preserving Earth's biodiversity is his top priority. The lead scientist for the Half-Earth Project, he races to identify and protect at least half the Earth's "hot spots"—where rare, diverse life intersects with life-supporting environments. His vision for an online Encyclopedia of Life includes an uploaded page for every species, estimated at 2.2 million living, with nearly two million logged as of 2021.

VIRUSES

Viruses have distinctive genetic material but are considered biological entities at the edge of life, incapable of reproduction outside of a host organism. They therefore exist outside the three domains. About 5,000 kinds have been identified, but given their widespread occurrence within living hosts, many millions are likely to exist.

DOMAIN ARCHAEA

Organisms in this domain superficially resemble bacteria—microscopic, with no organized nucleus within their cells—but their genetic makeup is so different from bacteria and all other forms of life that a new domain was designated in the late 1970s. Microbiologist Carl Woese and colleagues from the University of Illinois found these microbes living in hot-water springs at Yellowstone National Park and in deep-sea hydrothermal vents and determined that they are genetically and biochemically unique. Once thought to be rare and exotic organisms living only in limited high-temperature areas, Archaea are now known to be abundant in the open sea. They are also methane-producing inhabitants of environments ranging from the digestive tracts of cows and termites to extremely saline habitats, in anoxic muds of marshes, in deep-ocean vents, and in petroleum deposits deep underground.

DOMAIN BACTERIA

Bacteria are mostly extremely small organisms that perform the functions of life within individual cells that do not have an organized nucleus. As recently as the early 1990s, only about 4,000 bacteria had names, and few were thought to exist in the ocean. Now the actual number is thought to be many millions of distinctively different kinds, including a great many that exist in relatively small numbers in a dormant state. The greatest abundance and diversity of life are in this domain. As knowledge of the nature of bacteria has evolved, so has awareness of their comprehensive role in shaping the existence of the other domains of life.

At least a dozen major groups of bacteria exist, including the cyanobacteria, once classified as blue-green algae. These tiny blue-green creatures share with plants the ability to photosynthesize and have the distinction of having the longest known fossil record—3.5 billion years. Since early in Earth's history, cyanobacteria have consumed and displaced carbon dioxide during photosynthesis, generating much of the oxygen now in the atmosphere. And they are still at it. Another important attribute of these organisms is their ability to convert inert nitrogen from the atmosphere into an organic form used by photosynthetic organisms for their growth.

DOMAIN EUKARYOTA

All eukaryotes, from microscopic plankton to sponges to human scuba divers, have one thing in common: a membrane-bound nucleus within their cell or cells that contains genetic material organized into chromosomes. Five kingdoms are recognized here. Other classifications are recognized, and changes are likely as new biochemical techniques are used to gain better understanding of the relationships in this highly diverse domain.

VISIONARIES

Holger Jannasch
Small Is Powerful

....

Bologna sandwiches started it all. In 1968 the research submarine *Alvin* sank—without passengers aboard—off Massachusetts. When brought up a year later, the intended crew's lunch was waterlogged but otherwise in mint condition. This interested Woods Hole microbiologist Holger Jannasch. He determined that in the low-temperature and high-pressure sea, microbe metabolisms had slowed by perhaps 100 times, leaving the sandwiches "pristine." Over the next 30 years, microbes remained his mission: how they thrive in extreme deep-sea environments, especially around hydrothermal vents, and what they mean to life on Earth. By feeding on sulfur from the vents, he found, microbes were supporting extensive ecosystems. He came to believe that microbes are both central to all life on Earth and likely to represent the origin of life itself. In 1996 the genetic makeup of one microbe—later named *Methanococcus jannaschii* for him—was decoded, an early step in linking microbes to first life. His theories have prompted explorations of microbial communities beneath the seafloor and under the soil on Mars. "Small," wrote Jannasch, "is intrinsically powerful."

....

OPPOSITE ABOVE: Extremophile bacteria *Staphylothermus* thrives in sulfur-rich, superheated environments like black smoker vents. OPPOSITE BELOW: Corkscrew-shaped *Spirillum volutans* populate salt and fresh waters.

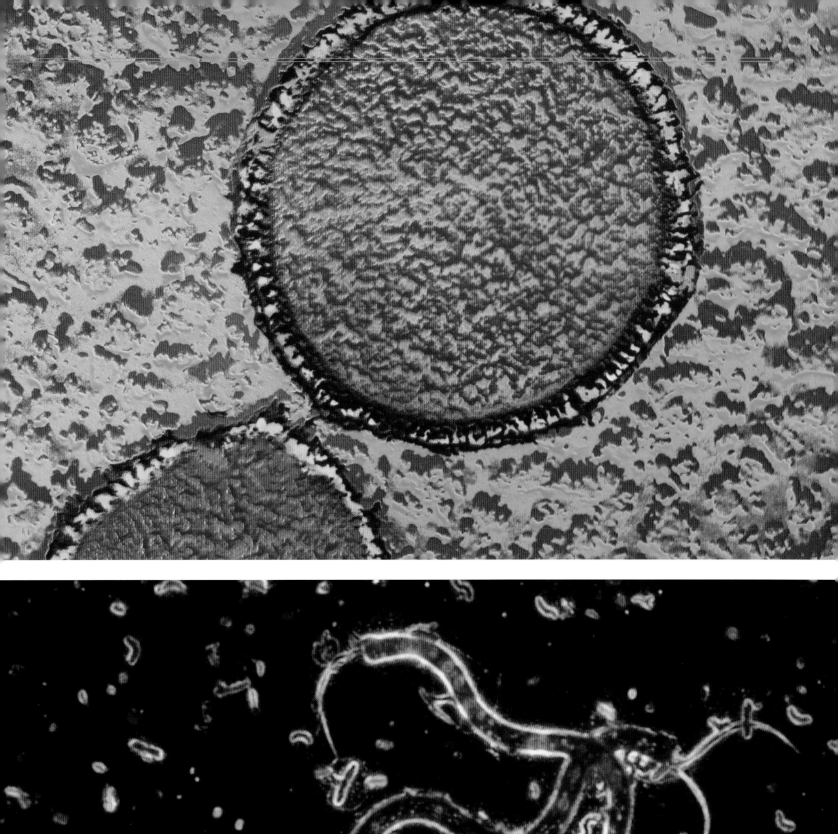
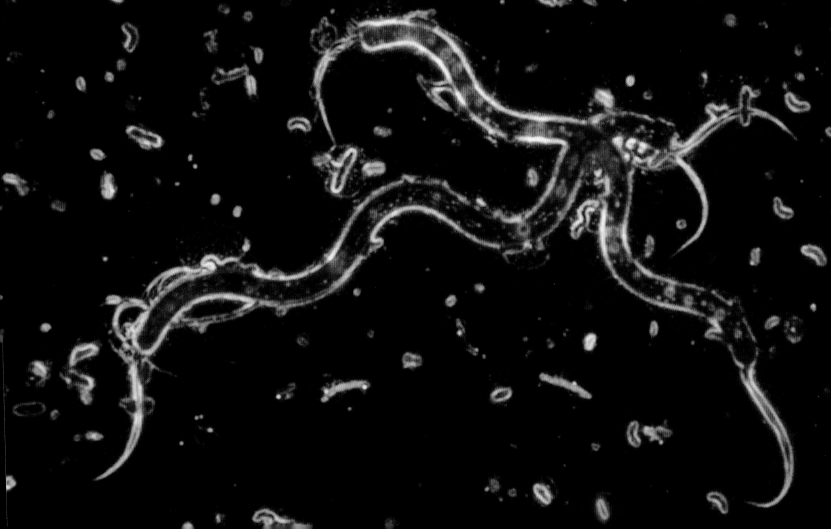

KINGDOM PROTISTA

The Kingdom Protista is a heterogeneous collection of organisms called protists that include small, single-celled protozoa, photosynthetic plankton, and many multicellular organisms, with cells that contain a nucleus with genetic coding. The photosynthetic members of this group, together with marine cyanobacteria, generate most of the oxygen in the atmosphere and capture the sun's energy through photosynthesis, thus serving as the base of the ocean's sunlit food webs and providing sustenance for a large portion of marine animals as well as numerous land dwellers with culinary connections to the sea.

Phylum Ciliata: Covering the surface of these single-celled but complex animals is a full or partial coating of short, dense, hair-like cilia that move for propulsion and for engulfing food. Common in both fresh and salt water, these small organisms grow rapidly in a wide range of environments.

Phylum Dinoflagellata: Single-celled but highly diverse, dinoflagellates are notorious as the organisms responsible for red tides and for creating some of the most brilliant displays of shallow-water bioluminescence. About half are photosynthetic; the rest consume other organisms or are parasitic. Typically they have a pair of flagella used for propulsion.

Phylum Foraminifera: Single-celled and shelled, foraminifera form intricate structures made of calcium carbonate organic compounds or sand grains cemented together. Enormous quantities of their shells form deep sediments on the seabed, and in places such as the Bahamas they contribute substantially to the characteristic white-sand beaches. The many species have different water temperature tolerances, so studying their shells has been an important tool in determining ancient climate conditions.

Phylum Chlorophyta: The green algae include numerous freshwater species, some terrestrial kinds, and a large number of ancient and unusual marine species. Most occur in the upper 30 meters of the ocean, but some are adapted through special light-gathering pigments in addition to two kinds of chlorophyll that enable them to thrive in depths greater than 200 meters. Several groups incorporate calcium carbonate into their cell walls, including species of *Halimeda* that are so abundant in some tropical areas that their remains make up the principal ingredient of the region's sand.

Phylum Choanoflagellata: Some regard the tiny, flagellated, single-celled choanoflagellates as the group "where it all began for animals." Resembling certain cells of sponges, these minute mobile organisms are entirely aquatic, feeding on even smaller creatures in marine and freshwater environments.

Phylum Euglenophyta: This group of minute, mobile, mostly one-celled photosynthetic organisms sometimes cause greening or reddening of salt ponds and saline lagoons with their prodigious numbers. Each cell bears a single whip-like flagellum.

Phylum Myxomycota: Referred to as "slime molds," these have common characteristics with fungi, animals, and other protists. One species, *Labyrinthula zosterae,* has been credited with killing off more than 90 percent of the seagrasses of the North Atlantic coast in the early 1930s.

Phylum Radiolaria: Greatly admired for the intricate symmetry of their glassy shells, the single-celled radiolarians are as well known to geologists

"IF GIANTS STILL SWIM UNSEEN . . . ACROSS THE VASTNESS OF THE SEA, WHAT OTHER SURPRISES AWAIT US IN THE SMALLER CREATURES SWARMING AMONG THEM?"

—EDWARD O. WILSON, *HALF-EARTH: OUR PLANET'S FIGHT FOR LIFE*

as they are to biologists, because their skeletons have endured in ocean sediments for millions of years and provide important clues concerning the nature of ancient seas.

Phylum Rhodophyta: The red algae are a highly diverse assemblage of more than 4,000 species, mostly marine, all with distinctive reddish-pink pigments and complex alternating life history phases. Some are thin, flat, or tubular; others grow in crusts, nodules, or upright clusters of branches and precipitate calcium carbonate from surrounding seawater and incorporate it into their cell walls. As much as 90 percent of the stony matrix of coral reefs is actually made up of coralline red algae. The red pigments effectively gather blue light in deep water, enabling some coralline species to thrive in depths well below 250 meters, carpeting the seafloor with stony pink nodules.

Magnified 1,000 times, this single-celled dinoflagellate uses two pair of flagella, hair-like extensions, for propulsion.

Phylum Labyrinthulomycetes: These small organisms with a big name have a major role in the cycling of nutrients in the sea. Once classified as fungi, these mostly parasitic or symbiotic protists produce a characteristic network of filaments that absorb nutrients from their hosts. They are ubiquitous in the ocean, likely as abundant as bacteria.

KINGDOM CHROMISTA

Sometimes lumped with the Kingdom Plantae, this varied group of organisms differs from them by having chlorophyll c—a form of chlorophyll found in certain marine algae that is lacking in plants—and by not storing their energy in the form of starch. They differ from protists in having distinctive pigments that give most of them their characteristic brown or golden color.

Phylum Bacillariophyta: Diatoms are renowned not only for the beauty of their intricate glassy shells, but also for their critical role as producers of food in the sea and in freshwater systems globally. Typically, two shells—one slightly smaller than the other—encase a single cell like a miniature jewel box. Well known in the fossil record, some rocks are formed almost entirely of diatom shells. Diatomaceous earth has been mined for many years and used for filtration and as a natural pesticide.

Phylum Chrysophyta: The golden-brown chrysophytes are microscopic, usually photosynthetic, sometimes colonizing organisms, some with silica, or glass, scales. Freshwater and marine, they are abundant in the fossil record and are still commonly found in the ocean's floating plankton layer or attached to a surface.

Phylum Ochrophyta (Phaeophyta): The brown algae are nearly all ocean dwellers, most luxuriant and conspicuously developed in cool or cold coastal waters. Some form wispy clusters of filaments, while others resemble delicate ribbons or leafy, golden-brown shrubs. In this phylum are kelps, the largest of all photosynthetic organisms in the sea. California's *Macrocystis pyrifera* can grow about a third of a meter a day and attain lengths as great as 30 meters.

DEEPER DIVE

Prochlorococcus

Its name in Latin, meaning "primitive green berry," is longer than the size of this kind of cyanobacterium—a single-cell microbe that creates energy through photosynthesis. Research finds that rafted together across the surface of the ocean, colonies of *Prochlorococcus* provide the key to Earth's atmosphere today—and to our future. From fossils some 3.5 billion years old—the oldest known evidence of life on Earth—scientists have traced the course of evolution and environment, of photosynthesis, and the addition of oxygen to the atmosphere. The tiniest cyanobacteria, *Prochlorococcus,* is perhaps the mightiest—a "superorganism" with "a story to tell," says its

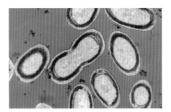

Prochlorococcus cyanobacteria

champion, MIT biological oceanographer Sallie W. "Penny" Chisholm. The most abundant photosynthesizing cell in the sea, *Prochlorococcus* may contribute a significant 10 percent of Earth's total photosynthesis. Chisholm and fellow researchers continue to raise the profile of this one-cell wonder. Its genetic complexity heftily outranks ours—some 80,000 genes are meted out across all *Prochlorococcus* species, compared with humans' 20,000. Identifying its precise role in originating and supporting life on land and sea, guiding evolution, filling the atmosphere with life-sustaining oxygen, and influencing climate is key to determining our future.

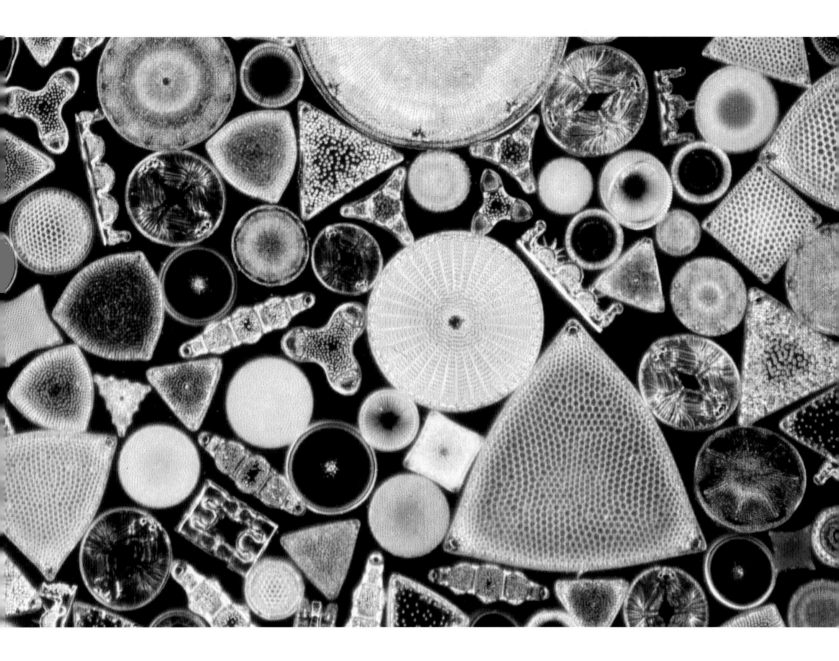

Phylum Haptophyta: Numerous species of this group are known as fossils, and about 500 living species now contribute significantly to photosynthetic activity in the sea, coccolithophorids among them. The delicate calcium-based shells of coccoliths have accumulated in enormous sedimentary deposits over broad areas of the seafloor and are a common component of chalk, including the famous White Cliffs of Dover on the southeast coast of England.

Natural stained glass, a fossil from Oamaru, New Zealand, holds microscopic diatoms, which build glassy shells from silica in the sea.

KINGDOM FUNGI

At least four phyla of fungi are recognized, including many common marine species that, owing to their cryptic nature, are rarely seen. Many are small, some are parasitic, and none photosynthesize. It is estimated that 1.5 to 5 million species of fungi exist, but so far only about 1,100 are known to be exclusively marine. According to the results of a 2018 workshop on marine fungi at the Marine Biological Laboratory in Woods Hole, Massachusetts,

"There is clear and ample evidence that fungi shape both biological and geochemical cycles at all levels of the ocean ecosystem, but there are vast gaps in our mechanistic understanding of fungal ecosystem function."

KINGDOM PLANTAE

Historically, almost any organism that photosynthesized was regarded as a "plant," but as currently defined, plants are multicellular, photosynthetic organisms with cell walls containing cellulose—a nucleus within the cells containing genetic material—and typically, a life cycle that involves a generation with a single set of chromosomes (haploid) and a generation involving a joined set or sets of chromosomes (diploid). Diploid embryos retained within female organs for development as seeds are characteristic of flowering plants, but methods of reproduction vary widely in other groups.

About a dozen phyla of organisms comprise about 250,000 mosses, ferns, liverworts, club mosses, horsetails, conifers, flowering plants (including most trees), and several extinct groups. A few ferns, shrubs, and marsh grasses live in brackish or salty water, but the best known and most widespread marine plants are 72 flowering plants in the Phylum Anthophyta, known as seagrasses, that typically occur as vast undersea meadows off every continent. Some seagrass beds as big as 400,000 football fields can be seen from space. All play a vital role in the health of coastal ecosystems. Not only do their strong roots and rhizomes literally shore up land along the sea by binding sediments, but they cleanse pathogens out of the water, shelter and feed fragile marine life, and act as carbon sinks storing overload the atmosphere has pumped into the ocean. A single square meter of seagrasses can fill the air with as much as 10 liters of oxygen a day.

Perhaps the most studied, eelgrass *(Zostera marina)* thrives throughout the Northern Hemisphere and is prevalent along the U.S. East Coast. Since 1984, environmental programs in the Chesapeake Bay have reduced toxic levels of nitrogen and phosphorus, increasing seagrass beds fourfold, which

······················· **DEEPER DIVE** ·······················

Superior Cephalopods

Smart, strong, adaptable. Among mollusks—soft-bodied invertebrates—marine-based cephalopods are superlative. Their name, meaning "head-foot" in Greek, reflects their makeup, with arms attached to their head. Squid, octopuses, cuttlefish, and the chambered nautilus have been called "unusually brainy." Their highly developed nervous system and sense organs promote complex and awe-inspiring behavior. Quick learners by example or trial and error, they can navigate mazes and have been shown to distinguish among human faces. Researchers observed one Indonesian octopus craft protective homes from coconut shells scattered

Chambered nautilus

across the coastal floor and also roll in them to travel. When danger threatens, a muscular tube beneath a cephalopod's eye shoots out water and propels it backward, sideways, or vertically, confusing predators as it escapes. Octopuses have only soft, fleshy, arms with underside suckers that attach to rocks and reefs and curl around prey. Cuttlefish and squid have arms and also blade-shaped, retractable tentacles for hunting. Color-changing cells called chromatophores enable these escape artists to hide in plain sight. Octopuses change texture, too, and even turn kaleidoscope colors in their sleep—perhaps dreaming about action in the ocean.

has helped restore depleted populations of blue crabs and bay oysters.

Along with seagrasses, about 80 species of mangrove occur along tropical and warm temperate coasts around the world, with the red mangrove *(Rhizopora mangle)* most prevalent. Hefty vertical roots and twisting horizontal rhizomes maintain fortress-like security, capture carbon, and form labyrinthine safe havens for many species of invertebrates, fish, birds, and mammals.

KINGDOM ANIMALAE (METAZOA)

This kingdom includes multicellular, nonphotosynthetic organisms with cells that lack walls and have genetic material contained within nuclei. Most form tissues organized into specialized organs and have diploid embryos, those with two sets of chromosomes. The phyla of animals are sometimes lumped or subdivided into more or fewer categories as new insights are

One of more than 5,000 species of sponges, yellow tube sponges adorn a Caribbean reef.

gained. No evolutionary ranking is intended in the order of the following list of phyla, but in general, those with relatively simple structures are followed by those with increasing complexity.

Phylum Porifera: More than 5,000 living species of sponges represent three categories of these ancient, mostly marine animals, their name meaning "hole-bearing." The Hexactinellida (glass sponges), the Calcarea, and the Demospongia have distinctive calcareous or siliceous spicules—calcium or glass spikes—held within a spongy or soft matrix. One family of sponges snares crustaceans with Velcro-like spicules, then digests them with special cells that migrate to the site of capture. Reefs of glass sponges were common in warm, shallow seas during the Jurassic period, but were thought to exist in modern times only as fossils—that is, until large living reefs were found on the western continental shelf of Canada in the late 1980s. As exploration of the deep sea has progressed, glass sponges have been found in canyons off the northern coast of Spain, in Antarctica, and in numerous other places.

Phylum Placozoa: One very small species of this category of animal (0.5 millimeter across) was discovered in the 1880s adhering to the glass of a marine aquarium. Beyond that, little is known about them.

Phylum Rhombozoa: Celebrated for having the smallest number of cells of any animal (between eight and 40), these tiny, entirely marine parasites live only in the kidneys of certain octopuses and squids.

Phylum Orthonectida: About 20 species of these small, ciliated marine parasites occur internally in various invertebrate animals.

Phylum Myxozoa: These tiny parasitic animals use marine invertebrates and some vertebrates as hosts. They have distinctive multicellular spores and are regarded as relatives of the phylum Cnidaria.

Phylum Cnidaria: Thousands of jellyfish, corals, anemones, hydroids, and sea pens are included in this group, as well as a small number of freshwater species, all known for specialized stinging cells embedded within their tentacles. Cnidarians are distinctive in having only two basic cell layers—ectoderm and endoderm—with a soft material, the mesoglea, in between that may make up the bulk of large jellies and anemones. Corals form calcium carbonate skeletons with cups for individual polyps. Thousands of interconnected polyps of numerous species form enormous stony structures.

Phylum Ctenophora: Eight distinctive bands of iridescent cilia adorn the sides of comb jellies, sea gooseberries, Venus girdles, and sea walnuts—usually translucent, bioluminescent, and always marine carnivores. Most live in the water column, but a few creep along the seafloor. Unlike the cnidarians that they superficially resemble, ctenophores have three distinctive cellular layers and have no stinging cells, although some have retractable tentacles armed with sticky cells useful in capturing prey. All consume other creatures, sometimes engulfing other jellies as large as they are.

Phylum Platyhelminthes: There are three groups of flatworm known, and two—the Cestoda and the Trematoda, or flukes—are parasitic, with complex life histories sometimes involving several hosts. Most notable in the ocean are the free-living Turbellaria, including some that are brilliantly colored and swim with the grace of aquatic butterflies. Most are a few millimeters long, but some can reach 15 centimeters.

A Pacific native called the fuchsia flatworm; part of its scientific name, *Pseudoceros ferrugineus*, means "rusty."

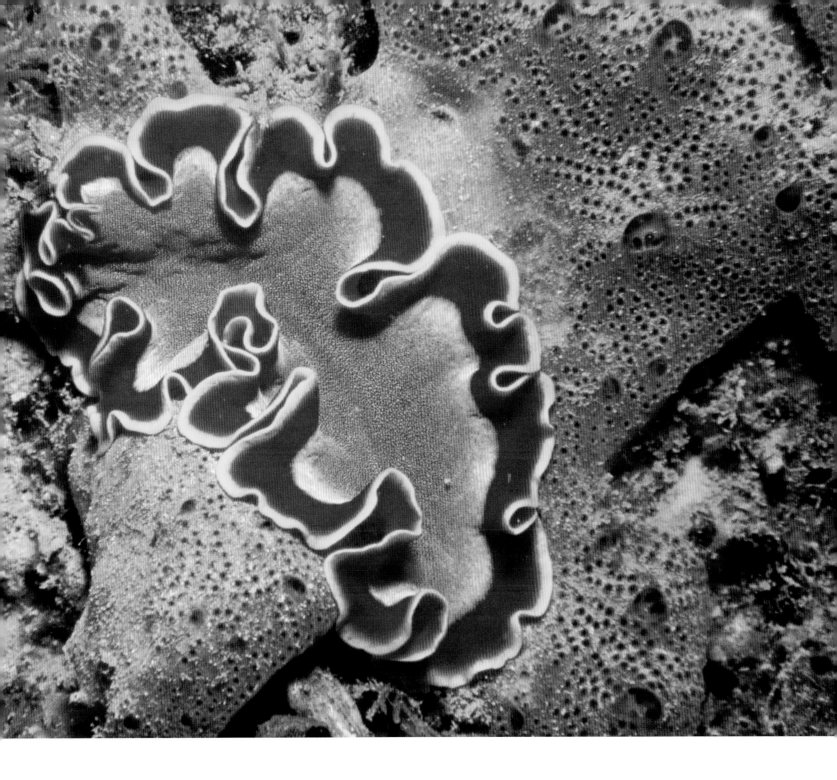

Saving Sharks: It's in Their eDNA

Sharks: sleek, charismatic predators. Through time their dominance in the food chain has balanced the marine web of life; but human activity, especially industrial-scale fishing, has reduced their numbers by as much as 90 percent, disrupting that balance. To restore and protect sharks, it is important to know how many there are of each species, where they live, and how far they travel. To do that, researchers have developed a method called metabarcoding, a technique that rapidly assesses species information using

Silky sharks in Cuba

DNA-based identification and DNA sequencing. By extracting DNA from Caribbean Sea and Pacific Coral Sea water samples, scientists have identified some 21 shark species, their habitats and ranges, and have found that their distribution patterns align with human activity. Not surprisingly, shark numbers decline where commercial fishing occurs but thrive where conservation is in full swing. Not only does metabarcoding help define diverse shark populations, but it also leads to better conservation strategies.

Phylum Nemertea: Nearly a thousand species of ribbon worms—soft, unsegmented animals that live throughout the ocean in sandy, muddy, and rocky places—use an extendable proboscis to gather food, sometimes wrapping prey with their sleek body. Most are a few centimeters long, but one species reaches 30 meters.

Phylum Rotifera: Of the 2,000 or so species of rotifers known, a few hundred live lightly attached to seaweed or other homes, or as floating plankton in the sea. Tiny (one to 2.5 millimeters long), translucent, and often occurring in great numbers, these predators have a circular cilia-covered organ that propels smaller organisms into their gullet.

Phylum Gastrotricha: About 400 kinds of gastrotrichs—animals less than one millimeter long with characteristic hairy undersides—have been discovered living in muddy sand in aquatic places worldwide.

Phylum Kinorhyncha: From coastal sands to the fronds of seaweed to the deep sea, the tiny, entirely marine 13-segmented, spiny, cylindrical, mostly transparent kinorhynchs are common but not often seen because of their inconspicuous size and cryptic habits.

Phylum Nematoda: The long and narrow shape of these organisms is suggested by their name, based on the Greek word *nema*, meaning "thread." Longitudinally aligned muscles enable the animal to bend side to side but not in other directions. Free-living species thrive on bacteria in moist soil, deep-sea mud, polar ice, hot springs—essentially every place where food and water are available. Parasitic forms live on or in almost every other kind of organism known, including a 13-meter-long species that inhabits the gut of sperm whales. About 15,000 species are recognized, but it is estimated there may be half a million yet to be discovered, thus rivaling arthropods—insects, spiders, and crustaceans—in diversity, and most likely exceeding them in sheer mass.

Phylum Nematomorpha: Most of the 240 species of long, slender horse-hair worms live in fresh water, some as parasites in marine crabs. Related to nematodes, with larvae that are invertebrate parasites, the adults do not feed, but invest their short lives in producing eggs and fertilizing them to ensure the future of their kind.

Phylum Acanthocephala: As adults, these prickly-headed worms are entirely parasitic in vertebrate animals, mostly freshwater fish. Their larval stages are parasitic in crustaceans or other arthropods that must be eaten by a vertebrate host for the larvae to mature.

Phylum Entoprocta: About 170 species of these tiny stalked animals occur widely in the oceans of the world; only one lives in fresh water. Some are solitary, but most link together in colonies, and all have a distinctive crown of tentacles with which they capture small prey.

Phylum Gnathostomulida: Entirely marine, about 100 kinds of these small, wormlike animals with paired jaws have been discovered living in shallow and deep mud and among sand grains worldwide.

Phylum Priapula: About 20 species of small, spiny cactus worms, entirely marine, constitute this distinctive and widespread but inconspicuous group of tubular-shaped organisms that appear to be a cross between a worm and a cactus. Mused one science writer, "If enlarged to human size, it could be the terrifying star of a science fiction movie."

"POSSESSION OF MITOCHONDRIA IS A *SINE QUA NON* OF THE EUKARYOTIC CONDITION."

—NICK LANE, *POWER, SEX, SUICIDE: MITOCHONDRIA AND THE MEANING OF LIFE*

With the grace of a dancer, a hula skirt siphonophore, a cnidarian relative of jellyfish, wields stinging tentacles.

Phylum Loricifera: These very small, translucent, barrel-shaped, totally marine creatures with a characteristic sheath of spines and scales were first discovered amid grains of sand in several locations in the Atlantic. About 50 species are now known to live in depths from 10 to at least 500 meters.

Phylum Cycliophora: Not discovered until 1994, just one species is known of these small animals. They adhere to bristles in the mouths of Norwegian lobsters and filter-feed with cilia surrounding their own mouths.

Phylum Sipunculida: About 330 species of peanut worms are known, all entirely marine. They have a muscular, bulbous, smooth, unsegmented body elongated at one end with a frill of tentacles surrounding a small mouth used to feed on detritus. Some are tiny (two millimeters or so long), but some are nearly as big as a baseball bat. Most live in sand or mud, but some burrow into rock or coral, or nestle among the holdfasts of large seaweeds.

Phylum Echiura: About 150 species of entirely marine tube-shaped spoon worms are widely distributed in the ocean, each with an extendable

A European fan worm spreads spiral feeding fans from its home, a leathery tube; when threatened, it pulls them back inside.

feeding proboscis at one end of their body and spiny hooks on the other. A Pacific species is known as "the innkeeper" because numerous small shrimp, crabs, polychaetes, and other creatures move into its burrow.

Phylum Annelida: About 9,000 kinds of marine annelids—segmented creatures—are known, mostly polychaetes—typically having many bristles or spines. Enormously diverse and widespread, polychaetes live in sand and mud and on and in sponges and corals, or free floating as plankton. Some create tubes of calcium carbonate or sand grains to live in. The tiny trochophore—or trochosphere—larvae are distinctively sphere- or pear-shaped and fringed with circles of cilia, which help them swim.

Phylum Pogonophora: Entirely marine, the roughly 80 species of bearded worms are mostly small animals with no mouth or stomach. They live in areas ranging from shallow coastal regions to the deep sea. The most spectacular representatives live in cold seeps and hot vents on the deep-sea floor, where they derive sustenance from symbiotic bacteria living in their distinctive, colorful, plumes. Some exceed two meters in length.

Formidable at 30 centimeters long, the South Pacific white-spotted hermit crab will spar with other crabs for a gastropod-shell residence.

Phylum Onychophora: Once abundant in ancient oceans, about 100 species of these small, caterpillar-like velvet worms today live in moist, land-based forest habitats, mostly in the Southern Hemisphere.

Phylum Tardigrada: Some 800 species of tardigrades are known, about half of them in marine places ranging from deep-sea sands to the surface of certain seaweeds. Some are predatory, but many simply suck the juices from other organisms. Resembling small, stubby-legged bears (and often called "water bears"), they are usually a millimeter long or less. These mostly translucent creatures share with certain microbes the ability to become dormant for very long periods. In 2019, some of these tiny animals were accidentally released on the moon when a spacecraft carrying them as a science experiment landed there.

Phylum Arthropoda: Arthropods have jointed legs, a segmented body, compound eyes, and an exoskeleton that is typically shed periodically during growth. In this most diverse group of animals, embracing about 80 percent of all known species, insects dominate, except in the ocean, where only a few have become established. In the sea, about 30,000 kinds of crustacea have been discovered, including species of crabs, shrimp, barnacles, copepods, ostracods, stomatopods, and many others.

Phylum Chelicerata: Closely related and sometimes lumped with arthropods, chelicerates—sea spiders, horseshoe crabs, spiders, and scorpions—are regarded here as a group distinguished by having four pairs of walking legs and two pairs of specialized mouthparts, including the distinctive claw-like chelicerae. An ancient group originating in the ocean, Chelicerata's marine species today include four or five kinds of horseshoe crabs

VISIONARIES

Edith "Edie" Widder
Communication in a Flash

....

So many deep-sea creatures generate light, Edith "Edie" Widder has said, that bioluminescence is likely the most common form of communication on Earth. A pioneering specialist in animal bioluminescence, Widder has dedicated her career to creating vehicles and instruments to study these animals' makeup, behavior, and distribution sustainably. Her inventions include the High-Intake Defined Excitation (HIDEX) bathyphotometer, which identifies creatures by the duration of their flashes, and the Low Light Auto-Radiometer (LoLAR), a light meter that determines patterns of animal distribution. In 2012 the Medusa camera—based on her remotely operated deep-sea camera, Eye-in-the-Sea (EITS)—filmed the first deep-sea video of the legendary and elusive giant squid. "To be good ocean stewards, we have to know how many animals there are, how they are distributed, and how they behave," she has said. Her long-term vision is to reverse "the trend of marine ecosystem degradation."

....

LEFT: The Pacific's spotted wonderpus octopus, from the German *wunder*, meaning "marvel," is a camouflage artist. OPPOSITE: The deep-sea jellyfish *Atolla wyvillei* in artificial light (below) and self-illuminated (above)

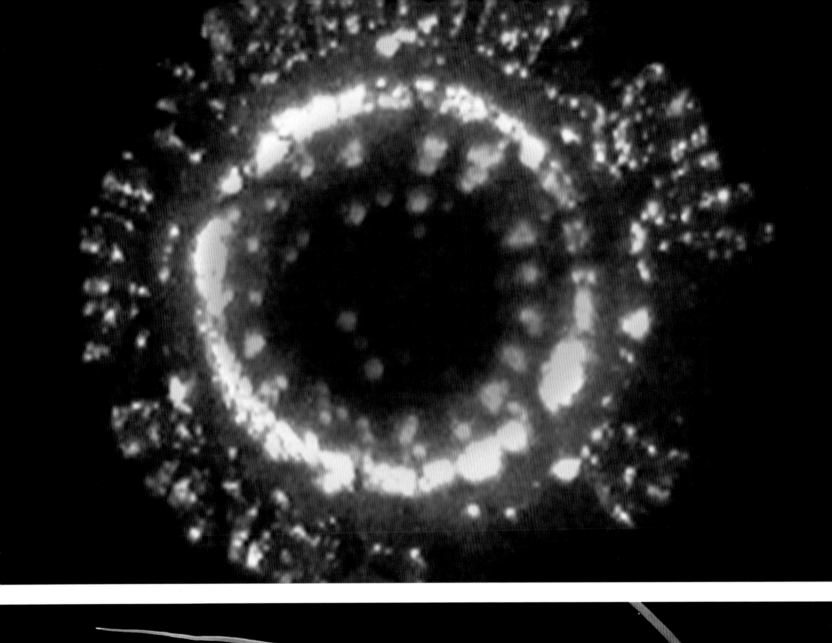
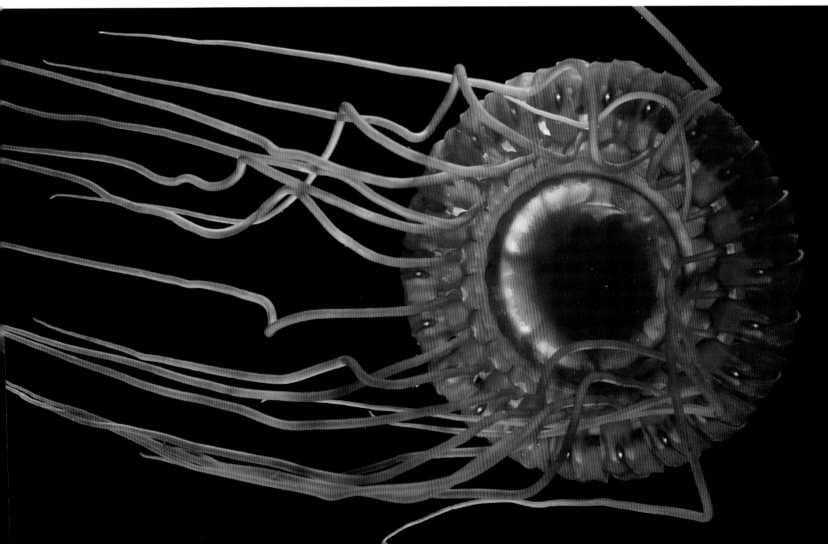

and about 600 kinds of sea spiders. Extinct chelicerates, the eurypterids, grew to be as large as grizzly bears, nearly three meters long.

Phylum Mollusca: Most of the nearly 50,000 kinds of mollusks known are marine, including more than 600 kinds of cephalopods (which include octopuses and squid), 700 tooth shells that resemble elephant tusks, 1,000 or so chitons that are oval with overlapping plates, and two kinds of monoplacophorans—with a single, cap-like shell—that are entirely marine. Also included are pelecypods (clams) and gastropods (snails) that are mostly marine. All are highly organized into tissues and organs, and most have an internal or external shell or shells. Cephalopods have especially well-developed sensory organs, eyes structured much like those of vertebrates, and a complex nervous system.

Phylum Phoronida: About 15 entirely marine species make up the horseshoe worms: small, widely distributed, tube-dwelling, filter-feeding animals.

Phylum Ectoprocta (Bryozoa): Far more abundant and diverse as fossils, about 5,000 filter-feeding bryozoan, or moss animal, species currently live in the sea. A few live in fresh water, where they typically grow as lacy colonial crusts, sometimes in upright clusters.

Phylum Brachiopoda: An ancient group of entirely marine filter-feeding animals, lamp shells superficially resemble clams, once including thousands of species so abundant that they paved the seafloor during the Paleozoic era. Now far less common and diverse, they mainly occur in cold, deep water.

Phylum Echinodermata: This entirely marine group of "spiny-skinned" animals—including starfish, crinoids, sand dollars, sea cucumbers, and brittle stars—have distinctive planktonic larvae and adults with calcareous plates embedded in their skin. All have five-part radial symmetry—even sea cucumbers when viewed in cross section—and all have tube feet powered by a water-based, very slow-moving system with a specialized suction effect.

Phylum Chaetognatha: About 125 species of arrow worms are known. They are small, entirely marine, usually transparent predators that often dominate planktonic communities, where they voraciously consume small crustaceans and larval fish—and other arrow worms.

Phylum Hemichordata: Entirely marine, sometimes classed with chordates but more closely related to echinoderms, these soft-bodied acorn worms now include several hundred species, mostly burrowers, but there are also many extinct forms.

Phylum Chordata: Chordates include some of the most conspicuous forms of life in the sea: about 17,000 kinds of fish; 70 or so species of dolphins, whales, seals, sea lions, and otters; eight to 10 kinds of sea turtles; several species of crocodilians; about 80 kinds of sea snakes; one semi-marine toad; and hundreds of species of seabirds. Less well known are numerous kinds of salps, sea squirts, and small leaf-shaped lancelets. One chordate species, *Homo sapiens,* is technically not marine, but owing to the dominant place humans occupy as predators in ocean ecosystems—profoundly altering the physical and chemical nature of the ocean—it is appropriate to include humankind here. Some humans have succeeded in becoming temporary marine mammals by diving, and a few have engaged technologies enabling them to live in underwater laboratories or encased in various submarines.

Like many deep-sea dwellers, this planktonic polychaete worm *(Tomopteris)* lights the darkness with bioluminescence.

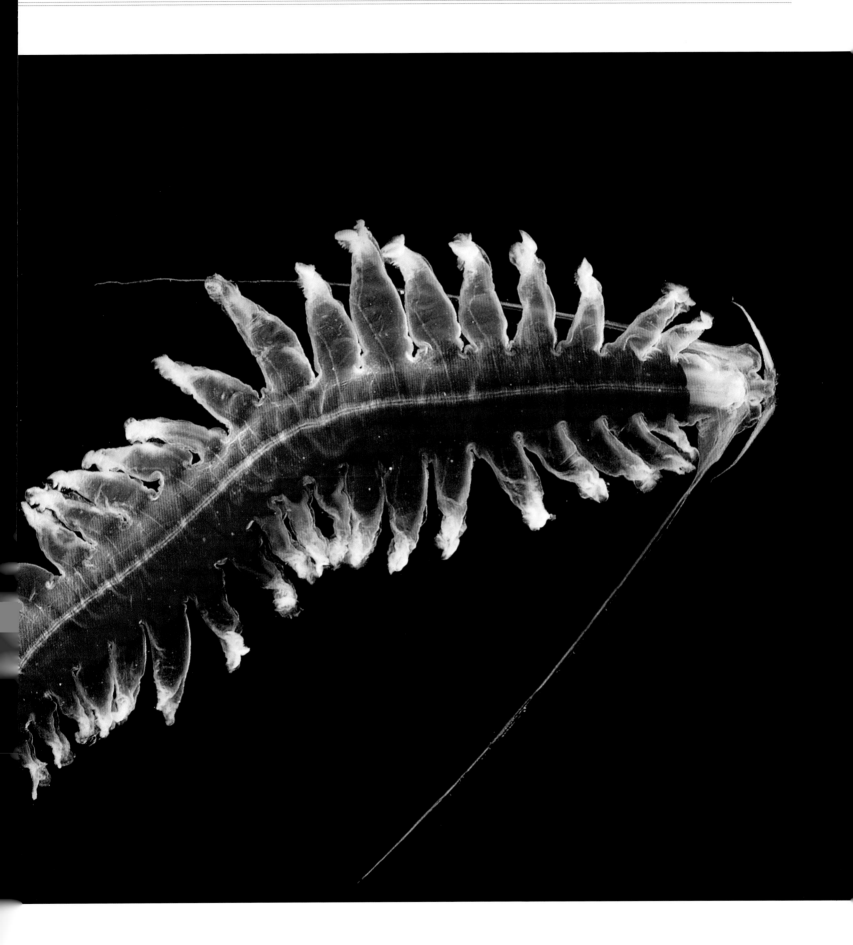

"YOU HAVE MADE YOUR WAY FROM WORM
TO MAN, BUT MUCH OF YOU IS STILL WORM."

—FRIEDRICH NIETZSCHE, *THUS SPOKE ZARATHUSTRA*

THE NATURE OF SEA LIFE

Humans have a lot in common with their fellow vertebrates. About 500 million years ago, during what is referred to as the Cambrian explosion, the emergence of organisms with a cranium and brain, a rod-like flexible notochord or spine, a hollow nerve cord, gill slits, and a tail marked the beginning of the amazingly diverse group of animals that include fish, amphibians, reptiles, birds, and mammals. For some, including humans, the notochord, gill slits, and tail occur only during embryonic stages of development, but they are there nonetheless, as clear evidence of our ancient origins and modern connections to our vertebrate cousins.

Connections seem obvious among mammals, birds, reptiles, and amphibians, all equipped with four appendages either as embryos or adults—arms, legs, wings, or flippers—and lungs needed to breathe air during some or all of their life. Fish seem very different—and in some ways, they are. Over thousands of millennia, fish have developed enhanced sensory capabilities, have come to occupy more spaces and more ecologically varied places than other vertebrates, and have developed greater diversity of species, anatomical features, and behaviors than all other vertebrate animals combined. Like other vertebrates, they have evolved communication systems involving sound, light, smell, and touch, as well as enhanced chemoreception and likely other senses that we have yet to discover. Still, for many, a fish is a fish is a fish. Restaurant menus frequently offer "catch of the day" or "fish chowder" or "fish and chips" without indicating what kind of the thousands of species of fish might be on your plate. Names such as "bass" or "trout" might be applied—or misapplied—to dozens of different species. Chickens are generally not referred to as "fried birds." Cows, pigs, and sheep are known for what they are rather than as "mammal burger" or "mammal steak." Like every other form of life, every fish is an individual, with distinctive features and behavior—a reality not evident when cooked or canned.

Humans have much in common with invertebrates, too. It is relatively easy to compare the anatomy of creatures as diverse as jellyfish, polychaete worms, starfish, shrimp, and squid, and find features that are counterparts of our own. Most have organs sensitive to light, and the majority have well-developed eyes. Nearly all have a place where food is taken in and another where feces are released. All have structures and strategies for reproduction, and most have at least two genders.

And consider the cephalopods. Zoologist Roger Hanlon describes octopuses, squids, cuttlefish, and nautiluses as "brainy, colorful, fast, sophisticated, inspiring," with "large brains, keen senses, and complex behaviors." Those seeking intelligent life elsewhere in the cosmos might consider connecting with the thriving universe of life beneath the surface of the sea.

> ## "WATER'S GIFT IS LIFE."
>
> —SANDRA POSTEL,
> GLOBAL WATER POLICY PROJECT

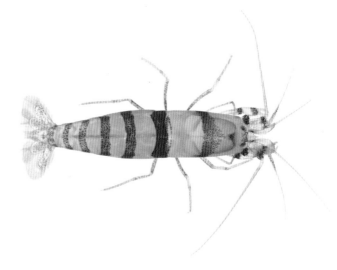

Like other snapping shrimp, this Galápagos species snaps its claw, stunning prey with shockwaves.

Phylum Platyhelminthes
Candy-striped flatworm, *Prostheceraeus vittatus*

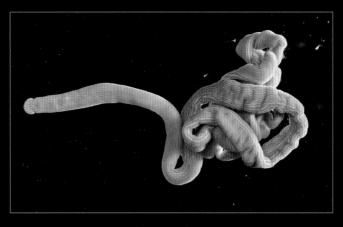

Phylum Nemertea
Ribbon worm

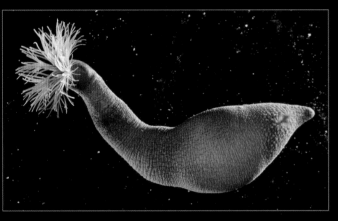

Phylum Sipunculida
Peanut worm

Phylum Echiura
Spoon worm, *Metabonellia haswelli*

Phylum Annelida
Feather duster worm, *Bispira* sp.

Phylum Annelida
Fireworm, *Hermodice* sp.

Phylum Annelida (Pogonophora)
Lamellibrachia sp.

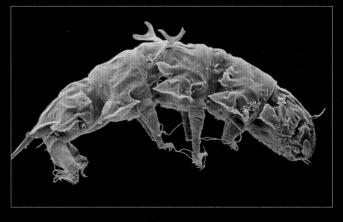

Phylum Tardigrada
Water bear, *Echiniscus testudo*

Phylum Porifera
Giant barrel sponge, *Xestospongia* sp.

Phylum Porifera
Glass sponge, *Euplectella aspergillum*

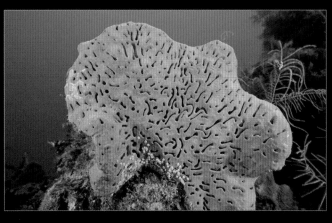

Phylum Porifera
Orange elephant ear sponge, *Agelas clathrodes*

Phylum Cnidaria
Gorgonian coral, *Gorgonia* sp.

Phylum Cnidaria
Stony coral, *Agaricia* sp.

Phylum Cnidaria
Crown jellyfish, *Netrostoma setouchianum*

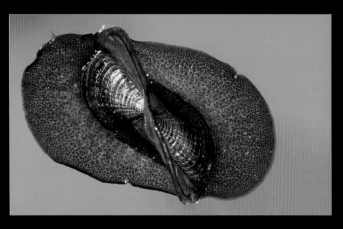

Phylum Cnidaria
By-the-wind sailor, *Velella velella*

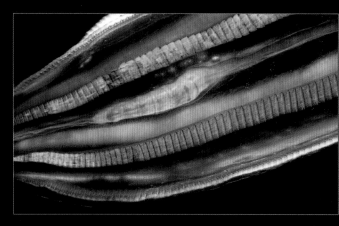

Phylum Ctenophora
Comb jelly, *Callianira antarctica*

Phylum Anthophyta
Turtle grass, *Thalassia testudinum*

Phylum Anthophyta
Turtle grass, *Thalassia testudinum*

Phylum Anthophyta
Turtle grass, *Thalassia testudinum*

Phylum Anthophyta
Glasswort, *Salicornia* sp.

Phylum Anthophyta
Cordgrass, *Spartina alterniflora*

Phylum Anthophyta
Red mangrove, *Rhizophora mangle*

Phylum Anthophyta
Black mangrove, *Avicennia germinans*

Phylum Anthophyta
Black mangrove, *Avicennia germinans*

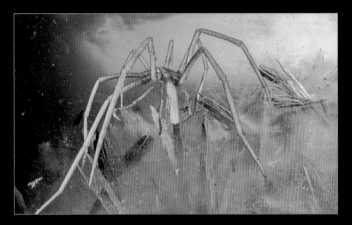

Phylum Chelicerata
Sea spider, *Colossendeis* sp.

Phylum Arthropoda
Spider crab, *Libinia* sp.

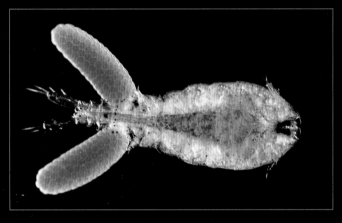

Phylum Arthropoda
Copepod, *Sapphirina sali*

Phylum Arthropoda
Giant isopod, *Bathynomus giganteus*

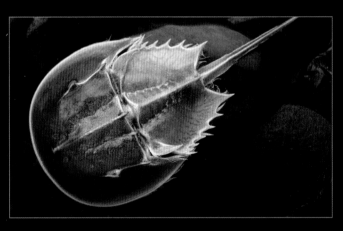

Phylum Chelicerata
Horseshoe crab, *Limulus polyphemus*

Phylum Mollusca
Giant clam, *Tridacna* sp.

Phylum Mollusca
Oriental sundial, *Architectonica perspectiva*

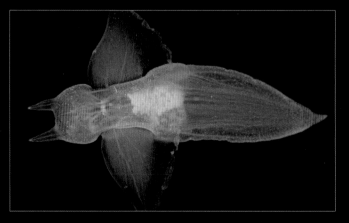

Phylum Mollusca
Thecosomata sp.

Phylum Mollusca
Pacific giant octopus, *Enteroctopus dofleini*

Phylum Mollusca
Broadclub cuttlefish, *Sepia latimanus*

Phylum Ectoprocta (Bryozoa)
Neptune bryozoan, *Reteporella couchii*

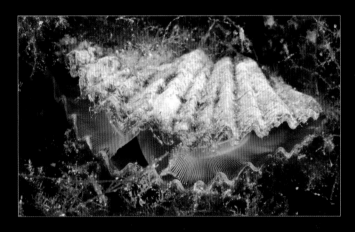

Phylum Brachiopoda
Lamp shell, *Coptothyrus adamsi*

Phylum Brachiopoda
Lamp shells, *Terebratulina septentrionalis*

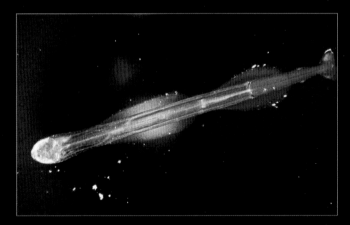

Phylum Chaetognatha
Arrow worm, *Sagitta* sp.

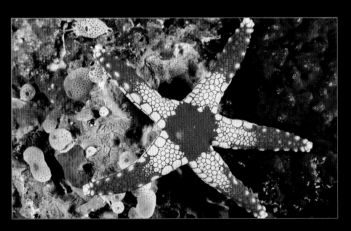

Phylum Echinodermata
Candy cane sea star, *Fromia monilis*

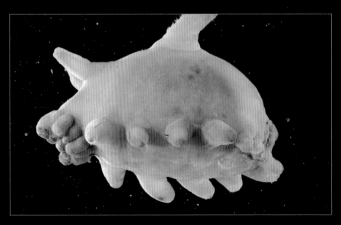

Phylum Echinodermata
Sea cucumber, *Scotoplanes globosa*

Phylum Rhodophyta
Coralline red alga, *Lithothamnion* sp.

Phylum Ochrophyta (Phaeophyta)
Brown alga, *Padina pavonica*

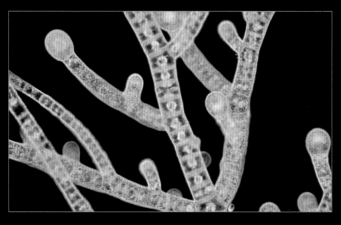

Phylum Ochrophyta (Phaeophyta)
Filamentous brown alga, *Haplospora globosa*

Phylum Ochrophyta (Phaeophyta)
Giant kelp, *Macrocystis pyrifera*

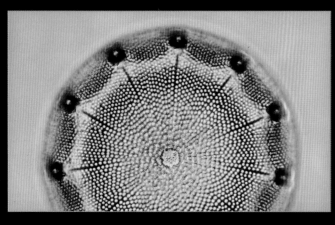

Phylum Bacillariophyta
Marine diatom

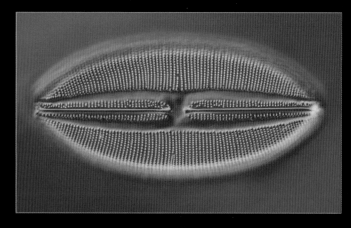

Phylum Bacillariophyta
Marine diatom, *Lyrella lyra*

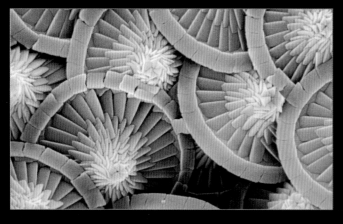

Phylum Haptophyta
Coccolith sp.

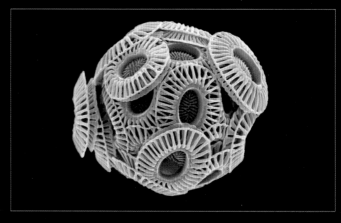

Phylum Haptophyta
Coccolithophorid, *Emiliania huxleyi*

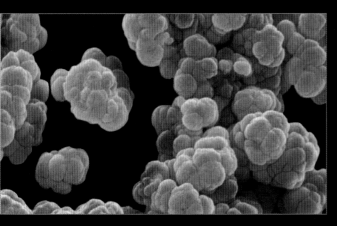

Phylum Euryarchaeota
Methanosarcina sp.

Phylum Cyanobacteria
Prochlorococcus marinus

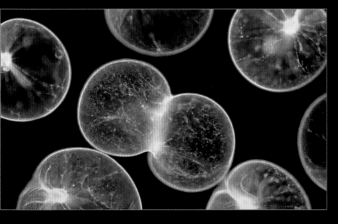

Phylum Dinoflagellata
Sea sparkle, *Noctiluca scintillans*

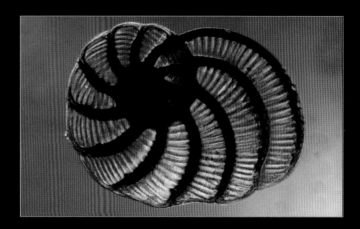

Phylum Foraminifera
Peneroplis pertusus

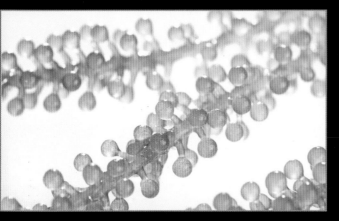

Phylum Chlorophyta
Green algae, *Caulerpa racemosa*

Phylum Chlorophyta
Green algae, *Halimeda* sp.

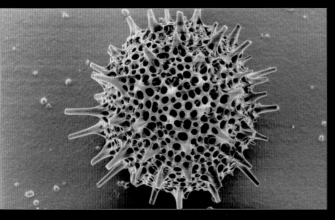

Phylum Sarcomastigophora
Radiolarian sp.

Phylum Rhodophyta
Red alga, *Halymenia* sp.

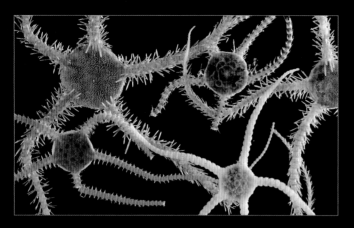

Phylum Echinodermata
Brittle stars, *Ophiacantha* sp.

Phylum Echinodermata
Slate pencil urchin, *Heterocentrotus mammillatus*

Phylum Hemichordata
Acorn worm, E*nteropneusta* sp.

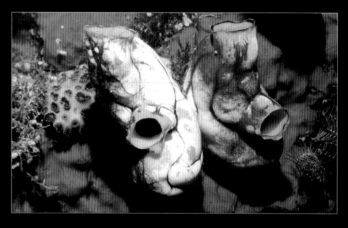

Phylum Chordata
Goldmouth sea squirt, *Polycarpa aurata*

Phylum Chordata
Mexican hogfish, *Bodianus diplotaenia*

Phylum Chordata
Green sea turtle, *Chelonia mydas*

Phylum Chordata
Humboldt penguin, *Spheniscus humboldti*

Phylum Chordata
Bottlenose dolphin, *Tursiops truncatus*

Duotone for camouflage, Australia's
blue dragon nudibranch swallows air
bubbles to float on the sea surface.

The Secret Life of Fish

I n his 2016 book *What a Fish Knows,* Jonathan Balcombe gives voice to the "anonymous trillions" of fish who have many attributes comparable to wild mammals and birds but have yet to be accorded the same respect, affection, empathy, understanding, and protection as their furred and feathered relatives. Like other vertebrates, including humans, fish have developed striking social behaviors including collaboration, elaborate courtship, parental care, use of tools, problem solving, and learned behavior that can be passed from one generation to the next.

There are well-documented examples of groupers teaming up with a moray eel or an octopus to go hunting, behavior initiated by the grouper. Biologist Robert Wicklund describes in *Eyes in the Sea* how he watched a Nassau grouper *(Epinephelus striatus)* swim back and forth in front of a mesh trap with bait inside but too small an opening for the grouper to get at it. After several minutes of what appeared to be contemplation, the fish flipped on its side and used its tail to waft the bait to a place it could reach. The action was repeated several times until the bait was completely consumed. Problem solving and the use of water as a tool—by a fish.

Learned pathways are well documented in birds and mammals, so it should not be surprising that fish, with a much longer history, should have developed comparable traditions. There are reports of "cod highways" in the North Atlantic Ocean that once were led by large, old fish migrating to feeding and breeding areas, with younger, less experienced fish tagging along. Bluefin tuna and great white sharks travel widely but are known to return to breeding and feeding areas.

Some fish, notably the sailfish *(Istiophorus platypterus)* hunt and feed in packs. Carine Strubin, a researcher from Switzerland's University of Neuchâtel, documented surprisingly stable relationships in hunting alliances of goatfishes in the Red Sea, each recognizable by distinctive facial stripes. Manta rays *(Mobula alfredi)* form enduring social bonds that correspond to friendship, according to a 2019 study by Rob Perryman, a researcher for the Marine Megafauna Foundation. Like many other vertebrates, various species of seahorses and butterflyfish pair for life, and a 2004 report by E. A. Whiteman and I. M. Côté notes at least 70 kinds of monogamous fish species.

Chimpanzees—once regarded as suitable for use as food, pets, entertainment, and subjects for medical research—have achieved growing respect, care, and protection. Historically, songbirds, eagles, owls, elephants, tigers, lions, bears, and many other wild animals have been killed for sport and trophies, for traditional medicine markets, and as bushmeat—and some still are, but there is widespread empathy and respect for them. Attitudes about fish and other sea creatures may change, too, as people get to know them on their own terms, each one a living miracle.

"FISH HAVE DIFFERENT WAYS OF SHOWING THEIR CURIOSITY. OFTEN WHILE SWIMMING ALONG WE WILL TURN BACK ABRUPTLY AND SEE THE MUZZLES OF ECHELONS OF CREATURES FOLLOWING US WITH AVID INTEREST."

—JACQUES COUSTEAU,
THE SILENT WORLD

Groupers gather to spawn in the Pacific's coral Fakarava Atoll, a UNESCO Biosphere Reserve sheltering a rectangular reef and enclosed lagoon.

Personalities of Fish

In 1970 an all-woman team of scientists and engineers lived and worked for two weeks in the undersea lab *Tektite II*. While NASA scientists on the surface were studying the behavioral activity of the "aquanauts," the real takeaway was the insight acquired concerning the behavior and personalities of fish. Each morning the *Tektite* team encountered five angelfish that stayed together as they grazed on the reef, then went separate ways at night; butterflyfish that swam in pairs; and a big green moray eel that occupied a certain section of the reef as its home base. During observations day

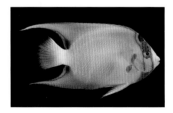

Caribbean queen angelfish

and night, the team got to know "that" snapper and "that" grouper, learning their different faces, spots, and shyness or aggressiveness. Fast-forward 50 years, and research shows that fish play, recognize humans, and use tools. Some even outperform chimps in solving problems. But because their faces don't move, their eyes don't blink, and they usually do not cry out in pain, they don't receive, as behavioral ecologist Culum Brown puts it, "the same compassion as warm-blooded vertebrates." Attitudes are changing as more people get to know fish alive, rather than swimming in lemon slices and butter.

Family Life Under the Sea

Since the first living cell divided into two, billions of years ago, astonishing variations and embellishments have developed to help ensure that species will endure. For bacteria and archaea, the process of reproduction appears to be fairly straightforward, with cell division replicating individuals in response to availability of nutrients and other favorable conditions. Among eukaryotes—organisms with genetic material organized into chromosomes contained within nuclei—the advent of sex (basically the exchange of genetic material between partners) changed everything. Combining different genetic codes enhances the potential for variability, and thus adaptability to a wide range of environmental conditions.

While there are great variations in terms of social structure, sexual orientation, courtship, fidelity, and length and conditions of pregnancy and caring for young, our furry or sleek air-breathing, warm-blooded cousins face challenges most humans can relate to: maintaining themselves while doing what it takes to ensure their genetic continuity— courtships, friendships, social bonding, parental care, and enduring family ties.

Many marine organisms, including sponges, corals, starfish, marine polychaete worms, lobsters, and numerous species of fish, release gametes—eggs and sperm—simultaneously into the surrounding sea, where fertilization occurs and the young begin a precarious existence, finding food to eat while doing their utmost to avoid being consumed by others.

An example of this kind of reproduction occurs at a specific evening hour once a year, determined by certain phases of the moon. Then, some species of corals, sponges, brittle stars, mollusks, polychaetes, and likely other less conspicuous coral reef residents respond to ancient rhythms, sending clouds of gametes into the ocean. The surrounding water serves both as a liquid nursery and a bountiful banquet for hungry egg-eaters.

The great majority of animals on Earth are invertebrates, encompassing more than 30 distinctive phyla living in the sea. Among this vast and varied population, it should not be surprising that a wide range of reproductive strategies have developed over time, but the habits of most simply are not known. Even well-studied, commercially sought-after species, such as clams, mussels, and oysters, are full of surprises.

What is known about squid suggests a "love them and leave them" approach to mating and egg care, with fertilized egg masses left unattended.

ABOVE: Pacific staghorn coral spawn twice yearly, spewing a blizzard of eggs and sperm to ensure fertilization. OPPOSITE: A mouth-brooding male cardinalfish babysits some 400 fertilized eggs.

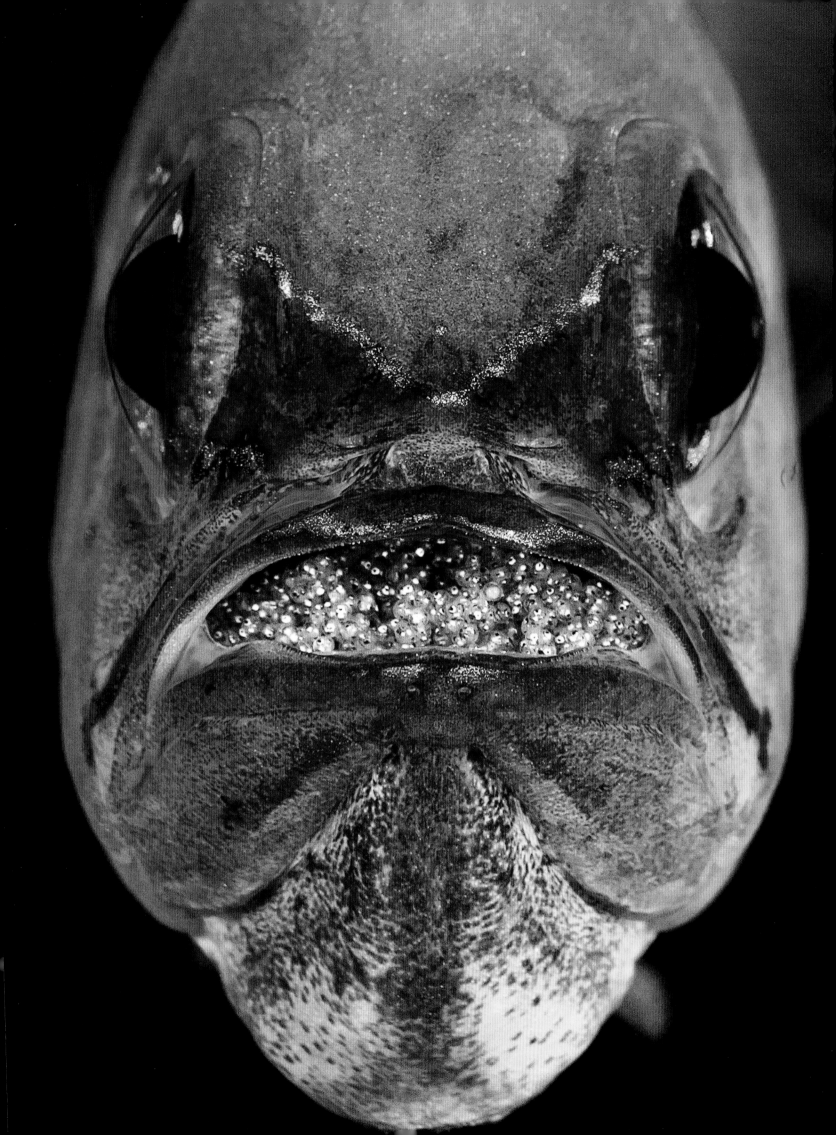

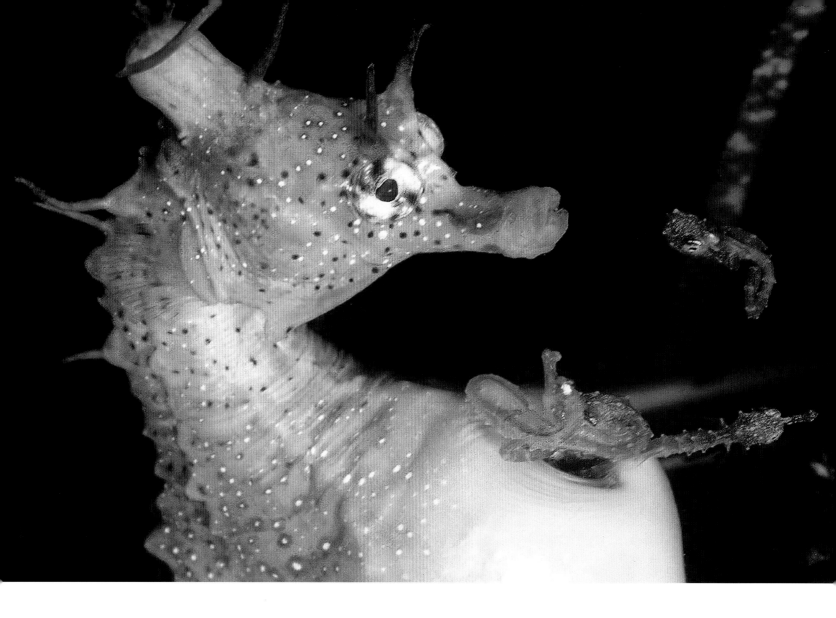

Octopus mothers tend to guard their eggs, but once fully formed, their hatchlings must fend for themselves. One of the "Top Ten Ocean Stories of 2018," reported by *Smithsonian* magazine, was the discovery of two locations where hundreds of octopus mothers, each firmly embracing her own clutch of eggs, gathered together in extensive octopus nurseries near areas of volcanic activity in the deep sea.

Crustaceans such as lobsters and crabs typically release eggs and sperm simultaneously into the sea, where the gametes fertilize and the young are hatched as larvae that may take weeks or months to develop, undergoing many stages of growth with different needs for food along the way. A male peacock mantis shrimp performs a courtship dance before retreating with a female into a burrow where mating takes place. Females hold their fertilized eggs on their abdomen and they remain sheltered there for weeks until hatching. Since barnacles are stuck in place as adults, it makes sense for them to be hermaphrodites in order to reproduce, and most of them are. But some also are equipped with the longest penis relative to their size of any animal. With surgical precision, these impressive organs can inject sperm into neighbors more than 20 barnacle lengths away.

During more than 400 million years, fish have figured out the most varied and ingenious ways to procreate of any of the vertebrates. Most exist as two sexes throughout their lifetime, but others change as they mature. Parrotfish may begin as females and later become males, or they may start out as

Offspring swarm as a male seahorse gives birth after his mate deposited her eggs in his teaspoon-size pouch to fertilize and incubate. One prolific father produced 1,572 young.

male and stay male. Certain individuals—whether initially female or always male—mature as "super-males" that have distinctive coloration and sometimes enhanced physical features.

Many fish are renowned for their elaborate courtship and nest-building behavior, but a recently discovered species—the finger-length white-spotted pufferfish *(Torquigener albomaculosus)*—may match the bowerbird for fanatic artistry. Males spend weeks creating an intricately patterned circle of sand more than two meters wide and decorated with shells and sand dollars. The result is a masterpiece of ichthyological engineering intended to impress a discriminating female who, if she likes what she sees, will swim to the center of the circle. The artist joins her there, gently biting her cheek while they both release gametes into the nest to mate.

Prior to the mutual release of eggs and sperm, Atlantic cod *(Gadus morhua)* engage in melodious drumming sounds and one-on-one caressing and dance displays that to human eyes appear deeply affectionate. All members of the family Sciaenidae—the 300 or so species of croakers or drums—are notorious for their distinctive vocalizations. But one in particular—the Gulf corvina *(Cynoscion othonopterus)*—produces thundering songs in the Gulf of California so loud that a gathering of thousands of lovelorn males can be heard kilometers away.

How animals in the darkness of the deep ocean find mates is largely unknown, but communication using sound, bioluminescent signals, and special chemical emissions is involved. The mysteries of deep-sea anglerfish sex have been pieced together, clue by clue. A small, strange growth discovered on the skin of female anglers turned out to be a male angler completely merged with the tissues of its mate. The male has a continuous source of sustenance, the female a ready supply of sperm to fertilize her eggs.

The life histories of most of the creatures who live in the deep sea—and even in mangrove forests, rocky shores, and tropical coral reefs—remain unknown to humans, but clearly, these animals know what, when, and how to do what it takes to ensure the continuity of their kind.

DEEPER DIVE

Mating on the Move

For speed-demon dolphins, mating means spooning on the move, male behind female. Others have more complex rituals. The hermaphrodite sea hare, a kind of sea slug with projections resembling bunny ears, takes to eelgrass beds for a community lovefest. Three at a time form mating chains so that the one in front turns female for the one behind it. Over several hours or days, driven by the pheromone attraction, the group may produce 80 million eggs. Seahorses mate for eight hours straight, dancing snout-to-snout, changing colors. The female fills her mate's brood pouch with eggs, which he fertilizes and then carries until they hatch. The argonaut, a

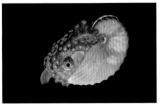

Female paper nautilus, an argonaut

distant relative to the chambered nautilus, crafts a paper-thin calcium-carbonate case that will hold 170,000 eggs per mating—a million over a female's lifetime. Into it, her 30-times-smaller mate detaches a sperm-filled arm called a hectocotylus ("hollow thing" in Greek). Once she's fertilized, the female stockpiles his arm—along with those of past mates—to use remaining sperm later. The male dies. Another femme fatale, the fertilized female octopus strangles and eats her smaller mate. Then, in James Bond–like drama, her own body begins to self-destruct through a kind of cellular suicide. Once the new generation hatches, she joins her octopedal ancestors.

Locomotion in the Ocean

Whether with paws, fins, flippers, suction cups, or sophisticated tubes just right for jet propulsion, locomotion in the ocean is tuned to the nature of water, a medium that in many ways behaves like air but is many times denser. To live is to move, sometimes at a pace marked in microns in a century, the lifestyle of microbes living in watery cracks deep beneath the ocean floor. Some life-forms, such as corals, sponges, barnacles, and most seaweeds, seagrasses, and mangroves, are stationary as adults but as juveniles are highly mobile, swimming or drifting with currents. Others start out with one mode of transportation—such as the young of flounders and halibut, which begin life as tiny, fish-shaped larvae swimming and drifting with currents and feeding on floating plankton, then later shift to a flat shape and an eye and fin arrangement better suited for life camouflaged in the sand as a bottom dweller.

The appendages of marine vertebrates that relate to the arms and legs of their terrestrial counterparts have developed into various sleek forms coupled with supple bodies that enable them to more effectively power their way through water. Hundreds of millions of years ago, the limbs of sea turtles morphed into flippers, and their bodies took on a more hydrodynamic shape than their land-based relatives. Most of the 50 or so species of sea snakes tend to have flattened, oar-like tails and spend much of their life submerged, surfacing occasionally to breathe. Marine iguanas power their way through the water by swinging their muscular tail back and forth, their land-friendly arms and legs held close to their body, thereby reducing drag.

Amphibious polar bears have typical bear-like arms and legs, but their unusual hollow hair provides buoyancy and warmth while fluffing their oversize paws with greater surface area, which is helpful in swimming. Seals, walruses, and sea lions retain their ability to lumber around on the land but are most at home in the sea, where their body shapes and powerful flipper-like appendages are most effective.

Among vertebrates, fish have evolved the most varied ways of getting around under the sea. Seahorses and pipefish tend to move slowly, propelled mostly by maneuvering their two pectoral fins, one on either side of the head behind the gills, with course adjustments made with back dorsal fin action and twists of their tails. Sea robins crawl on their spiny, finger-like pectoral fins. Flounders flap their tails as a primary power source and lift their top-facing pectoral fins as sails to determine direction. Most fish use

ABOVE: South America's southern king crab cruises the seafloor up to 1.6 kilometers a day when migrating. OPPOSITE ABOVE: Bodily contractions pulse the bigfin reef squid through Indo-Pacific waters. OPPOSITE BELOW: A banded sea krait ranges tropical Pacific coasts with its paddlelike tail.

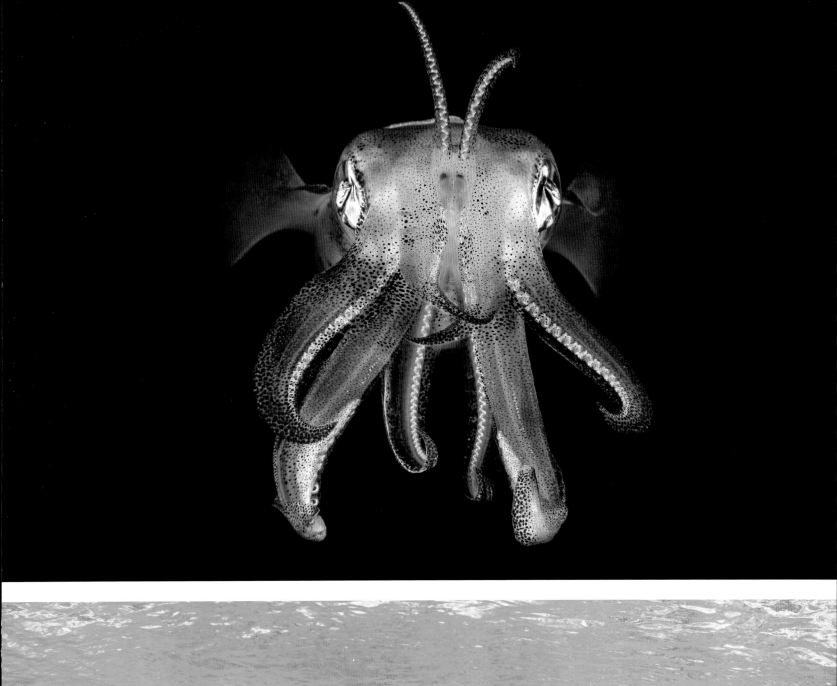

a combination of tail and body flexing for speed, with fins and spines applied as needed. Many are coated with special kinds of slime that reduce drag and make movement more efficient.

One of the fastest fish in the sea, the sleek, streamlined bluefin tuna, has remarkable features that enhance its speed and efficiency. The largest fins can retract into grooves, and a row of small rudder-like finlets along the top and bottom of its body breaks up the flow of water that would normally cause drag. Horizontal keels near the tail stabilize movement, enhance maneuverability, and minimize the resistance of water. Back-and-forth movement of the tail creates vortices—whirling eddies that push the fish forward. As biologist Carl Safina says in his *Song for the Blue Ocean,* "the fish is actually propelled, as if surfing on its own wake." Engineers at the Massachusetts Institute of Technology used creative techniques to replicate the tuna's legendary speed with a torpedo-shaped model, "robotuna," but they did not succeed in replicating the tuna's 97 percent efficiency or 65-kilometer-per-hour speed.

Spider crabs, lobsters, and numerous other crustaceans walk on masterfully articulated limbs, but some of the legs of the portunid, or swimming crab, are shaped like paddles. Sea urchins and sand dollars tiptoe along on spines coupled with numerous long-tubed suction cups that, like those of their starfish relatives, enable them to hold on to slick surfaces or, for starfish, also to pull apart prey. Their cousins, sea cucumbers, often are regarded as sluggish, bottom-dwelling creatures that look and generally behave like the vegetable for which they are named. There are deep-sea versions, however, that have winglike extensions and literally fly through the sea. Even among those that normally hug the bottom, some have the ability to flood their bodies with water until they are neutrally buoyant, so they can be carried to new locations by ocean currents.

Among invertebrates, none can match cephalopods for exceptional propulsion systems. Squid are notorious for their ability to jet in reverse at high speed, but they can also use their eight arms and two tentacles for precision maneuvering. Octopuses use water propulsion to move swiftly, and can also shape shift to squeeze through tight places or discreetly creep along the seafloor in slow motion. The chambered nautilus takes the prize for jet-propulsion efficiency, however. Graham Askew and Thomas Neil, scientists at Britain's University of Leeds, used lasers and high-speed cameras to demonstrate that as a nautilus sucks in water and spits it out, it is able to capture 30 to 75 percent of the energy it has transferred to the water to move. Squid tend to be about 40 to 50 percent efficient, and dome-shaped jellyfish that pulse their bells to squirt out water move at less than 50 percent efficiency.

In modern motorized ships, only 25 to 35 percent of the energy contained in the fuel is effectively used for propulsion based on designs that have evolved over a few thousand years. It may be useful for designers to consider systems that have their origin hundreds of millions of years ago.

From its land base a polar bear paddles up to six kilometers per hour through Arctic waters to hunt seals, propelled by large front paws.

Traveling Far, Wide & Deep

f the activities of all of the animals in the sea could be tracked and made visible, the image would be a seething, swirling network of moving trails, with patterns shaped by currents, temperature, availability of oxygen and food, and by the habits and capabilities of individual species and their varied social structures. Some creatures such as sponges, corals, and oysters are homebodies, permanently anchored in one place for their entire adult lives; others—seahorses, damselfish, conchs, starfish, sea urchins, and such—move around but stay close to a particular reef or lagoon or rocky shore. But along the coasts, across the broad surface of the sea, and deep within the ocean's dark interior, many animals are on the move, migrating over hundreds or thousands of kilometers in a season. Some are passive vagabonds, drifting with ocean currents, while others are ocean athletes, capable of powering their way over, on, or through the sea on very long-distance journeys.

Seabirds hold the record for traveling the farthest in a year, notably the feathered champion of champions, the Arctic tern *(Sterna paradisaea),* weighing just 100 grams. After nesting in the high Arctic, these marathon fliers follow the sun south, moving with favorable winds and pausing at feeding sites, eventually reaching the other end of the planet, where dining on krill enables them to have the energy to begin their long return trip north. Red knots *(Calidris canutus rufa),* after spending the warm winter months in Tierra del Fuego, migrate northward for summer breeding. As they stream by the thousands along the eastern seaboard of North America, they stop to refuel in coastal bays and inlets where horseshoe crabs *(Limulus polyphemus)* migrate from deep water to mate and lay eggs in wet beach sand. The timing is perfect, the result of ancient rhythms that coincide in favor of the birds that gorge on *Limulus* eggs before resuming their journey. Dozens of other seabird species feast on the eggs as well, but the seasonal transfusion of nutrients from the crabs to the red knots and the deposit of nutrients from the birds to the beach and into the sea form one of the longest, narrowest ecosystems on Earth.

Stanford University oceanographer Barbara Block and her students have succeeded in documenting the travels of various species by attaching markers on turtles, whales, sea lions, sharks, tuna, and even certain jellyfish to learn their migrations. After dining on deep-dwelling squid and fish, the sharks return to favored places in Mexico and California. Block has enlisted fishermen as allies in her Tag-A-Giant project to help attach instruments to bluefin tuna in the Atlantic and Pacific and thus gain better knowledge of the fishes' population size and range.

The tail markings of humpback whales have been photographed and used since the 1970s to verify the migration of specific animals over thousands of kilometers, from cold water feeding areas to tropical nurseries—and back. Now, computer programs use pattern recognition to monitor photographs of the distinctive spots on individual whale sharks or the

> "VERTICAL MIGRATION . . . OVERSHADOWS ALL THE MIGRATIONS ON THE PLANET."
>
> —BRUCE ROBISON, MONTEREY BAY AQUARIUM RESEARCH INSTITUTE

scales on the faces of different sea turtles to track their movements during their long-distance travels.

The ocean migration of birds and large animals has been witnessed and noted for centuries, but the greatest migration on Earth is less well known. It happens vertically, every night, over large areas of the ocean when trillions of small fish, squid, swimming worms, crustaceans, arrow worms, and various kinds of gelatinous animals ascend hundreds of meters from the depths to regions close to the surface. There, a banquet of phytoplankton and zooplankton is consumed, and, as the sun rises, the animals descend en masse back to deeper water.

Whether moving horizontally or vertically, over short distances or long, every living thing, from tiny crustaceans to giant whales, has contributed to the vibrant tapestry that shapes the nature of Earth in ways that favor our existence.

King of migrators, the Arctic tern flies 40,233.6 kilometers from Arctic breeding grounds to summer in Antarctica each year.

Synergies in the Sea

B iologist David Kirk declares, "It is doubtful whether there is an animal alive that does not have a symbiotic relationship with at least one other life form." Collaboration, not competition, has been the primary driving factor in shaping and achieving the success of life on Earth, according to Lynn Margulis, an eminent American evolutionary biologist who championed the concept of symbiogenesis—the idea that by cooperating, different species will survive and thrive. Her theory that critical elements of plant and animal cells including chloroplasts and mitochondria were once independent bacteria has been widely accepted and underscores a basic principle of collaboration in living systems. Merging of interests at the cellular level was preceded by merging of elements at the molecular level and followed inexorably from the development of cells to the formation of organelles, then organs, species, communities, ecosystems, and ultimately to the interdependent fabric of life that now exists.

Partnerships are everywhere in the sea. Some give benefits to both parties, symbiotic mutualism, where organisms are better off together than they would be alone. The bond between reef-building corals and numerous kinds of zooxanthellae, algae that live within their tissues, gives each an advantage. The algal cells have a home and a source of nutrients; the host corals gain a direct source of oxygen and nourishment generated by their photosynthetic residents. Giant clams, some kinds of anemones, and sponges have similar relationships with microscopic algal partners.

Alliances between tiny microbes and larger, multicellular hosts abound, from the bacteria that provide sustenance for giant pogonophorans, or plume worms, living near hydrothermal vents in the deep sea, to the complex microbial communities vital for breaking down and absorbing nutrients within the digestive tracks of oysters, fish, whales, and likely all creatures with stomachs and intestines, including scientists. In humans, microbial cells outnumber the other cells of our bodies by at least a hundred to one. Without them, we could not exist, and the same is true for other organisms that host an internal microbiome.

Numerous deep-sea animals harbor symbiotic bioluminescent bacteria within special organs that provide just the right environment for the microbes to prosper. Anglerfish have them in a specialized "lure" that is conveniently deployed in front of the fishes' large, gaping mouth. Other fish host bioluminescent bacteria under their eyes, on their bellies, down their sides, and in other strategic places useful for luring or finding prey, attracting mates, confusing predators, and likely other uses that scientists have not yet discovered.

Some fish line up like shoppers at a checkout—waiting one by one, head down, tail up, gills flared, and mouth open, while small, brightly colored fish dart into action, nipping away parasites and bits of torn skin. This is

> "OVER THE LONG TERM, SYMBIOSIS IS MORE USEFUL THAN PARASITISM. MORE FUN, TOO. ASK ANY MITOCHONDRIA."
>
> —LARRY WALL, COMPUTER PROGRAMMER

A scarlet skunk cleaner shrimp dines on dead skin and parasites meticulously picked from an Indo-Pacific yellow-edged moray eel.

"cleaning behavior," a kind of mutualism where a feast for one yields health benefits for the other. Cleaner shrimp casually clamber among the fearsome-looking teeth of moray eels, snacking on scraps while avoiding being snacks themselves. Several kinds of blind snapping shrimp share their burrows with bright-eyed goby fish, who serve as lookouts for predators and signal the shrimp when retreat is prudent.

There are small anemones whose only home is in the claws of tiny boxer or pom-pom crabs. Each may gain protection and food-acquiring advantages from the other. Numerous animals benefit from camouflage while the creatures providing it have movable anchorage, including sponge crabs, abalones, queen conchs, decorator crabs, and hermit crabs that carry a specific anemone on their borrowed shells.

Symbiotic relationships that favor one party without harming the other are known as commensal partnerships. It is easy to see advantages to the soft-bodied pinnotherid crabs that enter an oyster, sea urchin, sea cucumber, worm tube, or glass sponge as larvae and grow inside their host. There they dine, mate, and produce offspring that as planktonic larvae seek new homes of their own. The presence of the little crabs seems not to harm their hosts, but it is not likely to help them, either. Similarly, the small crabs that live on the underside of sand dollars and heart urchins, the brittle stars that are tightly wound within the branches of soft corals and sponges, and the barnacles that hitch a ride on the backs of sea turtles and whales are taking benefits without obvious favors in return. The fat innkeeper worm, a member of the phylum Echiurida, carves burrows into soft, muddy sand, providing a home for as many as 17 different species of crustaceans, annelids, and even small fish. There is no apparent benefit to the innkeeper since the tenants pay no rent, but they do provide interesting company.

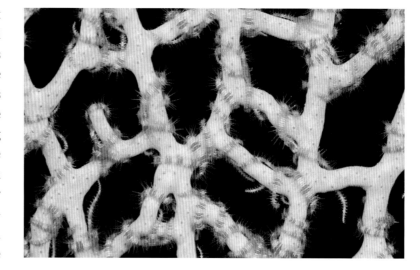

ABOVE: Brittle stars hug their gorgonian host. OPPOSITE: Immune to stings within the embrace of a bulb-tentacle sea anemone, a spine-cheek clownfish watches for predators in Indo-Pacific waters.

A different relationship, parasitism, occurs when one organism prospers at the expense of the other. In the sea, many crustaceans—copepods, barnacles, isopods, and others—have evolved as freeloading ectoparasites that cling to the gills or skin of fish or other hosts, biting or sucking sustenance. Endoparasites also take many forms, from protozoans and flatworms to roundworms and strangely modified barnacles that live within the organs and tissues of other animals. Successful parasites are rarely lethal, because if the host dies, so does the boarder, but they can reduce a host's vitality, causing it to be more vulnerable to stress, disease, and predation.

From bonds between individuals, to social gatherings, to ecosystems, and ultimately to the entire planet, all living things appear to be dependent one way or another on the existence of other living things, and all multicellular organisms have ties to the existence of microbes. As naturalist John Muir observed, "When we try to pick out anything by itself, we find it hitched to everything else in the universe."

Conservation

n 1804, when Captain Meriwether Lewis and Captain William Clark embarked on the Corps of Discovery expedition across North America to find the Northwest Passage, hundreds of millions of bison ranged over the continent, and billions of migrating passenger pigeons darkened the skies. During the following century, the advance of humans from east to west had so consumed American wildlife and wilderness that there were calls for protection. President Ulysses S. Grant designated the first national park, Yellowstone, in 1872. That same year the H.M.S. *Challenger* set out to explore the ocean on the first global oceanographic expedition.

By 1900 only a few hundred bison remained, mostly in private reserves, and the last wild passenger pigeon was shot in 1901. President Woodrow Wilson signed into law the consolidation of the nation's parklands as the National Park System in 1916, acknowledging the need to protect "natural, historic, and cultural" values. Two years later, the Migratory Bird Treaty was signed by the United States to protect songbirds and other migratory species that were in sharp decline. Subsequent national and international laws were enacted to maintain wildlands and wild species. By 2000 there were parks, protected areas, international laws, and hundreds of conservation organizations that recognized the importance of embracing Earth's natural living systems—mostly terrestrial—with care.

Meanwhile, well into the 20th century, the ocean seemed so vast, so resilient, that humans could do little harm, no matter how much wildlife, oil, gas, or minerals were taken out, nor how much waste was deliberately or inadvertently put in. Coastal marshes, mangroves, and seagrass meadows were considered to be trivial tradeoffs for engineered deep-water ports, inland waterways, and increased waterfront property. Little thought was given to the consequences of noise generated by growing ship traffic, seismic testing, or the use of sonar for mapping, locating underwater navigational hazards and shipwrecks, or fish finding. No one seemed concerned about using water as ballast for cargo vessels, unaware that hitchhiking organisms were being transported from places where they lived peacefully with their neighbors to new locations where they often became aggressive, invasive newcomers.

Until recent decades, we did not widely appreciate the connections between burning fossil fuels and the resulting impacts on planetary temperature, chemistry, and the ancient rhythms of life on land and in the sea. Even now, despite knowing the dire consequences of these actions, many current policies, laws, and attitudes reflect profound complacency.

One hundred years after the first national park came into being, and a century after the *Challenger*'s first global expedition launched, the United States passed legislation in 1972 authorizing the establishment of national marine sanctuaries. Australia created the Great Barrier Reef National Park Authority in 1975 to protect more than 1,500 kilometers of reefs along its eastern coast. By the year 2000, many nations had enacted policies aimed

WHY IT MATTERS

Losing Marine Life

While land and air pollution have loomed for decades, the ocean has seemed immense and untouched. No longer. Pollution ravages, waters acidify, marine habitats disintegrate. Life in once pristine waters suffers what a 2017 report from the National Academy of Sciences called "biological annihilation." As of 2020, fish species threatened with extinction totaled 2,849 on the International Union for Conservation of Nature (IUCN) Red List, joined by more than 3,000 other marine species. Hovering just above extinction are coelacanths, bluefin tuna, and some sea turtles.

Why does it matter if we lose marine species? Because they may produce more antibiotic, anti-cancer, and anti-inflammatory medicines than any land-based organism. Because they inspire aerodynamic, architectural, robotic, and biomedical designs. Because their role in the food web, in sustaining ocean nutrients, and producing life-giving oxygen helps maintain the balance of life on Earth. And because unknown species—and our opportunities to mine their wisdom—may be gone before we even discover them.

A fishing line drag means injury or drowning for an olive ridley sea turtle.

at sustaining populations of marine mammals, birds, fish, and other ocean wildlife, but by then only a fraction of one percent of the ocean had been proactively embraced as protected places.

At the time, large areas of the ocean were protected by their inaccessibility—especially in polar regions, the deep ocean, and the High Seas. By 2020, about 3 percent of the ocean had been safeguarded by laws aimed at full protection, and as much as 10 percent had some form of protection. However, the decline of ocean health has outpaced global action.

Human impacts are altering the nature of nature in ways that are undermining the capacity of Earth to support conditions favorable to life as it currently exists—including human life. The good news is that a great deal of new knowledge is widely available. The challenge now is to put into place vital actions soon enough to slow, stabilize, and reverse the current negative trends.

In a thoughtful article in *Science* magazine in 2019, marine ecologists Jane Lubchenco and Steven Gaines summarized the current state of ocean conservation, concluding, "The ocean is not too big to fail, nor is it too big to fix. It is too big to ignore."

Shark fishing in the Seychelles will be curbed with the archipelago's 2020 commitment to protect 30 percent of its waters—400,000 square kilometers—and support the comeback of hammerhead sharks, giant tortoises, seabirds, and others.

GREAT EXPLORATIONS

Census of Marine Life

This larva of a deepwater decapod—a crustacean—was found on a Census of Marine Life expedition.

Fred Grassle was frustrated. After pioneering research on deep-sea biodiversity at Woods Hole Oceanographic Institution, he had joined Rutgers University to start its Institute of Marine and Coastal Sciences. A 1995 report he'd compiled for the U.S. government on identifying marine biodiversity was languishing, and he wanted action. The world needed an organized archive of ocean life that could be shared by everyone. He huddled with colleagues including Woods Hole scientist Jesse Ausubel, and in 2000 they launched the visionary Census of Fishes—soon to become the Census of Marine Life (COML)—with the intent of documenting all sea life. "Everyone said we were crazy," Ausubel later wrote.

Over the next decade some 2,700 scientists from 80 nations snorkeling, scuba diving, on ships, in submersibles, and using satellite imagery gath-

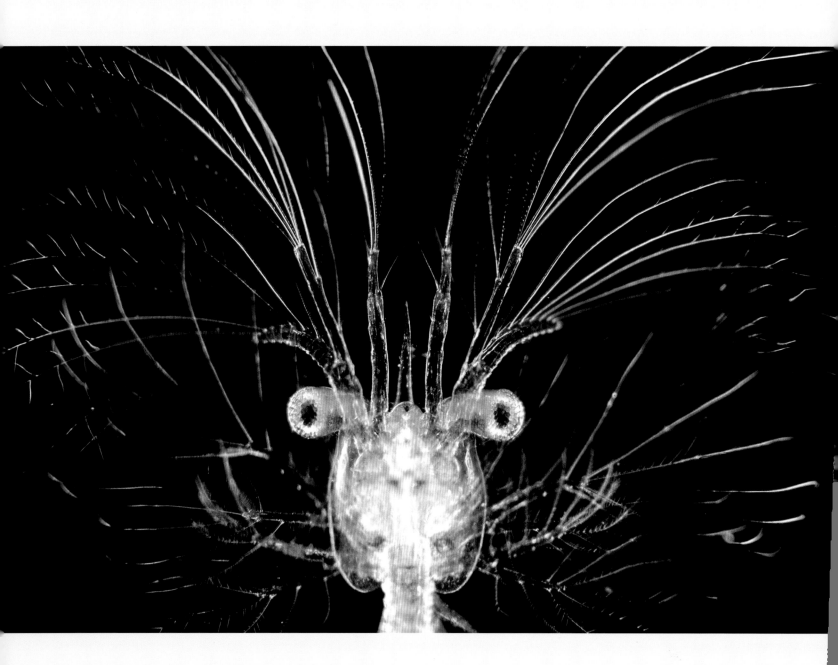

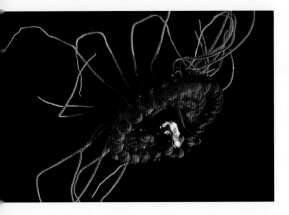

ABOVE: The Gulf of Maine's medusa jellyfish wields stinging tentacles to grab prey. RIGHT: Minerals from black smoker vents on the Mid-Atlantic Ridge sustain tubeworms, anemones, and shrimp.

ered data on habitats and animals from deep-sea hydrothermal vents in the mid-Pacific to waters along Antarctic ice shelves to determine: 1) who lived in the ocean past, 2) who lives there now, and 3) who is expected to inhabit it in the future. They also estimated the numbers of each species, how those numbers have changed over time, what role humans have played, and how we can reverse the damage by humans to ensure a healthy future.

COML discoveries have led to greater understanding of—and respect for—the ocean. What humans can compete with the awe factor of the Jurassic shrimp (*Neoglyphea neocaledonica*), considered extinct for 50 million years but found to thrive on an underwater peak in the Coral Sea off Australia? Or the 38,000 kinds of microbes that inhabit a single liter of water? Or the community bonds of a shoal of fish the size of Manhattan that swarms off the coast of New Jersey? Or the endurance of the sooty shearwater, whose annual migration around the planet takes it 70,000 kilometers, often flying 1,000 kilometers a day.

Through its tracking and documenting systems, COML and its partners, including the Alfred P. Sloan Foundation, the National Geographic Society, Galatée Films, and the multi-institutional Encyclopedia of Life, encouraged global collaboration. New technology included genetic barcoding to quickly identify species; electronic tagging to get real-time views of habitats and behaviors; and the publicly accessible Ocean Biogeographic Information System (OBIS), the world's largest database of marine life.

Over its decade, COML formally identified 1,200 new species and recorded information on 5,000 more. When it officially ended in 2010, its work had just begun. Estimates say there may be as many as 10 million ocean species to identify, and OBIS, now operated by the United Nations Educational, Scientific, and Cultural Organization's (UNESCO's) Oceanic Data and Information Exchange, with other key marine life databases, is growing fast. By 2021, its accepted species count had reached nearly 140,000.

Census Superlatives

1996: Fred Grassle and Jesse Ausubel start brainstorming the Census of Marine Life (COML).

2000: Scientists begin 14 field projects: polar oceans, mid-ocean ridges, vents and seeps, seamounts, abyssal plains, continental margins and shelves, coasts, coral reefs, regional ecosystems, microbes and zooplankton, and top predators.

2002: COML begins building digital database Ocean Biographic Information System (OBIS), with these sample entries:

- Yeti crab (*Kiwa hirsuta*, hairy-armed crab), 2,133 kilometers deep, Easter Island
- Atolla jellyfish (*Atolla wyvillei*), Japan, which "scream" for help via bioluminescence
- Bob Marley worm (*Bobmarleya gadensis*), Caribbean, named for its "dreadlock" tentacles
- Four-eyed fish (*Anableps anableps*), South America, which sees above and below the surface at once
- Jurassic shrimp (*Neoglyphea neocaledonica*), New Caledonia's Coral Sea, living relatives of 50 million+-year-old fossils
- Deepest recorded benthic comb jelly, Japan's Ryuku Trench, 6,000 meters deep
- Hottest, deepest known black smoker (a kind of hydrothermal vent), Atlantic's Ascension Island, swarming with shrimps, clams, bacteria
- Crustaceans, jellyfish, and single-celled creatures under 700 meters of Weddell Sea ice, Antarctica
- The manylight viperfish (*Chauliodus sloani*), which lives in a quarter of the world's ocean; called the deep ocean's "Everyman" by scientists

2010: COML formally ends; Encyclopedia of Life begins uploading 90,000 pages of species.

2021: OBIS holds some 65 million georeferenced records for instant access and has accepted some 140,000 species.

COASTAL
SEA
LIFE

Chapter Five

"THE OCEAN IS A WILDERNESS REACHING ROUND THE GLOBE, WILDER THAN A BENGAL JUNGLE AND FULLER OF MONSTERS, WASHING THE VERY WHARVES OF OUR CITIES AND THE GARDENS OF OUR SEASIDE RESIDENCES."

—HENRY DAVID THOREAU, CAPE COD, 1864

Gently rippling along a coastal marsh or in thundering waves crashing on rocky cliffs, dynamic forces are in play where land and sea converge. Living in any part of the 620,000 kilometers of the world's ocean coasts means coping with the certainty of uncertainty. Instability is normal where tidal rhythms, waves, and storms shape and reshape the contours of what inhabitants along the ocean's edges regard as home. Creatures must adapt to changes in temperature, tides, and periodic drying. From the high tide line to the rim of submerged land around continents and islands, coastal ocean life also is impacted by the nature of the land above and the flow of fresh water from rivers.

Adapting to shifts in salinity poses challenges at river mouths and where groundwater seeps from land into the ocean through subterranean channels. Globally, 165 major rivers and tens of thousands of smaller rivers and streams pour fresh water, sediment, and a cargo of other material from upstream into the sea. There, intricate marshes and broad deltas form highly productive feeding and breeding areas for wildlife from the land as well as the sea. Seabirds gather to feast on the metropolis of intertidal organisms: tens of thousands of small polychaete worms, crustaceans, mollusks, and other creatures that burrow into each cubic meter of healthy mud or sand. Giants are attracted to coastal beaches, too, as National Geographic Explorer Michael Fay discovered when he emerged from his epic Megatransect across the Congo Basin of Africa in 1999 to the coast of Gabon and witnessed elephants strolling the waterfront and hippopotamuses surfing within sight of singing humpback whales.

Canada is the country with the longest coastline, with 202,801 kilometers of rugged, mostly icy edges of high Arctic, Pacific, and Atlantic land and sea that also includes the world's largest estuary, the Gulf of Saint Lawrence. The next longest coastline borders Norway, 58,133 kilometers of filigreed shores ranging from the frozen waters of Svalbard to hundreds of temperate fjords that bring salty fingers of ocean water—and ocean wildlife—far inland. The country with the shortest coast is Monaco, a small but meaningful 4.1-kilometer gateway to the Mediterranean Sea and the blue world beyond.

RIGHT: In coastal waters off Chile's San Ambrosio Island, a starfish shares space with sea urchins, fellow echinoderms. PAGES 206-207: Trumpeting threats through their inflatable proboscises, elephant seal bulls in their prime—up to four meters long and two metric tons each—battle for breeding rights on a North Pacific shore.

Living Coastlines

merican poet Sarah Kay suggests that "there's nothing more beautiful than the way the ocean refuses to stop kissing the shoreline, no matter how many times it's sent away." Smooth and sloping, steep and ragged, or fringed with reefs or trees, the edges where land and sea converge are exuberant with life. Seabirds gather along coasts for food, shelter, and nesting sites; sea turtles travel thousands of kilometers to return to favored beaches to mate and lay eggs; seals, sea lions, sea otters, walruses, whales, and dolphins range widely but are lured back to known coastal havens to mate and give birth. Coral reefs flourish in shallow, tropical seas, as do seagrass meadows, marshes, and mangroves that rival rainforests as highly productive regions of diversity. But coasts have irresistible appeal to people, too. A disproportionate number—about 60 percent of the world's human population—currently live within 100 kilometers of an ocean shore.

More than half of the world's coasts are rocky, including high, steep cliffs such as those bordering much of the coast of Chile and western North America, where vertical relief on the land is matched by deep water that hugs the coastline. European shores, the west coast of New Zealand, and the southern and eastern shores of Australia are examples of rocky coasts that extend seaward along wide, flat continental shelves. These rugged habitats are home to legions of life above and below the edge of the sea.

Recently developed satellite technologies have made possible a global survey of shorelines and have determined that about one-third of the ice-free coasts are sandy or gravelly. Based on Landsat records going back to 1984, NASA and the U.S. Geological Survey have demonstrated that some coasts have been eroding or subsiding while others have been growing. Africa has 66 percent of the world's sandy beaches, but most are gradually eroding. Marshes, beaches, and barrier islands in the Gulf of Mexico along the coast of Louisiana and Mississippi are shrinking, with more than 5,000 square kilometers lost since 1930.

The highest rate of growth is taking place in Asia, in part due to the process euphemistically called "reclamation," where sand and mud are taken from the ocean to create land called "fill" for farms, communities, and industrial complexes. Worldwide, natural coastal systems are being displaced to make way for various human uses. Along the coast of Dubai, in the United Arab Emirates, elaborate landfill islands built in the shapes of spreading palm trees and a map of Earth support resort housing, as well as an extensive market and restaurant complex. While the amount of linear coast is increased, these man-made structures have replaced wildlife that prospered there until just a few decades ago.

Starting in the mid-1800s and continuing far into the 20th century, much of San Francisco Bay, including vast expanses of seagrass, was converted to land where tall buildings now stand. Similarly, Florida's Tampa Bay, Miami Beach, and numerous other harbor cities across the globe have

A canopy of stars illuminates Tasmania's Sandy Bay from above, while light shines below from a galaxy of bioluminescent plankton.

made changes of geological magnitude to the configuration of their coasts.

In 2019 geologist Robert Sayre and several co-authors reported in the *Journal of Operational Oceanography* about a new 30-meter spatial-resolution global shoreline vector (GSV) study based on Landsat data. They acknowledge that "management-scaled inventories of global and coastal ecosystems are not yet available." Still, there is a sense of urgency in filling this need, owing to rapid impacts brought about by industrialized activity and the magnified importance of healthy coastal areas to the prosperity of life, human and otherwise.

Humans & the Coasts

Viewed from space, the margins of continents and islands clearly show the role of humans in shaping the geography of places where land and sea converge. If an overlay could be imposed of current coastlines on those of a thousand years ago, remarkable changes would be evident virtually everywhere people have established a presence. In the past 100 years, and especially since the 1950s, about half of the world's coastal forests, marshlands, seagrass meadows, and coral reefs have been displaced to accommodate human structures. They have been consumed by mining for sand, offshore drilling operations, pollution, aquaculture, and destructive fishing and shipping activities. Since 90 percent of goods traded globally are currently transported by sea, port cities have expanded rapidly to accommodate large ships and the associated facilities required to support commerce. Harbors for naval operations also have grown significantly across the world in recent decades.

Some wild fjords remain in Norway where seabirds, dolphins, orcas, and other whales come to feast on seasonal surges of herring. But starting in the 1970s, fjords in Norway, Chile, and New Zealand, and coastal embayments in the United States, U.K., Canada, Australia, and Russia, have been converted for use as fish farms, mostly to produce large quantities of Atlantic salmon *(Salmo salmar),* a species domesticated by Norwegian entrepreneurs. In the wild, this species lives for years at sea and returns to spawn, sometimes repeatedly, far upstream in specific rivers of North America and Europe. In the same way that agriculture has displaced forests on the land, salmon farming has replaced and disrupted extensive native ecosystems.

Similarly, large areas of mangroves and marshes in Thailand, China, Brazil, Ecuador, Mexico, and parts of the U.S. Gulf of Mexico have been carved into shallow ponds where several species of shrimp are farmed, mostly for export to global markets that have developed in recent decades. About half of the mangrove forests globally have been lost to farming and other coastal uses that likely seemed justified at the time, but new awareness of the importance of mangrove systems to the health and well-being of people is spurring protection and restoration projects globally. Sri Lanka, with coasts heavily impacted by the 2004 Indian Ocean tsunami, became the first nation to comprehensively protect all of its mangrove forests.

Where the Mississippi River meets the Gulf of Mexico, a lacy, 300-kilometer-wide, roughly triangular-shaped delta often likened to a giant bird's foot shifts in size and shape depending on the volume of water and sediments that are continuously being swept into the sea. Here thrived

ABOVE: Called "flowers of the sea," anemones, related to corals, often anchor along rocky shores. OPPOSITE: Seawater flowing into the rivers of Norway's fjords shelters marine life, in some places deep-sea corals.

brown pelicans, clapper rails, seaside sparrows, and gulls along the shore; blue crabs, shrimp, oysters, and paddlefish in shallow waters; and sperm whales offshore. That flow has been greatly reduced over the years with resulting changes in salinity, addition of toxic chemicals and fertilizer, and depletion of material needed to maintain coastal wetlands. According to a study called *Drawing Louisiana's New Map,* 62 square kilometers of wetland were lost each year between 1990 and 2000, owing to processes that continue to reshape and alter the nature of both land and sea. Some of the most notable coastal changes caused by human endeavors start far upstream in rivers where more than 800,000 dams of different sizes have been built to harness or divert waterways for agriculture, hydroelectric power, and various other industrial and domestic uses. For instance, the flow of the Mississippi River into the Gulf of Mexico has been boxed in by more than 5,000 kilometers of levees and a complex system of locks and dams intended to control water flow throughout the river's network of tributaries.

The world's largest delta is the Ganges-Brahmaputra, shaped by the tidal ebb and flow of the Bay of Bengal in India and Bangladesh. Here prosper wetlands with dozens of species of mangrove and other flora, and marine life including shellfish, saltwater crocodiles, and the endangered olive ridley sea turtle. It is also a fertile agricultural area where people and wildlife are vulnerable to flooding and increasing pressure from human uses of the coastal wetlands. Construction of the Aswan Dam in Egypt in 1960 reduced annual flooding in the Nile Delta, but also caused loss of silt and nutrients along the riverbanks and has resulted in a noticeable shrinking of the delta where the Nile flows into the Mediterranean Sea.

Even more dramatic coastal shrinkage has occurred to the delta and estuary in the upper reaches of the Gulf of California, where, over thousands of millennia, the Colorado River brought fresh water and nutrient-rich silt to the sea, feeding what ocean explorer Jacques Cousteau called "the world's aquarium." The Gulf is home to more than 900 kinds of fish, thousands of invertebrates, and 33 species of dolphins and whales including the smallest, the vaquita, and the largest animal ever, the blue whale. Now the Colorado River ends several kilometers from where it once flowed into the Gulf.

OPPOSITE ABOVE: Dovekies, puffin relatives, flock to nest on a rocky Norway coast after feeding on ocean plankton and crustaceans. OPPOSITE BELOW: Shrimp, crab, and oysters thrive in Mississippi Delta waters; its shores shelter seabirds.

MISSION BLUE

Ocean Color From Space

Since 1978, when NASA launched its Coastal Zone Color Scanner (CZCS) experiment, satellites have been determining coastal ocean health through remote sensing instruments that measure color. Yellow, for instance, may signal suspended particles of runoff from rivers. As of 2020, with six "color-coding" NASA satellites circling the globe—including the Geostationary Ocean Color Imager (GOCI)—scientists are focusing on the color of phytoplankton populations. Single-celled phytoplankton may be the size of a pinhead, but when masses

Black Sea plankton

bloom, their chlorophyll—a green pigment—turns coastal waters such vibrant shades of blue-green that satellites can easily measure the extent from hundreds of kilometers above Earth. The bluer the water, the less dense the bloom; the greener the water, the more dense. Using carbon dioxide for photosynthesis, big populations pull more carbon from the atmosphere. If sustained over time, such populations could significantly lower carbon dioxide levels, and, in turn, lower average ocean temperatures, helping combat global warming.

Finding Their Way

How do migrating fish find their way as they travel ocean highways? By following coastlines? Detecting scents? Responding to Earth's magnetic field? To help discover answers and draw attention to the many kinds of fish that migrate over long distances, April 21 has been designated World Fish Migration Day. There are many kinds of fish to consider.

Atlantic and Pacific salmon leave the coast after hatching in freshwater rivers and, with remarkable fidelity, return to those same rivers years later to begin the next generation. In reverse, American and European eels begin life in the Sargasso Sea but are drawn to the coastal areas from which their parents departed several years prior. How they find their way across hundreds of kilometers of open sea to places they have never seen remains a tantalizing mystery.

In a torrent of sleek bodies and flashing fins, billions of sardines turn the cool waters along the South African coast into a moving banquet for thousands of seabirds, dolphins, whales, larger fish, and millions of small animals that thrive on the scraps. Following the shore and running counter to the warm, southbound Agulhas Current, the migrating *Sardinops sagax* annually provide one of the ocean's greatest wildlife spectacles. It is the food web in action: plankton being converted into sardines, sardines becoming the energy powering a frenzy of creatures with fins, feathers, fur, scales, tentacles, and teeth.

Hugging both coasts of Florida, southbound mullet *(Mugil cephalus)* darken the sea in columns many kilometers long while en route to Gulf Stream breeding grounds. The 80 or so species in the family of mullets are typically coastal, whether along major landmasses such as eastern Australia or along island edges as in the Galápagos.

Other sea creatures are long-distance swimmers as well. For sea turtles, returning to shore is an essential phase of a mostly oceanic lifestyle. All eight species range thousands of kilometers at sea but eventually return to the beaches of their birth to mate and build nests close to where they began their life's journey. Leatherback sea turtles are renowned for their long migrations at sea, where they are largely at the mercy of ocean currents. But with uncanny navigation capabilities, using the Earth's magnetic field, they find their way back to the place of their origin.

From his office window at the Scripps Institution of Oceanography, ichthyologist Carl Hubbs counted gray whales as they passed by, migrating from feeding areas in the Bering and Chukchi Seas to warm, shallow lagoons along the western side of Baja California thousands of kilometers to the south. There, parent whales give newborns their first swimming lessons. In 1946 Hubbs helped lead efforts to save the few whales that remained, and in 1947 killing gray whales was prohibited by the International Whaling Commission. Commercial whaling of all species was banned in 1982, but whales and other wildlife are increasingly in conflict with people.

Off the coast of Sri Lanka, green sea turtles recently hatched on a sandy shore try out their new ocean home.

A Whale's World Record

Varvara's 22,511-kilometer swim across the Pacific turned out to be the world record for any mammal migration. Starting from her summer feeding grounds off Russia's Sakhalin Island, the nine-year-old female western north Pacific gray whale swam across the Bering Strait, then south along the North American coast to Baja California in Mexico. Next, she joined her "cousins"—eastern north Pacific gray whales traveling from north of Alaska to Mexico. Until recently, scientists believed that the western cohort was isolated across the Pacific,

Gray whale, Vancouver

following their own pattern of migration. All that changed after scientists led by Bruce Mate, director of Oregon State University's Marine Mammal Institute, electronically tagged and tracked Varvara from 2011 to 2014. Perhaps the western whales aren't a different group at all, but an extension of the eastern cohort. Mate also says that Varvara may be a Mexican, not a Russian, native. Knowing the annual migration habits helps Mate and others pursue ways to protect and grow endangered populations on both sides of the Pacific.

When Is the Ocean a Sea?

What's in a name? Ocean basins are fringed with features known variously as bay, gulf, lagoon, cove, sound, bight, and, most often, sea. "Ocean" and "sea" are terms often used interchangeably, but geographers define a sea as a portion of the ocean that is partially surrounded by land. Size does not define a sea. The Mediterranean Sea is more than 2.9 million square kilometers in area, while to its northeast the petite Sea of Marmara, which connects the Aegean Sea and the Black Sea, is less than 13,000 square kilometers. The Caribbean Sea covers 2.7 million square kilometers, and it in turn is connected to the ninth largest body of water on Earth, the Gulf of Mexico.

There are 50 or so places on Earth referred to as "seas." The Arctic Ocean embraces six sizable seas that range in area from the Barents Sea—1.4 million square kilometers—to the 57,000-square-kilometer Wandel Sea. Those two, together with the Kara, Laptev, Chukchi, and Beaufort Seas, as well as Baffin Bay, border the lands at the top of the world and are notably cold, partly or totally frozen for at least a portion of the year. Hudson Bay, although south of the Arctic Circle, is regarded by the International Hydrographic Organization as part of the Arctic Ocean. Canada regards the Hudson Bay as an internal body of water, with outlets to both the Atlantic and the Arctic Oceans. On his fourth voyage to the region, British explorer Sir Henry Hudson was set adrift in a small boat when the crew of his ship, *Discovery,* mutinied in 1611. He likely would not care whether those icy waters that would later bear his name were considered a bay or a sea.

In the North Pacific, just south of the Arctic Circle, the Bering Sea is bordered by the North American continent on the east and by Asia on the west. Native people know these waters by many names, but it officially carries the name of Vitus Bering, a Danish navigator and officer in the Russian Navy and the first European to systematically explore the region. In 1741, on the last of Bering's 16 years of voyaging, the young naturalist on board, Georg Wilhelm Steller, managed to collect or describe 143 species of plants as well as many kinds of birds, mammals, and fish during just 10 hours ashore in Alaska. The Steller's jay is named for him, and so is the Steller's sea cow, a giant, gentle manatee last known to exist just 27 years after Steller identified it. The Southern Ocean has cold-to-frozen seas along the Antarctic continent, notably the Ross, Weddell, Amundsen, and Bellingshausen Seas, named for early explorers from Britain, Scotland, Norway, and Russia.

Most of the Earth's embayments occur around the planet's warm-to-temperate midsection—the Red Sea, Persian Gulf, Arabian Sea, Andaman Sea, and the Bay of Bengal. More than a dozen major warm-water seas embellish the coasts of Australia, China, Japan, the Philippines, and Indonesia. In the Atlantic, numerous bays, sounds, channels, gulfs, and seas carve the coasts.

Ironically, the largest body of water known as a sea defies the definition. Occupying almost two-thirds of the North Atlantic Ocean, the loosely defined, slowly rotating Sargasso Sea is bound by liquid currents—not land.

After a plunge dive, a brown pelican, said to have a beak that can hold more than its belly can, captures fish in Turtle Bay, off Santa Cruz Island in the Galápagos.

The Abundance of Coastal Habitats

Emerald mangrove forests and salt marshes embroider the margins of tropical and warm temperate seas. More than 60 species of seagrasses—plants with true flowers and seeds—thrive from warm tropical to cool temperate regions in depths to about 30 meters. Phytoplankton and benthic algae go deeper, using special pigments to capture the low level of sunlight that powers photosynthesis in depths beyond 200 meters.

Even in polar regions, where enough sunlight penetrates ice shelves and frigid water to support photosynthesis, rich communities of life occur. Masses of diatoms and other microbial forms grow on the underside of pack ice as well as in the surrounding water in numbers so rich that the sea at times turns the color of a grassy lawn. Crustaceans swarm to feed on them, then they become food for fish that in turn become meals for other animals

Evergreen forests on the land give way to forests of kelp in the sea along broad areas of the cold temperate coasts in both the Northern and Southern Hemispheres. California's giant coastal redwood trees *(Sequoia sempervirens)* may exceed 100 meters in height and live for more than a thousand years—the fastest growing cone-bearing tree. But for speed in attaining height, no land plant matches the "redwoods of the sea," the giant kelp *(Macrocystis pyrifera)*. Under the right conditions, kelp grows taller by more than 30 centimeters a day. Although individuals may reach the height of coastal redwood trees, kelp remains anchored for only a few years and can be replaced in a season, not centuries. Widespread along the western coasts of North and South America, *Macrocystis* was observed along the rocky shores of Tierra del Fuego by Charles Darwin, who noted kelp's cushioning effect against shore-bound waves.

Several species of the humble oyster, en masse, can also provide storm protection, as well as maintain a productive coastal reef habitat. In his book *The Big Oyster,* Mark Kurlansky describes how billions of *Crassostrea virginica,* the Eastern or American oyster, formed such massive mounds in New York Harbor before the 20th century that they were hazards to navigation. Consumed by native populations for millennia, oysters became a wildly popular culinary attraction for European settlers in the 1600s and eventually became the target for insatiable American and European markets. It takes about five years for an oyster to reach market size, and, left in place, individuals may live as long as humans.

By 2000 nearly all of New York's oysters had succumbed, done in by either pollution or their popularity as food. Now the nonprofit Billion Oyster Project is giving *Crassostrea* a break. Founder Peter Malinowski is optimistic. By reintroducing thousands of oysters to suitable locations in the New York Bay, reef restoration is under way. The recovery in New York of oysters, valued not only as tasty morsels but even more for their role in filtering enormous quantities of not-so-pure water, is inspiring promising partnerships between man and mollusk along other coastal environments.

ABOVE: Mangrove forests line tidal inlets of Australia's Kimberley Coast, a protected marine park. OPPOSITE: Mangrove roots shelter silversides in coastal Cuba.

The Power of Seagrass

Plant seagrass. The simple solution surprised Caribbean communities who'd been pouring millions into expensive concrete walls and other treatments to save their economy-dependent tourist beaches. The finding was published in a 2019 issue of *BioScience* by engineers and scientists responding to the red alert that the beaches were disappearing to buildings, complicated by rising sea levels and increasing storms due to climate change. A beach with both healthy seagrass beds and calcifying algae "is a resilient and sustainable option in coastal defense," reported lead author Rebecca James from the Netherlands University of Groningen and the Royal Netherlands

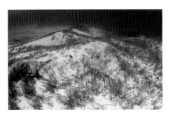

Turtle grass, Bahamas

Institute for Sea Research. Turtle grass, the most abundant seagrass along Caribbean coastlines, is anchored by an armor-strong horizontal root system that holds sand and silt in place. The roots strengthen beaches and keep coastal waters clear so coral communities—with resident algae that require sunlight to photosynthesize—can thrive. In turn, strong barrier reefs become a frontline defense against storms, ensuring the calm waters most seagrasses need to grow. While protecting one another, this duo also supports each other's communities—bacteria, fish, and manatees in seagrass; polyps, fish, and octopus in reefs—contributing to a healthy ocean system.

HOPE SPOT

Chilean Fjords and Islands

Along its 6,400 kilometers, Chile's coastline is a maze of breathtaking cragged and creviced fjords and offshore islands. Whales, dolphins, seals, and other marine mammals thrive along the coast, and the fjords are home to unique species of fish, coral, sponges, and other animals. The Chilean government has established more than 40 percent of its coastal waters for protection from human impacts. Waters around Rapa Nui (Easter Island) are open only to local fishermen using traditional methods of capture.

....

A basket star forages with lacy arms in a Chilean fjord.

Sculpting the Shore

The carpenters of Norwich, England, likely had new respect for chalk after the distinguished geologist Thomas Huxley spoke to them in 1868. "A great chapter in the history of the world is written in the chalk," Huxley observed, referring to the fossilized remains of single-celled planktonic coccolithophorids and other shelled creatures that are the principal ingredients for the chalky deposits underpinning broad areas of Northern European coasts and limestone formations throughout the world.

Volcanic rock, sandstone, and even granite are shaped by wind and water into cliffs, stacks, and arches in coastal regions globally. In the Galápagos Islands, Darwin's Arch and Kicker Rock are among the iconic features born of volcanic action, now providing perches for seabirds above the ocean's surface and homes for marine creatures in the depths below. Volcanic outpourings provide the foundation for many oceanic islands and rugged coastlines. The spectacular Giant's Causeway along the north coast of Northern Ireland, resembling elegantly carved steps leading to the sea, consists of about 40,000 interlocking basalt columns derived from the cooling of volcanic material. According to Icelandic folklore, the enormous basalt stacks lining the coast of Reynisfjara were once trolls who tried to lure ships to the rocky shore at night; they ventured too far and in the light of dawn were turned to stone. Actually, the rocks themselves are gradually being turned to black sand, the result of persistent erosion by wind and water.

Like dunes in a desert, coastal mounds of sand are shaped and moved by wind, sculpted into restless slopes, joined by rippled troughs. Held steady in places by vegetation—sea oats, beach grasses, sandburs, ground-hugging morning glory vines, and other dune-loving plants—coastal dunes may resist the relentless sweep of wind, at least for a while. Storm surges and winds eventually bite into these ephemeral mountains, sweeping sand to new destinations, building offshore sandbars, and creating new beaches kilometers away.

Animals living within the dynamic interface between land and sea have developed nimble strategies for survival tuned to the predictable rhythms of change, including periodic strong winds, exceptional high tides, and storm surges. Reproductive cycles of many—from burrowing ghost crabs to beach-spawning fish such as California's grunion (*Leuresthes tenuis*)—have adapted to tidal and seasonal pulses and can weather occasional highs and lows of wind and water.

The Scottish architect Ian McHarg in his insightful book *Design with Nature* strongly advises against building structures on a beach or dunes, not only recognizing the folly of seeking permanence in places that are naturally unstable but also respecting the vital protection that beaches and dunes provide. His wisdom has been vindicated in coastal communities that have kept their natural dunes and beaches intact and have homes still standing after great storms.

Legend says that giants built it, but Giant's Causeway, a UNESCO World Heritage site in Northern Ireland, formed from volcanic action 60 million years ago.

Reefs:
Some Like It Hot

G lowing like opals scattered across warm, shallow waters around the midsection of Earth, coral reefs have beguiled humans throughout history. It is little wonder that people once regarded corals as plants ("zoophytes"), given the colorful, flowery, bushy, and tree-like appearance of many. Whether forming upright branches, mushroom shapes, or lumpy mounds, colonies of individual coral polyps resemble small anemones held within a stony matrix. Not until the 18th century did the scientist William Herschel, using an early version of a microscope, verify their true animal nature. And not until well into the 20th century did scientists, using advanced versions of microscopes and careful experimental methods, determine that, in fact, shallow-water, reef-building corals live in close partnership with microorganisms that photosynthesize. Although most corals use their crown of tentacles to capture plankton, they are, in fact, well-orchestrated, animal-vegetable-mineral combos.

As a graduate student in the late 1950s at the University of California, Berkeley, Leonard Muscatine demonstrated that "anemones and probably reef-building corals" derive significant nutrition from their algal tenants. In the decades since, Muscatine, his students, and other researchers have confirmed the close nutrient and oxygen connections between many shallow-water corals, anemones, and certain mollusks and sponges and the creatures who dwell interdependently within them, called their endosymbionts. So close is the relationship that if the algal partners, collectively called zooxanthellae, disappear, the host usually dies. Prolonged changes in temperature or salinity or changes in light and water quality are factors that can cause the loss of zooxanthellae and thereby the loss of color. Coral bleaching is the result, with living corals appearing snowy white, an increasingly common phenomenon closely correlated with a global rise in sea temperature.

On his round-the-world voyage aboard H.M.S. *Beagle,* Charles Darwin spent 12 days in 1836 exploring the coral reefs and atolls of the Cocos (Keeling) Islands, 36 square kilometers of land in the open sea of the eastern Indian Ocean. There he found evidence to support his theory that gradual subsidence of volcanic islands would lead to the progressive formation of fringing reefs, barrier reefs, and coral atolls. Darwin's ideas have been generally substantiated by deep-drilling operations on various Pacific atolls where volcanic rock has been found hundreds to thousands of meters below layers of coral growth. There are variations, however, including some reefs maintaining pace or exceeding changes in sea level and growth on various materials from granite to shipwrecks—and even glass bottles and beer cans.

Corals have graced the planet in one form or another for at least 500 million years, and reef-building corals appear in the fossil record dating more than 400 million years back. Species have come and gone, but throughout

VISIONARIES

John "Charlie" Veron
Godfather of Coral

. . . .

T he "godfather of coral" to many, marine biologist and Australian native John Edward Norwood "Charlie" Veron has spent more than 7,000 hours underwater in coral regions around the world, identifying some 20 percent of Earth's coral species. In a 2018 interview he recalled his first reef dive in the 1960s as immersing himself in "a metropolis . . . humming and buzzing." A 45-year love affair ensued, leading him around the globe to identify and explore reefs' vital role in the ocean ecosystem as chief scientist for the Australian Institute of Marine Science. Now Veron sounds "an air raid siren," as he puts it, to protect reefs. He warns that rising ocean temperatures and resulting bleaching have already killed half of Earth's corals, and the rate is picking up. Described by wildlife luminary David Attenborough as "one of the world's great authorities on coral," Veron continues to prophesize and proselytize about reef futures and for their survival.

. . . .

Called the "soft coral capital of the world" by Jacques Cousteau, Fiji's warm waters host some 80 species.

the ages, corals and coral reef systems have persisted, nearly extinguished at times but flourishing when favorable conditions arise. For most reef-building species, favorable means clear, sunlit water in depths to about 50 meters, temperatures above 18°C and below 28°C, salinity between 30 and 35 parts per thousand, and pH above 7.6. Some tolerate and even thrive in wider physical ranges and varying amounts of nutrients and sediment. Recently discovered cold-water corals far from coastal waters thrive in the open ocean at depths up to 6,000 meters and at temperatures as low as 4°C.

The same factors that apply to warm-water reefs apply to the often over-looked but critically important species of calcareous red and green algae that may dominate the carbonate structure of them. Some species give rocks the appearance of being painted pink; others form dark red or pink crusts or small, lime-infused upright branches. The 35 or so species of the calcareous green algal genus *(Halimeda)* may occur in such abundance that sand in some reef areas is predominantly composed of their decaying segments. According to algal expert Walter Adey, their overall contribution to production of calcium carbonate likely exceeds that of corals in tropical seas.

Favorable conditions also mean that the diverse components of a reef system are essentially intact. In addition to microscopic planktonic life in the water embracing Australia's iconic Great Barrier Reef, the system is home to more than 600 kinds of corals, 23 kinds of reptiles including six species of sea turtles, 215 species of birds, more than 30 kinds of marine mammals, at least 1,500 species of bony fish, and 134 kinds of sharks and rays, as well as at least 800 kinds of echinoderms, 3,000 mollusks, and 5,000 or so kinds of sponges. More than 30 other phyla of animals, as well as at least 500 species of algae, 39 kinds of mangroves, and 15 species of sea-grasses, also live here. No one has yet enumerated the bacteria, archaea, and other microbes at home there, but it is fair to say that every member of this diverse population has a piece of the action that enables the reef to function as a system.

A comprehensive study of coral reefs by biologist Jeremy Jackson and 18 co-authors, published in *Science* in 2001, awakened many to the inter-dependence of corals, fish, and other reef dwellers based on paleontological, archaeological, historical, and ecological evidence. Extraction by humans of large numbers of grouper, sharks, snappers, sea turtles, manatees, lobsters, pink conch, and other species, coupled with habitat destruction by fishing, clearly has had adverse consequences. Human actions have quickly unraveled hundreds of millions of years invested in developing tightly coupled nutrient links, life cycles, space allocations, food webs, predator-prey interactions, cooperative behaviors, soundscapes, chemical signals, and numerous other subtle but significant elements vital to the health of reef systems.

Parrotfish and surgeonfish, notable for their diet of algae and once considered unpalatable, became market favorites in the 21st century, reaching consumers thousands of kilometers from where the fish live. As their popularity in restaurants increased, parrotfish decreased, and fast-growing species of filamentous and fleshy algae rapidly began displacing and smothering corals, altering both the chemistry and the physical nature of the ecosystem. Katie Cramer and Richard Norris, researchers from Scripps

VISIONARIES

Paola Rodríguez-Troncoso
Restoring Reefs

....

Corals are her passion. From the University of Guadalajara in Puerto Vallarta, Mexico, where National Geographic Emerging Explorer Paola Rodríguez-Troncoso researches coral reef biology and ecology, her fieldwork takes her to the south Mexican Pacific coast to explore subtropical coral communities. Her focus: How do hermatypic—hard-skeletoned, reef-building—coral reproduce, and how do they cope with the stress of warming ocean temperatures at organism, population, and community levels? Her specialty is the genus *Pocillopora*, also called cauliflower coral for its wart-like surface growths and branching colonies. With other researchers and Mexico's National Commission of Natural Protected Areas, she is working to forecast how global climate change, including increased El Niño events, will put destructive pressure on coral reef ecosystems. Their restoration program, supported by National Geographic and the Waitt Foundation, targets and determines solutions for hard-hit areas. So far, a gradual increase in coral coverage at Mexico's Islas Marietas National Park, Jalisco's south coast, and Punta de Mita brings hope.

....

Institution of Oceanography, verified the apparent cause-and-effect relationship between reef health and parrotfish with current and fossil evidence spanning 3,000 years in the Caribbean.

John E. "Charlie" Veron has immersed himself in coral reef research for decades and has personally witnessed how the Great Barrier Reef has declined—a system that originated half a billion years ago, with modern reefs established about 8,000 years ago. In a decade the decline has been as much as 90 percent over broad areas. Veron and other experts, including Ove Hoegh-Guldberg, director of the Global Change Institute at the University of Queensland, predict that global warming and ocean acidification, exacerbated by declining water quality, overfishing, and loss of diversity, will drive coral reefs to functional collapse in this century.

An estimated 1,500 fish species swim the 2,500 individual reefs of Australia's Great Barrier Reef, a UNESCO World Heritage site.

Shelves: Living Along the Edge

I f the ocean could be drained away, it would be obvious that in some places high cliffs along the shore continue in steep descent into the depths below. It would also be apparent that along most continents and many islands, the terrain slopes gradually from the shore to a break at about 200 meters depth, where it abruptly drops off to very deep water. The shelf is an extension of the land, and over time much or all of it has been submerged by global sea-level rise from ice melting since the height of the last ice age 18,000 years ago. Along parts of the California coast, the shelf is less than a kilometer wide, but Siberia's continental shelf extends seaward almost 1,300 kilometers. Globally, the average width of the continental shelf is 65 kilometers, much of it once dry land.

A legal definition of a continental shelf is different from the geological one. According to the United Nations Convention on the Law of the Sea (UNCLOS), every nation may have an exclusive economic zone extending 200 nautical miles (nm) from the nation's coastline, giving it rights to the economic resources of the water column and continental shelf below. In some cases, if a nation can demonstrate that the geologic edge of its continental margin extends beyond 200 nm, it can petition for jurisdiction over the resources of the seafloor (but not the water above) up to 350 nm offshore. Overall, nations might increase their jurisdiction by about 10 percent by mapping and proving the geological definition of the continental shelf. Just by using the 200 nm extension seaward, Australia and the United States more than double their area of control over resources, and Costa Rica's jurisdiction expands 10 times. The area of ocean jurisdiction by island nations may exceed 100 times the landmass.

Sunlight powers photosynthesis in life-forms floating in the water column above continental shelf areas, but owing to a shelf's relatively shallow nature—average depth 150 meters—large areas of attached vegetation thrive there as well. In depths to about 30 meters, vast meadows of 70 or so species of seagrasses provide food and shelter for numerous kinds of fish and invertebrates, as well as critical grazing grounds for sea turtles, dugongs, and manatees. Benthic algae grow deeper, with red algal nodules studding the seafloor in depths greater than 100 meters. Some algal species thrive below 250 meters.

A 2012 report led by biologist James Fourqurean at Florida International University provides evidence that seagrass meadows can capture and hold as much as twice the amount of carbon per unit area as temperate and tropical forests. Evelyn Gonzalez, also at Florida International University, confirms in a 2017 study that carbon dioxide exhaled and excreted by the abundant animals living among seagrasses can offset the amount of carbon captured, but the animals also consume and store carbon when they eat, when they are eaten, or when, like seagrass leaves, they die and become part of the carbon-storing seafloor sediment.

WHY IT MATTERS

Keeping Kelp Healthy

Charles Darwin in 1835 predicted that if kelp forests—which he called more important than land forests—were lost, many of Earth's species would follow. A forest of kelp, a kind of algae, is an ecosystem of thousands of species of other algae as well as mammals, fish, and inverte-

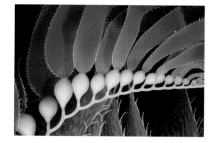

brates. The plant itself produces alginates, used to bind toothpaste and ice cream. More important, kelp fights climate change. Not only does it protect coasts from increasing storm surges and erosion; it also absorbs carbon and produces oxygen. When a heat wave swept California's coast in 2014, researchers at the University of California, Santa Barbara, discovered its extraordinary resilience and continued protection of coastal ecosystems. Recently, a blight killing starfish—predators that eat kelp-devouring sea urchins—has sent scientists searching for the cause and a cure. Keeping kelp ecosystems in balance is a priority.

ABOVE: The flotation bulbs of a giant kelp, *Macrocystis pyrifera*. OPPOSITE ABOVE: A flower hat jelly pulses in deep Pacific waters. OPPOSITE BELOW: At Cortes Bank, California, an orange sheephead navigates towering kelp.

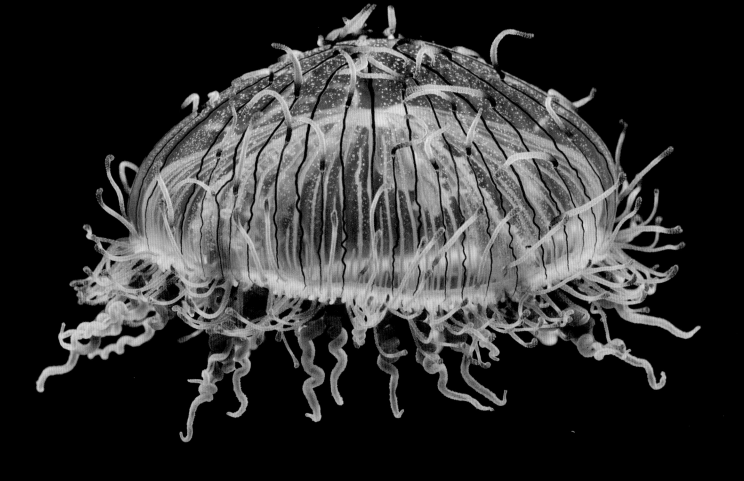

Numerous steep-sided canyons slice across continental shelves, created through erosion of ancient river mouths during ice ages when sea level was much lower than it is now. For thousands of years, people living in today's central California were well aware of the mountains to the east, and to the west the near-shore marshes, seagrass, and kelp forests, as well as sea otters, abalone, whales, and enormous populations of sardines and squid. But, until well into the 20th century, the undersea Monterey Canyon, a few kilometers off California, was unknown. It is a chasm steeper and deeper than Arizona's Grand Canyon. Concealed from view until recently are carpets of brittle stars, deep-sea corals, sponges, crinoids, and shrimp that live along the canyon's margins and in the great crevices themselves, hundreds of meters down.

The existence of hard and soft corals and other invertebrates living at the edge of light and beyond has been known since the early 1800s, but use of manned and remotely operated vehicles has made possible discovery of numerous previously unknown reefs. In 1984, hundreds of square kilometers of ancient living reefs primarily formed by glass sponges were found off the coast of British Columbia. Vast fields of similar species were later discovered about 40 kilometers beyond the shores of Washington State. Improved technologies are making possible spectacular new finds of sponges, corals, and numerous other heretofore undetected creatures in Antarctica, the deep fjords of Norway and New Zealand, canyons along the eastern coast of the United States, and the northern coast of Spain and Portugal. In the absence of photosynthesizers, many of these deep-sea communities have close associations with chemosynthetic bacteria that are considered to be sources of sustenance.

The consistent flow of materials from rivers, groundwater, and storm runoff from the land to the sea is well known. What has not been taken into account until recently is movement in the other direction. Salmon moving upstream to spawn after years at sea bring the essence of the ocean with them. When they die or are consumed by eagles, bears, insects, and other diners, nitrogen and other vital ocean-derived nutrients are transported to ecosystems far inland. In a 2002 study, biologists Morgan Hocking and Thomas Reimchen from the University of Victoria suggest, "This salmon nutrient subsidy extends from aquatic habitats into riparian forests, and is thought to be ecologically equivalent to the migration of the wildebeest on the Serengeti." Noting the severe depletion or extermination of bears and salmon in many of their historic areas, biologists Scott Gende and Thomas Quinn describe how state agencies are transporting salmon carcasses by truck and helicopter and depositing them in the forest in an effort to mimic natural processes, an endeavor they call the "Nutrient Express" in a 2006 report in *Scientific American*.

Nutrients derived from the sea often travel by air. Seabirds foraging on fish not only bring food to nestlings on shore but also paint the land with excrement rich in phosphorus, nitrogen, and various other elements vital to land vegetation. Off the coast of Peru, three tiny Chincha Islands have served as home base for millions of cormorants, pelicans, and boobies. Their droppings, over millennia, accumulated as rock-like guano as much as 60 meters thick. Most deposits have been stripped away for use as fertilizer.

VISIONARIES

Jonatha Giddens
Practicing the Art of Science

....

With a background in natural and social sciences, as well as training in art and traditional story-telling, ocean ecologist and National Geographic Fellow Jonatha Giddens is not your usual techie. While working with National Geographic's Exploration Technology Lab to develop the Deep-Sea Research Project for exploring ocean environments, she keeps a bigger picture in mind. By identifying indicators of ecosystem health at these sites, she intends to enhance local awareness of the ocean and inspire residents to support protection of threatened sites. Creative vision and communication are key—along with science. A deep-sea camera system helps identify and count deep-ocean species; with that information Giddens determines a mathematical formula modeling the relationship between animals and their environment. She then translates that back into a colorful, consistent conservation message that inspires others to champion similar ecosystems around the world. "I see science, technology, and art together allowing us . . . to bring back the wonders we have found," she says, "to bridge a relationship between people and nature."

....

The world's biggest fish, the gentle whale shark cruises Galápagos waters attended by hundreds of creole wrasse.

GREAT EXPLORATIONS

Picturing the Reef

As if posing for photographer David Doubilet, cardinalfish gleam against a Gorgonian coral.

n 1889 biologist Louis Boutan planted a sign against a bed of corals off southwestern France and clicked the shutter of a camera with a protective covering. The sign read *"Photographie Sous-Marine"* ("Underwater Photography"). Black and white and murky, the photograph was the first of a coral reef, setting the stage for 125 years of reef discovery.

In 1926 *National Geographic* magazine featured the first color photograph of reef life—a hogfish—by Charles Martin and ichthyologist William H. Longley from Florida's Dry Tortugas. The technology then developed gradually—more protective camera housings, faster film, better lighting.

Austrian scientist Hans Hass in 1940 captured a neighborhood of brain coral, sea turtles, and sharks in a 15-minute, black-and-white film off the Caribbean's Curaçao. By 1950 Hass had developed an electronic flash and

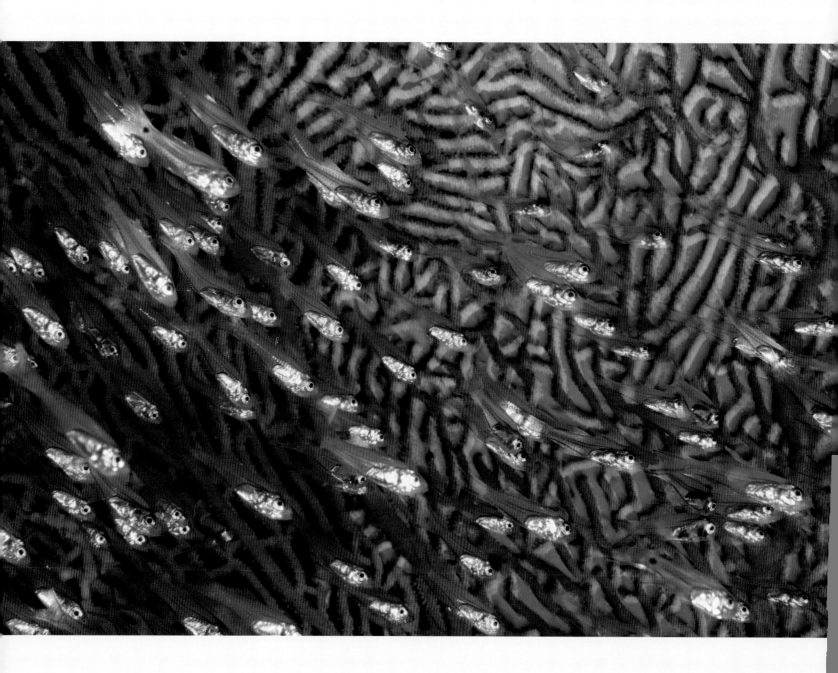

ABOVE: National Geographic photographer Brian Skerry documents a coral wasteland. RIGHT: Diving with Jacques Cousteau, Luis Marden photographed butterflyfish scraping corals for food.

protective housing for a Leica camera, which yielded clear, crisp imagery. Four years later his stereo cameras delivered 3D motion of his mermaid-esque wife Lotte sweeping among Caribbean reef ecosystems.

Through the lens of National Geographic photographer Luis Marden, who documented the mid-1950s adventures of pioneering French diver Jacques Cousteau, reef communities and their vast treasures came to the everyman. Tourists, recreational divers, and fishermen followed for commercial gain. But so did scientists and concerned citizens supporting reef protection.

Since then, milestones in reef photography have opened new worlds of understanding. In the 1970s National Geographic photographer Emory Kristof pioneered deep-diving color cameras to image hydrothermal vent communities; shipwrecks, including the *Titanic* in 1983; and myriad unique species. Since the 1970s, Phil Dustan, ocean ecologist at South Carolina's College of Charleston, has tracked reef health through spectrometry—measuring light emitted by a living body—and worked with NASA's Landsat imagery to pinpoint trouble spots.

Today crisp, digital images evoke deeper understanding, concern, and action. Smithsonian science photographer Owen Sherwood captures living corals from the inside out through his cross-section photography. Naturalist and filmmaker-photographer Kip Evans documents the life and struggle of reefs around the globe through National Geographic, NOAA's Sustainable Seas Expeditions, and Sylvia Earle's Mission Blue.

Forty-year National Geographic veteran photographer David Doubilet—arguably the world's foremost reef photographer—sees reefs threatened as never before and feels "an underlying urgency" to protect them through photography. "Pictures have the power to celebrate, educate, honor, humiliate, and illuminate," he says. "They can be a direct catalyst for conservation of a species or entire ecosystem. I'm especially proud of my colleagues Brian Skerry, Paul Nicklen, Thomas Peschak, and Laurent Ballesta—all on the frontlines of a battle to protect and preserve a vanishing world."

TIME LINE

Milestones in Ocean Photography

1899: French marine biologist Louis Boutan makes first clear reef image.

1926: Photographer Charles Martin and ichthyologist William Longley light image of reef hogfish.

1937: MIT's Harold "Doc" Edgerton designs first successful camera for underwater research; later teams with Jacques Cousteau.

1939–1954: Hans Hass makes underwater photographic leaps and develops a stereo camera, allowing for reef 3D images.

1942: Jacques Cousteau makes film *18 Mètres de Fond (18 Meters Deep)*.

1945: Ernest Brooks, Sr., photography instructor for World War II bomber pilots, founds California's Brooks Institute of Photography, featuring underwater classes.

1956: Photographer Luis Marden documents Jacques Cousteau aboard *Calypso,* making largest underwater photograph archive to date (1,200 images).

1960s: Photographer Bates Littlehales enhances reef photography with OceanEye Plexiglas camera housing.

1970s–1990s: Photographer Emory Kristof designs deep-diving cameras that photograph hydrothermal vents; shipwrecks including *Titanic;* submersibles; myriad species.

2008: Smithsonian with other institutions launches online Encyclopedia of Life to catalog photographs of all Earth's species; National Geographic's Enric Sala launches Pristine Seas photographic archive.

2009: Sylvia Earle's Mission Blue targets Hope Spots, a network of areas critical to ocean health and partners with Environmental Systems Research Institute (ESRI) to build image archive and database to tell stories, track change.

2015: David Gruber films biofluorescence of a hawksbill sea turtle.

2020: Photographers David Doubilet and Jennifer Hayes launch global coral reef baseline survey of reef health.

OPEN
OCEAN
LIFE

....

Chapter Six

"YES, GIVE ME THIS GLORIOUS OCEAN LIFE, THIS SALT-SEA LIFE, THIS BRINEY, FOAMY LIFE, AND YOU BREATHE THE VERY BREATH THAT THE GREAT WHALES RESPIRE!"

—HERMAN MELVILLE, *REDBURN*

The largest, deepest, widest, wildest, and least explored part of the planet is the open sea, the vast expanse of water beyond the coastal regions of islands and continents known as the pelagic zone, a term derived from the Greek word *pelagikos* meaning "sea or ocean." Politically defined, it includes the High Seas, the ocean beyond the exclusive economic zones claimed by coastal countries. Most of life on Earth, from microbes to whales, exists without touching the shore or the seafloor, suspended in the living liquid that enfolds most of the planet.

Sailors who venture across the open ocean may witness slick-calm days, when sky and sea merge into shimmering shades of blueness, mirroring clouds with a surface silky enough to allow oceanic water striders *(Halobates)* to skate freely, as if on a woodland pond. At such times, purple sea snails *(Janthina)* may sail on their nests of bubbles, and clusters of paper nautiluses *(Argonauta)* may bob along, embracing their precious cargoes of eggs. No one knows what happens to these open ocean voyagers when winds shift and the sea surface churns with mountains of frothing waves towering tens of meters high. Sailors have experienced such days. For some, stark images of a storm-tossed sea may mark their last memories.

Oceanographer and National Geographic Explorer Robert Ballard has spent decades locating and exploring some of the most famous ships that were swallowed by the sea during storms, during wartime battles, or—as happened to the H.M.S. *Titanic*—fatally torn and sunk by a giant iceberg. Recent technologies have made navigation more accurate, weather more predictable, and the occurrence of reefs, subsurface peaks, and icebergs mapped more completely, yielding far greater safety at sea than ever before—at least for humans. Since 90 percent of goods are now carried by more than 53,000 merchant ships crisscrossing the ocean—sailing along with 4.6 million fishing boats, many more millions of recreational vessels, more than 10,000 superyachts, and thousands of naval ships and submarines—it should be no surprise that the port reached by hundreds of vessels annually is at the bottom of the sea.

RIGHT: A cargo ship navigates stormy open seas, Earth's deepest, most mysterious realm, home for most of life on Earth. PAGES 236-237: In the waters of Mexico's Yucatán Peninsula, a sailfish circles a ball of sardines.

Deep & Going Deeper in the Open Ocean

Nowhere in the ocean are there straight lines defining constant conditions. Temperature, salinity, water clarity, and chemistry are constantly shifting vertically and horizontally depending on currents, the weather, the season, and the creatures moving within the water column, stirring and mixing as they go. Increasingly, humans are altering the natural rhythms and distributions by putting a new regime of sound and motion, as well as billions of tons of unnatural substances, into the ocean—and removing billions of tons of oil, gas, minerals, and wildlife from it. Natural zonation of terrestrial systems has been modified over thousands of years of human impact, but most of the changes in the sea have occurred since 1800. Nonetheless, it is convenient to note the traditionally recognized zones that characterize the open ocean environment.

Sunlit, Photic, or Epipelagic Zone (sea surface to 200 meters): Sunlight supporting most of the photosynthesis in the sea penetrates to 200 meters depth, generally known as the sunlit or photic zone. Light diminishes rapidly with increasing depth. Red light travels farthest in air, but least far in the sea. Within the first 20 meters or so, red and yellow light disappear, with green and blue light penetrating deepest. The quantity of light varies depending on the roughness of the ocean surface, the season, time of day, cloud cover, and the amount of plankton and other floating objects. Even in broad daylight it can seem like late evening under massive schools of fish or under the hull of a large ship. This is the realm where tuna, sharks, and whales power their way through the sea, often assisted by currents that transport ocean drifters.

Twilight, Mesophotic, or Mesopelagic Zone (200 to 1,000 meters): The ocean between 200 and 1,000 meters depth is generally thought to be below where photosynthesis can occur, but there are numerous records of green and red algae living below 200 meters. During the voyage of the H.M.S. *Challenger* from 1872 to 1876, fine mesh nets deployed in the mesopelagic realm captured an astonishing array of bizarre-looking fish, mostly black with various organs that emit an eerie blue light known as bioluminescence. Zoologist William Beebe, peering through the port of the *Bathysphere* and seeing the flash, sparkle, and glow of light emitted by lanternfish, hatchetfish, squid, and small shrimp, remarked that "the only other place comparable to these marvelous nether regions, must surely be naked space itself, out far beyond atmosphere, between the stars . . . where the blackness of space, the shining planets, comets, suns and stars must really be closely akin to the world of life as it appears to the eyes of an awed human being, in the open ocean, one half mile down."

Midnight or Bathypelagic Zone (1,000 to 4,000 meters): Below 1,000 meters and extending to 4,000 meters depth is the deep, dark heart of the sea, a region that remains largely unexplored but is known to be inhabited

"THE SEA, SO IMMENSE, SO BREATHTAKINGLY IMMENSE, WAS SETTLING INTO A SMOOTH AND STEADY MOTION."

—YANN MARTEL, *LIFE OF PI*

OPPOSITE ABOVE: Orcas come to Norway's Andfjord's waters in winter to feast on herring, first dazing them with thunderous tail slaps. OPPOSITE BELOW: A medusa jellyfish pulses through the deep Southern Ocean.

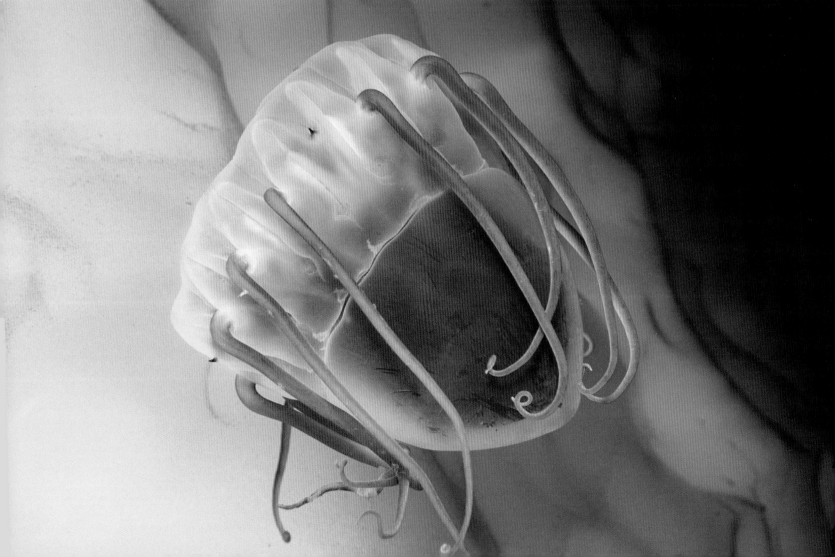

by myriad animals. Taken out of the sea, gelatinous creatures dissolve; long, fragile legs and antennae buckle; and soft bodies of squid, fish, and pelagic sea cucumbers collapse. Many are deep-sea drifters, adapted to exist in an eat-and-be-eaten realm that functions in darkness, where bioluminescence takes on magnified significance for finding food while avoiding being food for others. On—and especially within—the seafloor at these depths, communities of life can be as rich and diverse as anywhere on Earth. In the 1960s, biologist Howard Sanders and colleagues from the Woods Hole Oceanographic Institution were stunned to discover not monotonous uniformity but rather patchy assemblages of burrowing invertebrates as varied as life in a shallow tropical lagoon.

Abyssal or Abyssopelagic Zone (4,000 to 6,000 meters): The ocean below 4,000 meters to the rim of deep trenches in the seafloor maintains a temperature at about 4°C, with increasingly high pressure. This is the quintessential realm of darkness except for sparkles of light emitted by bioluminescent animals and the glow of thermoluminescence where superheated water jets through vents in the seafloor. Over broad areas of this mostly unexplored part of the planet, three kinds of mineral-laden regions are attracting the attention of industry: areas where nodules of minerals (manganese, iron, and various other metals) carpet the seafloor; crusts of minerals that are deposited around hydrothermal vents; and minerals adhering to the surface of seamounts. Though often characterized as regions where life is sparse or nonexistent, the abyssal zone abounds in microbial life and unique ecosystems where more than 80 percent of the organisms encountered are new to science.

Hadal or Hadalpelagic Zone (6,000 to 11,000 meters): Deep ocean trenches are long, narrow clefts produced by the subduction of oceanic plates, resulting in cold, dark, high-pressure environments that occupy about 3 percent of the ocean—an area about the size of Australia. Creatures living in these deep, steep chasms experience as much as 600 times the surface pressure, an increase from 14.7 pounds per square inch (psi) to 16,000 psi at the bottom of the Pacific Ocean's Mariana Trench, the deepest place in the sea, about 11,000 meters down. The deep ocean has often been characterized as a realm inhospitable for life, but observations confirm otherwise. During the first expedition in a submersible to the bottom of the Mariana Trench, Swiss explorer Jacques Piccard and Lt. Don Walsh of the U.S. Navy observed a "flounderlike fish." Later descents there by National Geographic Explorer and film director James Cameron in 2012 and by Texas-based businessman and explorer Victor Vescovo in 2019 and 2020 have documented the occurrence of numerous forms of life at these depths. Several remotely operated vehicles with cameras and various instruments have returned with evidence of crustaceans, echinoderms, various worms, and a wide array of microbial forms. Since all but a tiny fraction of all deep-sea trenches and the life they contain—from the 3,200-kilometer-long Aleutian Trench to the 1,750-kilometer Puerto Rico Trench—has yet to be seen, let alone explored, most of this unique part of the planet remains a mystery.

Sallie W. "Penny" Chisholm
On a Microbe Mission

In her 2018 TED Talk, titled "The tiny creature that secretly powers the planet," MIT biological oceanographer and ecologist Sallie W. "Penny" Chisholm spotlights her lifelong muse, the microbe *Prochlorococcus*. The mighty microbe also enjoys top billing on her Chisholm Lab website, described as "the dominant primary producer in the ocean, the smallest known phototroph, and the most abundant photosynthetic cell on the planet." No wonder Chisholm has chosen to spend her life with this compelling creature. As a scientist, she has always been fascinated by the behind-the-scenes work of microorganisms in shaping ocean ecosystems, and when she and colleagues discovered *Prochlorococcus* in 1986, the door opened for a new adventure. Her commitment to developing this microbe as a model for understanding the ecology and evolution of ocean microbes has a bigger mission—to lead to our understanding of how all life works in interdependent ways in an integrated, living planet. Reaching future scientists is hands-on. Not only does she engage her MIT students in her work but she inspires younger students through her award-winning illustrated children's books.

Small but fearsome, fangtooth viperfish are well equipped to dine several kilometers deep.

Life in the Open Ocean

To appreciate the nature of life in the open sea, it helps to think small, to envision the multitudes of microscopic creatures that populate the largest living space on Earth. It is also important to think large, as in half-ton leatherback sea turtles and blue whales, the largest animals ever to exist. Thinking fast matters, too, to imagine what it takes to power tuna, mako sharks, and billfish—the fastest fish in the sea—over long-distance migrations across entire ocean basins thousands of kilometers wide. Thinking slow is necessary as well, to account for the pace of jellyfish and the daily vertical migrations across hundreds of meters by minute crustaceans, fish, and squid under the cover of darkness.

Since all the ocean is connected, it is not surprising that some forms of life have managed to range globally. The open ocean hosts all eight species of sea turtles, as well as most of the great whales and High Seas sharks, including oceanic whitetips, makos, blue sharks, whale sharks, and great whites. Leatherback sea turtles may in their lifetime travel thousands of kilometers within the Atlantic, Pacific, or Indian Ocean, descending more than a kilometer deep to feed on jellyfish and other deep-sea life, and returning from time to time over many decades to the nesting sites where they began their life as hatchlings.

Sperm whales, orcas, humpback, blue, minke, fin, and right whales may migrate within a single year from tropical to polar regions, from feeding to breeding areas, descending along the way to dine on a movable mid-water feast of animals living as much as 1,000 meters deep. While these species may travel widely, some populations, such as the sperm whales of Dominica in the Caribbean Sea and the orcas of Puget Sound, Washington, tend to stay localized, developing unique behaviors and communication dialects.

In all open-sea regions, there are floating islands of debris. Mats of drifting vegetation from the land or detached blades of seagrasses and uprooted seaweeds have in recent years been interlaced with trash, largely plastics and discarded fishing gear. Hordes of young fish and swimming crabs find shelter under these drifting masses, while migrating songbirds pause to rest topside. Insects, spiders, even small mammals and lizards may be swept from one place to another aboard these vagabond islands.

Two species of brown algae, *Sargassum natans* and *Sargassum fluitans*, form golden floating forests in the Sargasso Sea, a slowly swirling, three-million-square-kilometer gyre in the North Atlantic Ocean, bounded on the west by the Gulf Stream. Most of the more than 50 species of *Sargassum* grow on rocks, but the two that live closely intertwined in the Sargasso Sea, the Gulf of Mexico, and the Caribbean Sea are true open-sea drifters, mostly reproducing by fragmentation. More than 100 kinds of fish and nearly 200 kinds of invertebrates rely on rafts of *Sargassum* for food and shelter. For about a dozen species, it is their only home.

WHY IT MATTERS

Shifting Migrations

Where have all the flounder gone? Although the fishing community of Virginia's Wachapreague—called the Flounder Capital of the World—is still booming, scientists are recording flounder populations moving north to cooler waters off New Jersey. American lobster, black sea bass, red hake, and more than a hundred other marine species are making similar northbound migrations. They are descending deeper, too, apparently

to find cooler temperatures in which to feed, spawn, and thrive. Such migrations, echoed worldwide, are prompting legal battles over fishing quotas between states and nations—and they're endangering species as well, since a new environment may require behavioral changes to which animals can't adapt. There are other dangers, too: North Atlantic right whales following krill migrations north have been struck by freighters in the Saint Lawrence shipping lanes. Damage caused by disruptions to ancient rhythms is compounded by the speed of change.

Eyes up, a sand flounder scouts for prey on the Pacific floor north of New Zealand.

SARGASSO SEA

Bordered by five currents, the Sargasso Sea is a three-million-square-kilometer gyre in the North Atlantic sustaining an ecosystem based on *Sargassum* weed. A clump may shelter thousands of organisms, from seahorses to hatchling turtles.

ABOVE: In tropical waters worldwide whitetip sharks are slow to mature and breed, a concern as numbers dwindle.

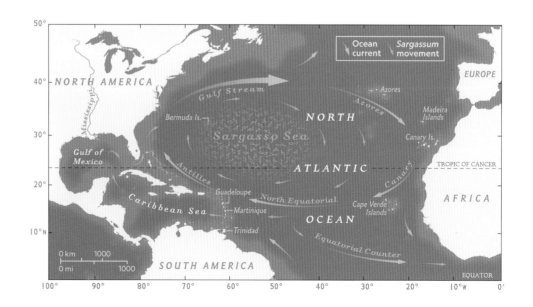

In July 1969, when the first footprints were being planted on the barren, lifeless surface of the moon, Swiss explorer Jacques Piccard led a crew of five men drifting beneath the Gulf Stream in the *Ben Franklin,* a mesoscaphe designed to descend to 600 meters. There they found abundant life, numerous fish, and transparent planktonic creatures called salps illuminated by the sub's lights "capering in the water, often making complete loops, flocking like butterflies, which they strongly suggest." During their 30-day, 2,300-kilometer journey, described in Piccard's book *The Sun Beneath the Sea,* the team personally experienced the impact of internal waves and eddies that bounced and turned the sub as it glided with the current under the western edge of the Sargasso Sea. Fifty-one years later, researchers at the Woods Hole Oceanographic Institution and the University of Washington discovered that tagged white sharks and blue sharks use such warm core eddies "like high speed tunnels" to travel into the depths to feed on squid and small fish that abound 200 to 1,000 meters below the surface.

Some have explored the open sea by temporarily becoming planktonic creatures themselves. In 1971, zoologists William and Peggy Hamner discovered a remarkable diversity of gelatinous drifters, sea turtles, and pelagic fish in the clear blue water offshore of the Bimini Islands in the Bahamas. They and a team of student researchers using scuba tanks suspended themselves at various depths under small boats in the Gulf Stream, drifting with the plankton, observing, documenting, and capturing in jars intact diaphanous "jellies" that are reduced to mush when scooped in nets. One of William Hamner's students, Laurence Madin—now a seasoned scientist at the Woods Hole Oceanographic Institution with thousands of hours of diving in the open sea—noted in a 2006 report that, while there, "You can't see the bottom. You can't see anything but blue water all around you, even up, and you have a sense of being suspended in infinity." Currently, it is popular among sport divers and scientists alike to participate in "black water dives," drifting in the open sea at night to witness the splendor of creatures that appear to be made of glass and rainbows, drifting against a dark void thousands of kilometers deep.

DEEPER DIVE

Light Show in the Deep

Twinkle, twinkle little . . . fish. From the ocean surface to the seafloor, more than 1,500 known kinds of marine creatures—bacteria, jellyfish, shrimp, squid, sharks, and more—glow with the chemical light of bioluminescence. Flashes of light can signal potential mates, lure prey, or confuse would-be predators. The flashlight fish *(Anomalops katoptron)* has bioluminescent bacteria living in special glowing organs below its eyes. To become invisible, or to signal by blinking, the fish can raise or lower a lid over the light. Marine biologists and National Geographic Explorers David Gruber and Brennan Phillips encoun-

A one-fin flashlight fish

tered a blue-glowing school of these fish that streamed "like a carpet of lights" from an underwater cave in the South Pacific. Fish use sight to form schools in the illuminated, upper part of the ocean, but this is the first report of bioluminescence assisting gatherings in the dark. After 350,000 video observations off the California coast, marine scientists Séverine Martini and Steven Haddock from Monterey Bay Aquarium Research Institute determined that 76 percent of the animals they viewed from surface to seafloor were bioluminescent. Those stats indicate that there is a major light show going on in the deep.

Through years of biweekly day and night plankton tows, scientists at the Bermuda Institute of Ocean Sciences (BIOS) have used nets to collect specimens in the top 200 meters at the Bermuda Atlantic Time-series Study (BATS) site in the Sargasso Sea. Resulting data have linked plankton dynamics to geochemical cycling, food web interactions, and changes in climate.

Every ocean animal, from single-celled protozoans to whales, consumes other organisms and produces excrement containing elements vital for phytoplankton. These in turn feed zooplankton, and so on, through the shifting, flowing web of life. Zooplankton may include more of the major divisions of animal life than are present in all terrestrial environments combined. Together, the vertical movements of the small, medium, and large forms of life in the open sea create the biological pump of ingredients that cause the ocean to be not just salt water, but rather a living liquid endlessly cycling vertically and horizontally within a medium of organic and inorganic compounds.

Deep-sea cutlass fish feed vertically on detritus floating from above. Multitudes of bristlemouth, lanternfish, and numerous other animals migrate from deep water toward the surface to feed in masses so thick that they appear as a "false bottom" when sensed with sonar.

The Deep Scattering Layer

"Phantom bottom" is the term physicists working for the U.S. Navy used in 1942 to describe what they found when they experimented with sonar as a means of detecting submarines in deep water near San Diego, California. They discovered mysterious "deep scattering layers" near the surface at night that moved to about 300 meters depth by day. The sound dispersed upon contact, creating "soft" images, compared with the hard, clear signals when the beams touched the seafloor. Martin W. Johnson, a Scripps Institution of Oceanography zoologist, surmised that the sonar echoes must be returning from masses of vertically migrating marine animals. Nets deployed to sample the layers returned with black finger-length fish with glowing spots along their sides, mirror-bright fish with large eyes and sharp teeth, plum-colored jellyfish, bright red shrimp, and translucent squid. In 1962, oceanographer Robert Dietz observed in *Scientific American* that research by physicists "presented the biologist with a powerful new ecological tool for understanding the mass distribution of life in the sea." He might have said life on Earth, because the multitudes of small fish, especially lanternfish and bristlemouths, or lightfish, appear to be present in the billions, possibly trillions, throughout the ocean.

The existence of animals that are at home in mid-water depths has long been known. Beachcombers near the Strait of Messina in the Mediterranean Sea often find along the shore windrows of deep-sea creatures thrust to the surface by strong currents and fierce upwelling. Scientists aboard H.M.S. *Challenger* in the 1870s described numerous kinds of lanternfish, squid, shrimp, and various gelatinous animals netted in depths greater than 100 meters, especially at night. When zoologist William Beebe and his engineer colleague, Otis Barton, descended nearly 1,000 meters into the Sargasso Sea in the 1930s, they witnessed large gatherings of euphausiid shrimp and squid that emitted bioluminescent ink when startled. In the 1960s, scientists discovered that Pacific fur seals were not dining on shallow-water fish but rather were diving into the deep scattering layer, chowing down on lanternfish. But until the U.S. Navy's discoveries in the 1940s, the abundance and prevalence of these mesopelagic creatures did not begin to come into focus. Even now, few realize that some of Earth's most numerous animals are thriving in the sea at the edge of light.

Fleets of ships from several Asian countries and Norway have mobilized to capture these heretofore "underexploited" sources of wildlife to be ground into feed for shrimp and salmon farms, for pet food, as well as for fertilizer, fish oil, and other products. At the same time, scientists are racing to discover the nature of these biologically vital regions, concerned that large-scale extraction may have disastrous consequences for seabirds, whales, seals, sharks, tuna, and sea turtles; disrupt nutrient cycles; and impact oxygen production, carbon capture, and planetary processes including global climate.

VISIONARIES

Larry Madin
Mesobot *as Avatar*

Woods Hole Oceanographic Institution (WHOI) biologist Laurence Madin calls his laboratory the largest living space in the solar system. His love is life in the open ocean, how it moves and behaves, from worms and jellyfish near the surface to the twilight zone's bioluminescent squid. To "open a world we rarely see," he has scoured the deep, first by scuba, then with submarines, remote-controlled vehicles, and most recently with an autonomous underwater vehicle. Completed in 2019, the *Mesobot* AUV, developed at WHOI by a team led by Dana Yoerger, started with Madin's wish for something that could go deeper and longer. The *Mesobot* can follow twilight zone species for hours, capturing behaviors via high-definition stereo cameras. For instance, Salps, gelatinous invertebrates, "jet-propel" themselves by contracting and pumping water, and also help remove carbon dioxide from the upper ocean and atmosphere. What more is out there? With *Mesobot* "as my avatar," Larry Madin will keep searching.

Pelagic squid, like the diamond squid, named for its fin shape, frequent deep scattering layers.

Getting Around

Early in the 20th century, the migration routes of whales were determined by implanting a "harpoon tag," retrieved when the whale was killed, thus yielding start-and-finish locations—once. A more effective way of tracking whales proved to be getting to know them as individuals. In the 1960s, biologist Roger Payne recognized numerous southern right whales by their distinctive facial markings and created a gallery of "mug shots" that enabled him to follow the actions of individuals. During more than 40 years of observations near Sarasota, Florida, Mote Marine Laboratory zoologist Randall Wells has come to recognize bottlenose dolphins *(Tursiops truncatus)* by their distinctive faces, fins, and behaviors.

In the 1970s, marine scientist Steven Katona initiated a catalog of distinctive humpback whale fluke shots in the North Atlantic, an approach that has been developed and used to track whales globally. Biologist and teacher Charles Jurasz documented humpback whales named "Notchfin" and "Spot" feeding together in chilly Alaskan waters in 1978 that turned up several weeks and thousands of kilometers later, splashing around in the warm waters off Hawaii. Soon thereafter, each swam with a baby whale at her side. Another whale, "Old Timer," first documented by Jurasz off Alaska in 1971, appeared in Hawaii in 1990 and was again photographed off Alaska in 2015.

In the 1980s, marine biologist Greg Marshall and colleagues at National Geographic devised ways to attach breakaway cameras and sensors to penguins, sea turtles, sea lions, whales, and even great white sharks. Called "Crittercam," the backpack cameras strapped to penguins helped document action for the popular 2005 film *March of the Penguins,* and a deep-sea version revealed social sounds and behaviors of sperm whales far below where human divers can venture.

During the 10-year Census of Marine Life from 2000 to 2010, U.S. scientists Barbara Block, Dan Costa, and Steven Bograd led an international team to observe wide-ranging sea animals, a project called Tagging of Pacific Predators (TOPP). A group of 90 scientists used sophisticated electronic tags tracked by satellite to follow 23 types of predators—tuna, sharks, squid, seals, sea lions, sea turtles, and seabirds. Data acquired made it possible to sort out the "homebodies" from the far-ranging animals and to collect and correlate physical ocean data collected by the tags with the animal migrations.

Since all living organisms, regardless of size, leave distinctive DNA traces in the sea from feces, urine, bits of skin, or mucus, there is now a more generalized but noninvasive method of determining what kinds of creatures live where in the sea. Many mysteries remain, but there is growing use of eDNA (environmental DNA) techniques, natural markings, clamp-on and suction-cup tags, as well as low-impact implanted tags coupled with satellite sensing. These methods are yielding a wealth of new insights about the habits of open ocean animals and are making traditional "catch, count, and discard" fish-counting techniques obsolete.

Playful Atlantic spotted dolphins range coasts, deep waters off continental shelves, and open ocean. Gulf of Mexico inhabitants migrate annually.

Animal GPS

The adage is true: You can set your watch by their travel. Gazing off the coast of California between March and June, you can see the massive spouts of gray whales and their calves moving north to summer feeding grounds, their built-in navigation systems—a primordial GPS— telling them where to go. Fish, sea turtles, and seabirds may navigate via an internal clock that syncs with the position of the sun, moon, or stars. Echolocation guides dolphins as they ping cues from the topog-

Gray whale calf off Baja

raphy of the ocean floor. Smell helps salmon find their breeding grounds. Genetically encoded migration routes is one possibility, while instinct, which drives the gray whales along California waters, is another. The most mysterious: Earth's magnetic field. While humans need a compass to detect it, many marine animals are born with sensors that tap into it as a steady guide. In addition to nature's inborn gifts, there is evidence that pathways in the sea may be communicated through generations.

HOPE SPOT

Sargasso Sea

Some call it the "golden rainforest of the sea." Bordered by North Atlantic Ocean currents, the Sargasso Sea harbors a unique ecosystem of free-floating brown macroalgae called *Sargassum*. Here bluefin tuna and the endangered Bermuda petrel and Porbeagle shark thrive. Mission Blue helped catalyze the Bermuda-led Sargasso Sea Alliance in 2010 to bring nations together to protect this region largely within the High Seas. So far, 10 nations have committed to the collaboration.

....

A sperm whale glides under a golden canopy of *Sargassum*.

Seamounts & Deep-Sea Corals

t was surprising to scientists in the late 1800s to find abundant and diverse organisms wherever nets, hooks, and dredges were deployed in the North Atlantic Ocean in depths as great as 7,300 meters—and later, in even deeper waters, during the global voyage of H.M.S. *Challenger.* It still surprises many to discover that corals, sponges, echinoderms, crustaceans, mollusks, and various other invertebrate animals, as well as numerous kinds of fish, are living in deep reefs offshore of North Carolina, near the edge of Antarctica's ice shelf, in New Zealand's Kermadec Trench, and adorning the slopes of seamounts wherever sampled.

Hawaii's Bishop Museum's intrepid diving scientist, Richard Pyle, uses a mix of gases in systems that enables him and his technical diving colleagues to explore depths as deep as 200 meters. In a 2017 *Science* article, Pyle notes, "Compared to what we know about shallow coral reefs, everything in the deep coral reefs is a big question mark." He has already discovered more than 100 new species of fish—13 on one very productive dive. And he predicts there are at least 2,000 more, as well as many thousands of heretofore unknown and unnamed corals, sponges, and other invertebrates in mesophotic, or twilight zone, reef systems. He points out the importance of the deep reefs as possible refuges for species that have declined in shallow systems owing to pollution, overfishing, and climate shifts.

Of the vast areas of the ocean that have yet to be explored, isolated submarine mountains—seamounts—that rise from the ocean floor to an elevation of 1,000 meters are accorded high priority for exploration and care. As undersea "islands," they have specialized communities of high diversity with unique species that are vulnerable to exploitation as commercially sought-after fish, crustaceans, and minerals. Only about a thousand seamounts have been named, but it is estimated that tens of thousands exist in the Pacific Ocean alone. Fewer than 100 have been surveyed for biological diversity, but as soon as they are identified, most get targeted for industrial fishing.

Some undersea mountains come close enough to the surface in cold water to host forests of kelp, and in warmer areas to be crowned with hard corals and communities of life similar to those typical of shallow, coastal coral reefs. Cashes Ledge, a seamount 128 kilometers offshore of Portland, Maine, is a Mission Blue Hope Spot that has been repeatedly proposed for full protection owing to the unusual richness of marine life there. It contains Ammen Rock, a peak that comes to within 10 meters of the sea surface and disrupts the Gulf of Maine current, forcing an upwelling of deep, nutrient-rich water. The area has the largest stands of kelp, mostly *Saccharina latissima,* along the New England coast, as well as exceptional blooms of plankton vital for young fish and a refuge for large cod *(Gadus morhua),* which have been greatly depleted by fishing. In the 1980s, Brown University biologist Jon Witman began researching the area, documenting cod as big

Richard Pyle
Fish Finder

....

"What an honor it is to be in the presence of a true naturalist," legendary biologist E. O. Wilson once praised fellow dinner guest Richard Pyle. Director of the Center for the Exploration of Coral Reef Ecosystems (XCoRE) at Hawaii's Bishop Museum, Richard Pyle began his career at age 19 after a close shave with death, which he recalls as "the best day of my life." After a 75-meter-deep dive chasing a fish he'd never seen before, he surfaced too quickly and suffered the bends. He soon championed safe deep-diving. Pyle has discovered hundreds of marine creatures new to science and is building an online, open-access catalog of deep-sea coral communities before overfishing, pollution, and climate change destroy them. The colorful reef "basslet" he identified in Hawaii's Papahānaumokuākea Marine National Monument—expanded by President Barack Obama in 2016—now officially goes by the name *Tosanoides obama.*

....

The submersible *DeepSee* enters the vent of an ancient seamount, Las Gemelas, off Costa Rica.

as the divers who came to observe them, as well as an abundance of lobsters, squid, cunner, pollock, blue sharks, dolphins, humpback whales, and large gatherings of seabirds. What he witnessed resembled accounts he had read by European settlers in New England from when they arrived in the 1600s. But Witman also watched catastrophic loss at Cashes Ledge, when fishermen using trawls, lines, lobster traps, and other destructive gear so decimated the system that by the late 1990s the New England Fishery Management Council temporarily banned bottom dredging and trawling. While not fully recovered, Cashes Ledge has rebounded; and if given full protection, it may regain its former robustness and serve as an oasis of renewal for nearby depleted systems.

In Hawaii, Terry Kerby, University of Hawaii marine scientist and *Pisces* submersible pilot, has observed the before-and-after consequences of dredging to obtain red, black, and gold corals used in making jewelry and the annihilation of deep-sea sharks that once populated seamounts near Honolulu. "It is heartbreaking," he says. "The sharks taken are likely to have been more than a century old, and the corals more than a millennium." Gold corals *(Kulamanamana haumaae),* also known as *Gerardia,* and black antipatharian corals may live thousands of years, among the oldest living animals on Earth.

Equally vulnerable are recently identified reefs dominated by glass sponges, a group of mostly deep-sea organisms that have existed relatively unchanged since their origin hundreds of millions of years ago. Individuals may be 11,000 years old. Fossilized sponge reefs are well known, but when living sponge reefs were identified at 150 to 250 meters deep offshore British Columbia in 1984, paleontologist Manfred Krautter of the University of Stuttgart said that the discovery "was like finding a living dinosaur." Glass sponges have been known to exist in deep water globally since the early 1800s, but shallower sponge reefs are now known to occur near Asturias, Spain; in deep, cold waters off Antarctica and New Zealand; and in deep water offshore of Washington State in the United States.

Richard Pyle and other scientists working with Conservation International have identified at least 29 seamounts in the open ocean of the north-central Pacific Ocean that come to within 300 meters or less of the surface. None of these has yet been surveyed visually, but deep coral reef ecosystems are likely to exist on their slopes. Efforts are under way to explore, document, and seek protection for them before they are damaged or destroyed by fishing and mining interests that are now expanding globally. Industries are gearing up to extract cobalt, nickel, lithium, and various rare Earth minerals that coat seamounts and exist as crusts around hydrothermal vents and in nodules on the deep seabed.

In the Gulf of Mexico, thousands of kilometers of pipelines, hundreds of deep-water oil wells, and fleets of industrial and coastal fishing operations have smothered or displaced deep coral reefs observed using the Delta submersibles in depths to 300 meters in the 1970s; the 1,000-meter Johnson Sea Link submersibles from 1990 to 2010; and the one-person, 700- to 1,000-meter subs used during National Geographic's Sustainable Seas Expeditions from 1998 to 2003. Since the 1960s, in waters near Australia, New Zealand, and the Arctic, industrial fishers using large trawls and nets have

WHY IT MATTERS

Clarion-Clipperton Controversy

....

Fist-size lumps of rock called manganese nodules carpet the seafloor between Hawaii and Mexico in an area the size of the United States—about 4.5 million square kilometers. Called the Clarion-Clipperton Zone, this ancient geologically and biologically active

region is also the site of controversy. Industries from several countries have invested in heavy equipment designed to mine the nodules to obtain the minerals they contain: nickel, manganese, copper, zinc, and cobalt. Here's the rub: Not only have these nodules taken millions of years to form by the action of still growing microbes, but they are vital habitats for fish, corals, sponges, and Casper, the ghost octopus—a new species discovered in 2016 that lays its eggs on sponge stalks anchored to the nodules. Scientists estimate that 90 percent of the species found here are new to science. Industrial mining will inevitably cause damage to these ancient living formations, gaining short-term financial returns but incurring losses of enduring natural and economic value.

....

A manganese nodule, in cross section, forms as bacteria deposit various minerals around a shark's tooth or bit of shell.

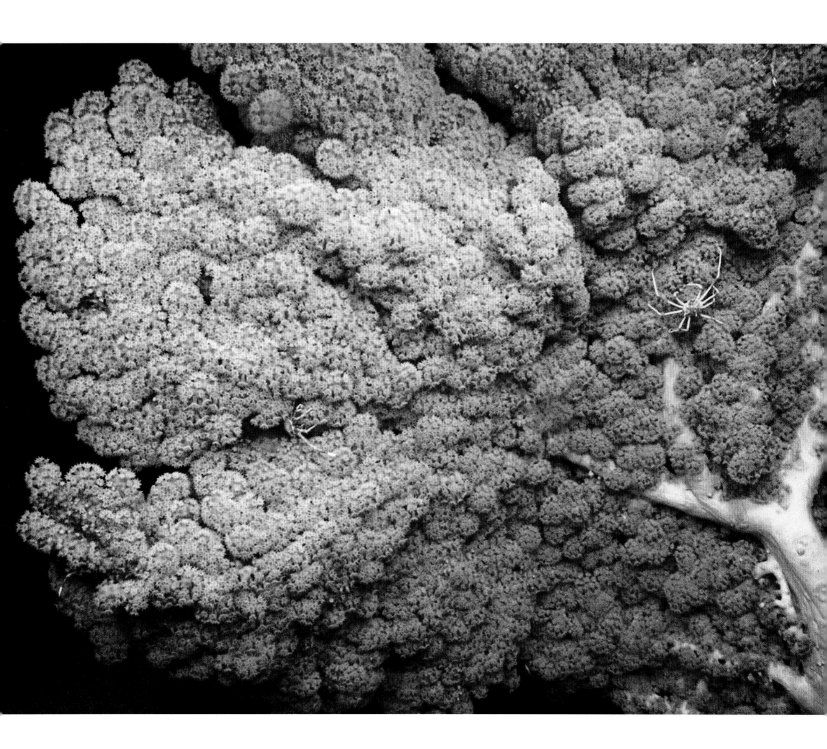

greatly depleted populations of century-old, late-maturing fish including orange roughy or slimeheads *(Hoplostethus atlanticus)*, oreo or oxeye *(Oreosoma atlanticum)*, Chilean sea bass or Patagonian toothfish *(Dissostichus* spp.), hoki or rattails *(Macruronus novaezelandiae)*, and other deep-sea fish. Fast-food chains as well as high-end markets and restaurants feature these animals in "fish" sandwiches, "fish" and chips, "fish" chowder, and other dishes with names that obscure their identity and origin. Industrial fishing has also wreaked havoc on forests of corals more than 5,000 years old. As concerns grow about loss of biodiversity and the need for maintaining resilience of ocean systems in light of climate change and other forces, there is increasing momentum to expand protection of the open ocean beyond current levels.

A colony of bubblegum coral *(Paragorgia arborea)* hundreds of years old prospers in Baltimore Canyon, a deep-sea coral protection area on the Virginia coast. It shelters shrimp, starfish, and shark eggs.

Deep-Sea Vents & Their Riches

The scientists aboard the Woods Hole Oceanographic Institution's research vessel *Knorr* were delighted but not surprised when images of shimmering hot water appeared in photographs recorded by the towed vehicle *ANGUS* (Acoustically Navigated Geophysical Underwater System), deployed 2,500 meters deep in the Pacific's Galápagos Rift, 420 kilometers west of the South American coast in February 1977. They had predicted that seafloor crevices and cracks would allow water to percolate down to the hot magma below and reemerge in ways similar to the geothermal springs of Iceland and the hot springs of Yellowstone National Park. What no one expected, though, were tall chimneys of minerals streaming hot water, surrounded by red-plumed tubeworms two meters tall, clams, pale crabs, plump eels, and pink, feathery clumps of something no one could identify thriving in the dark. It had long been believed that life in the deep sea existed on crumbs of sustenance drizzling down through a gantlet of creatures powered by photosynthesis. Now an alternative energy source involving microbes—chemosynthesis—was recognized. Geologist Peter Rona—then with NOAA—predicted vents in the Atlantic Ocean as well, and he proved it during a 1985 expedition. Instead of giant tubeworms, he discovered vents smothered with thousands of pink, eyeless shrimp as well as distinctive mollusks, echinoderms, fish, and other creatures.

Hundreds of hydrothermal vents and unique communities of life now have been found, mostly in the Pacific, but also in the Atlantic, Indian, Arctic, and Southern Oceans, and recognized as a critical part of the ocean's global plumbing system. At regions of seafloor spreading, water drains through the ocean's leaky bottom, is heated by hot magma, dissolves minerals from surrounding rocks, and returns to the deep sea in seeps and plumes that transport heat and chemicals from the interior of the Earth, continuously contributing elements used by ocean life and in shaping planetary chemistry. These vents also accumulate elements—cobalt, nickel, copper, and even silver and gold—in quantities that have potential commercial value.

In 2018, an international team of scientists discovered a new hydrothermal vent field near the Gigante Seamount in the Azores, a cluster of islands in the mid-Atlantic Ocean. Located at only 570 meters depth, 100 kilometers from Portugal's Faial Island, its relative ease of access presents an opportunity for regular monitoring. In 2018, on the science-technology site *Phys.org*, one of the expedition's leaders, Professor Murray Roberts from the University of Edinburgh, summed up the importance of such research, commenting: "Hydrothermal vents not only form oases of life in the deep ocean, but research over the last 20 years has shown the minerals they release also have important consequences for life throughout the ocean. As plans to mine deep-sea minerals are developed around the world, it's absolutely essential we understand these relationships to protect the oceans and the support functions they provide to all life on Earth."

VISIONARIES

Salomé Buglass
Galápagos Seamounts

Landing her dream job at the Charles Darwin Research Station to study the Galápagos Marine Reserve, National Geographic Early Career Grantee and marine ecologist Salomé Buglass took a deep breath and dived in. She joined a small team of scientists using novel technologies to survey and describe largely unexplored deep-sea communities on seamounts in the Galápagos. In 2018 Buglass and her crew headed to an open ocean area presumed to have subsurface volcanoes. Using high-definition cameras on an ROV, they took real-time video footage of the marine communities living in these tiered ecosystems as deep as 280 meters, including sponges, coral, and fish that beamed from the dark to the onboard monitor. Most fascinating was the discovery of extensive kelp forests. Buglass is from the island nation Trinidad and Tobago, surrounded by the deep sea. Her vision is to share her passion for seamounts with other islanders and inspire them to care about these little-known deep marine ecosystems.

At 2,980 meters deep, a black smoker sustains hundreds of species with minerals from inner Earth.

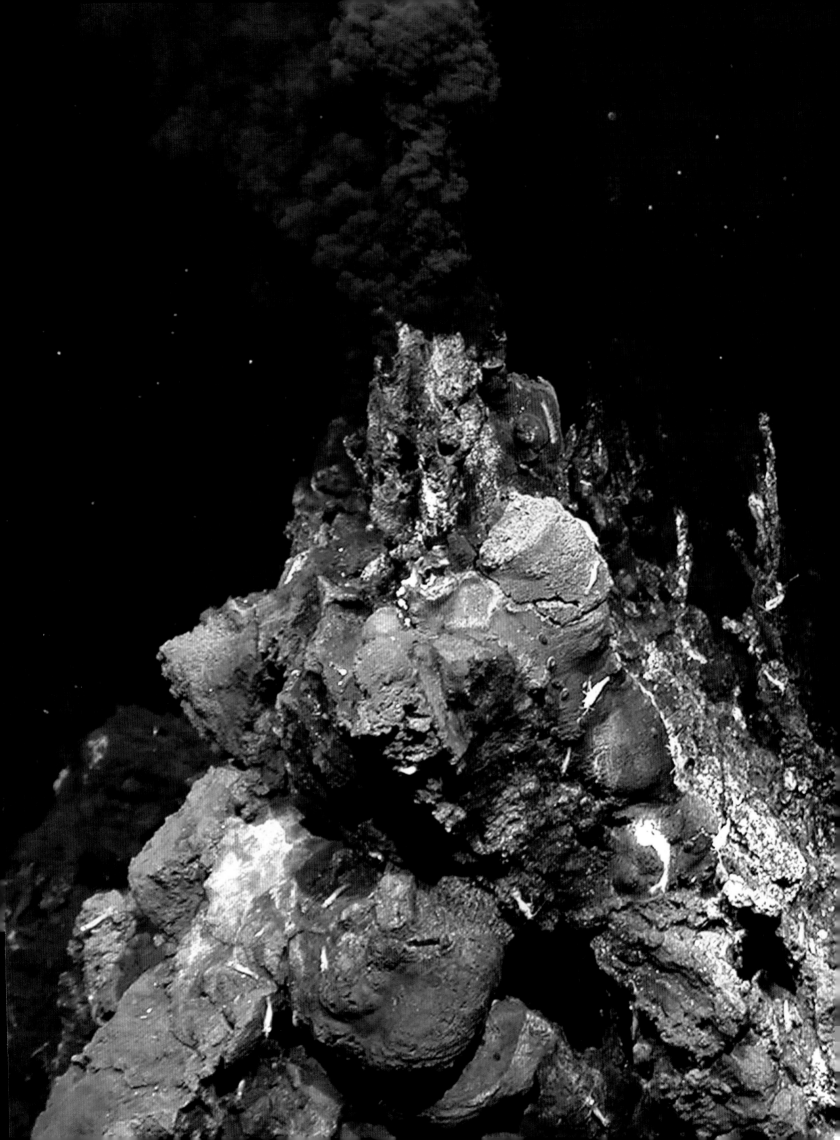

Discovering Life in the Polar Deep

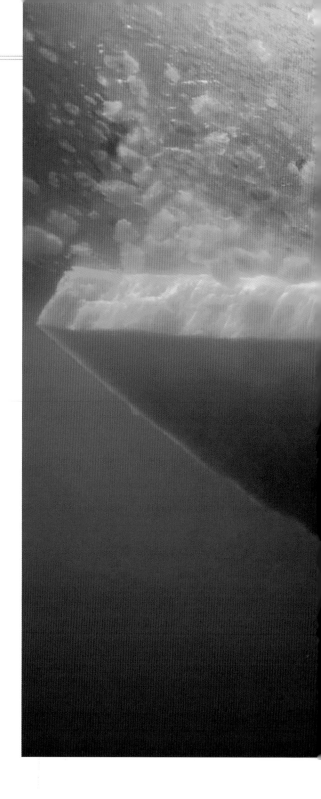

The frigid or frozen open sea in polar regions poses special challenges—and opportunities—for those who live there permanently or seasonally. Although pitch-black year-round below 1,000 meters, the long periods of continuous summer light and continuous winter darkness define special conditions for life in the high Arctic Ocean and in the rollicking, fast-moving Southern Ocean currents wrapped around the Antarctic continent.

Unknown to polar bears, narwhals, bowhead whales, belugas, and numerous birds and seals who have occupied Arctic land and sea for millions of years, humans currently claim territory deep into Arctic waters. With recent planetary warming and the corresponding decline of polar ice, new shipping lanes are opening up for fishing, transportation, and Arctic oil and gas drilling operations in areas heretofore inaccessible. Mindful of the potential harm unregulated fishing could impose in a High Seas area never before accessed by commercial interests, in 2019 10 countries signed a 16-year precautionary, collaborative moratorium that protects 2.8 million square kilometers, roughly the size of the Mediterranean Sea.

People have ventured far into the high Arctic for thousands of years, but to reach the North Pole means trekking hundreds of kilometers across sea ice, a feat recorded for the first time by Adm. Robert Peary, Matthew Henson, and four Inuit companions in 1909. But the first human access to the real North Pole, on the seafloor, occurred in 2007. Two Russian submarines, *Mir I* and *Mir II,* each carrying three explorers, descended and returned through a crack in the ice. On the bottom, they found mounds, trails, and burrows of numerous organisms lacing the soft sediment on the seafloor.

A 2019 report from NOAA describes efforts by an international team of researchers aboard the Norwegian icebreaker RV *Kronprins Haakon* to define the role of the Arctic as a pathway between the Atlantic and Pacific Ocean systems. Exploration of recently discovered hydrothermal vents under the blanket of sea ice is expected to yield new insights about relationships among vent organisms globally. Sampling organisms along the Gakkel Ridge, a slowly spreading rift system that is one of the three major mountainous formations cutting across the Arctic Basin, is a research priority aimed at correlating species with environmental data and acquiring detailed maps of the terrain.

Exceptionally wide continental shelves border the Arctic Ocean, mostly in depths less than 200 meters. Phytoplankton abounds there during summer months, including masses of diatoms that coat the underside of the pack ice. Vast numbers of amphipods, copepods, and other zooplankton devour the greenery, and they, in turn, are food for the notably abundant Arctic cod *(Boreogadus saida)* and more than 200 other species of fish sought by seabirds, marine mammals, and, recently, fishermen.

An emperor penguin speeds toward a diver beneath Antarctic ice, where golden clumps of microalgae flourish.

Paul Nicklen, a National Geographic Fellow and photographer, often behaves like a marine mammal himself, spending hours observing creatures that abound within scuba-diving depth. In *Bear: Spirit of the Wild* he writes, "After an hour under the ice, I emerge numb from the cold and am humbled by the experience." One elusive Arctic mammal species documented by Nicklen, the mysterious narwhal *(Monodon monoceros)*, dives to 1,800 meters, well beyond the greatest depths human athletes can attain on a single breath of air.

Mammals abound in Antarctic waters too, including some of the same species that flourish in the Arctic Ocean. During the 10-year Census of Marine Life, researchers documented more than 200 species of animals that occur in both polar regions, from orcas and humpback whales that do not migrate pole to pole to diminutive Arctic terns *(Sterna paradisaea)* that do.

A species of large king crab *(Paralomis birsteini),* typically an Arctic-area resident, has recently also been discovered in the Southern Ocean, where crabs and other decapod crustaceans are notably rare. Warming water may favor the advance of this newcomer, notorious for its appetite for mollusks, echinoderms, and other species that in Antarctica have no natural defenses against this robust predator.

The most formidable consumer of Antarctic wildlife arrived following the first sighting of the continent two centuries ago by Russian explorer Fabian Gottlieb von Bellingshausen. Whalers, sealers, and, more recently, industrial fishing fleets from faraway countries have made deep inroads into the abundance and diversity of Antarctic ecosystems. Of particular concern is the depletion since the 1980s of Antarctic krill *(Euphausia superba).* While enormously abundant, large quantities of these small crustaceans are essential to the well-being of the entire Antarctic ecosystem. Their health is vital to the existence of creatures small (Arctic terns) to large (whales), with implications for carbon cycling, climate, and, ultimately, humankind.

Antarctica's permanent ice shelves are shrinking faster than at any time in recorded history. As large sections of the ice shelves break away, places shielded by multiyear ice for as long as 120,000 years are opening to sunlight and the ocean beyond. As one large iceberg known as A-68 moves away from the Larsen C ice shelf, scientists from the British Antarctic Survey are moving in. Susan Grant, one of the researchers, regards this as a "fantastic, unknown area . . . We know very little about what might or might not be living in these areas and especially how they might change over time."

A discovery in 2015 confirms that they should expect the unexpected. Scientists thought they might find microbes on the seafloor under nearly a kilometer of ice on the Ross Ice Shelf, 850 kilometers from the open ocean. They did, but there was more. A camera lowered through a drilled hole returned images of a translucent pink fish, crustaceans, and a jellyfish—multicellular marine animals that were making a living in a dark, very cold realm far from the open sea. It is increasingly clear that the greatest era of exploration is just beginning, with every new discovery leading to more questions than answers.

MISSION BLUE

Monitoring Deep Reefs

The effects of increasing plastics, carbon, and acidification take their toll on all levels of the ocean. Monitoring deep-sea coral health is a frontline global mission carried out by government agencies, NGOs, scientists, and other individuals around the world. NASA, ESA (European Space Agency), and other international space agencies gather coral reef data from satellites and low-ranging aircraft and drones with remote-sensing cameras; their infrared imaging spectroscopy forms precise images based on colors assigned to thermal wavelengths. UNESCO's Global Reef Monitoring Network helps nations share monitoring information. In Australia, coral geneticist Madeleine van Oppen is experimenting with a new breed of coral that can withstand greater-than-usual heat. To help developing countries digitally record and share key information efficiently, technologies such as the Wildlife Conservation Society's MERMAID are helping grassroots groups upload and sync data in seconds to cloud-based databanks. This information is then shared with scientists working on coral-saving solutions.

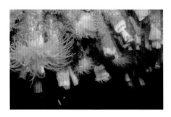

Deep-sea cockscomb cup coral

"WHEN WE EMBARK ON THE GREAT
OCEAN OF DISCOVERY, THE HORIZON OF
THE UNKNOWN ADVANCES WITH US
AND SURROUNDS US WHEREVER WE GO.
THE MORE WE KNOW, THE GREATER
WE FIND IS OUR IGNORANCE."

—GARDINER G. HUBBARD,
NATIONAL GEOGRAPHIC, VOLUME 1, 1888

"Unicorns of the sea," tusked
male narwhals surface between
800-meter dives for Arctic cod
in Canada's Lancaster Sound,
their summer grounds.

Life Under the Seafloor

Just under the surface of the seafloor of continental slopes and margins, at 500- to 2,000-meter depths, large amounts of carbon are contained in massive layers of frozen, biologically derived methane that has accumulated over millions of years. Methane combines with water under low temperature and high pressure to form solid, icy methane gas hydrates or clathrates. In some places, "ice worms"—plump, finger-length polychaetes—colonize the clathrates, dining on bacteria that in turn metabolize the methane. So widespread are masses of frozen methane in the deep sea that they may together hold as much hydrocarbon as all of Earth's oil, coal, and gas combined. Icy methane slush can block pipelines, one of the problems responsible for difficulties in capping the *Deepwater Horizon* oil well blowout in the Gulf of Mexico in 2010. As ocean temperature rises, there is concern that the frozen methane could melt and escape into the atmosphere. Already, methane released from human activity and from natural sources accounts for about 20 percent of the current rise in global warming.

Also hidden from view *beneath* the seafloor dwell microbial bacteria and archaea that may exceed all of the rest of the biomass of life on Earth. Here is a teeming world of microbes with complexity and diversity that rivals all other forms of life. However, their metabolic rate is exceedingly slow. Many bacteria double their number every few minutes, but sub-seabed microbes may divide once in a few centuries. Some may live millions of years, redefining what it means to be old and alive. They are extreme in other ways, too, with some prospering in temperatures as high as 121.6°C. Conditions just right for humans would be extreme for them.

At the 2018 American Geophysical Union meeting in Washington, D.C., a report from the Deep Carbon Observatory concluded that about 70 percent of Earth's bacteria and archaea live underground, a massive deep biosphere that is mostly under the ocean. The oceanic crust of volcanic-derived basalt rock lies below the sedimentary seabed, covering about two-thirds of the Earth's surface. On average, it is seven kilometers thick. In a March 2013 article in *Science,* Andy Fisher, professor of earth and planetary sciences at the University of California, Santa Cruz, noted, "The fact that you can get viable microbes out of those rock samples—that's just tremendously exciting."

In 2010, a team of scientists aboard the research drilling ship *JOIDES Resolution* drilled holes 530 meters into the ocean floor, then lowered in instruments that support long-term chemical and biological sampling and environmental monitoring of this heretofore unexplored netherworld. Fisher adds that the experiments will not only help determine how microorganisms may have developed on Earth but also "offer insight into how life may develop on other planets."

OPPOSITE ABOVE: Sailing over a field of manganese nodules with its two-meter-long tentacles, the *Relicanthus daphneae* anemone was first described in 2008.
OPPOSITE BELOW: From its manganese nodule base, a brisingid starfish will lift its arms to capture tiny crustaceans floating by.

GREAT EXPLORATIONS

Incredible Polar Photos

Before pressing the shutter, Amos Nachoum held the gaze of a mother polar bear swimming above him, cubs in tow.

Desolate beauty. Since 1910, when British photographer Herbert Ponting made the first polar images of the Antarctic seascape on Robert Falcon Scott's Terra Nova expedition to the South Pole, photographers have captured polar worlds and their waters.

Four years after Ponting, British photographer Frank Hurley braved Ernest Shackleton's ill-fated Antarctic journey, camera and photographs intact after their ship was crushed by ice and the crew spent two years finding their way home.

By the 1950s, underwater photography was pouring in from warm waters around the world, but cold waters remained mysterious. Venturing beneath the ice of Wisconsin's frozen Upper Nemahbin Lake to document the catch of an ice fisherman, National Geographic photographer

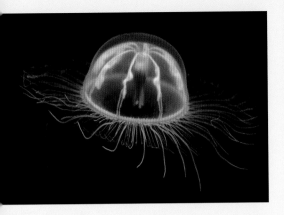

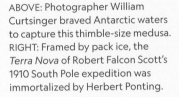

ABOVE: Photographer William Curtsinger braved Antarctic waters to capture this thimble-size medusa. RIGHT: Framed by pack ice, the *Terra Nova* of Robert Falcon Scott's 1910 South Pole expedition was immortalized by Herbert Ponting.

Thomas Abercrombie emerged to claim his episode "the act of a madman."

Soon after, madmen—and women—set flippers into the deep, cold sea. In the 1970s and '80s marine biologist Charles "Flip" Nicklin began building what National Geographic has termed "a majestic" portfolio of whale photographs, taken not only from waters off Hawaii and North America but also from the Arctic and Southern Oceans.

By the early 1990s William Curtsinger, a former U.S. Navy photographer, was making his trademark images of emperor penguins and "beautiful drifters" like the helmet jellyfish (*Periphylla periphylla*) gliding beneath Antarctic sea ice 2.5 meters thick.

From the bath-warm Red Sea to the frigid Arctic, Israeli photographer Amos Nachoum works to connect with his subjects, holding their gaze as he presses the shutter. From a 2016 Arctic expedition, his image of a languidly swimming mother polar bear and her cubs regarding him from above has become an icon.

Canadian biologist and photographer Paul Nicklen, raised on Canada's Baffin Island, has celebrated the beauty of extreme cold since childhood. His underwater photographic wonders include a doting female leopard seal who "adopted" him in frigid waters off Anvers Island in Antarctica. Brian Skerry spent months in Arctic waters documenting the plight of baby harp seals losing their habitat due to climate change. Laurent Ballesta dived a record 70 meters below Antarctic ice to find giant starfish, sea spiders, and coral reefs with the "color and exuberance" of their tropical counterparts.

ROVs, too, play a role, using pressurized, temperature-controlled cameras to capture high-definition imagery at 3,000 meters deep.

Still, human presence brings unparalleled art—and insight. In 2018 Viktor Lyagushkin, in the Arctic Ocean's White Sea, compared entering a hole in the ice to Alice descending the rabbit hole into Wonderland. His cameras captured carnivorous starfish, stalked jellyfish, sea anemones as graceful as flowers, and mating ephemeral sea slugs—in a "dance of the sea angels."

TIME LINE

Images From the Ice

1908-09: Explorers Robert Peary, Matthew Henson, and Donald MacMillan photograph the Arctic Ocean.

1910-12: Explorer Robert Falcon Scott and photographer Herbert Ponting make first known images of the Southern Ocean.

1932: Geologist Lauge Koch photographs Greenland glaciers, beginning an archive to document disappearing Arctic ice.

1964: First satellite maps of polar sea ice are made from 40,000 photographs taken by NASA's Nimbus 1 satellite.

1970s-present: Ex-Navy photographer William Curtsinger pioneers underwater photography in Antarctica; Charles "Flip" Nicklin makes legendary whale photographs.

1986: Phantom® ROV obtains first Antarctic under-ice images of giant sponges and other life.

1990s-present: Paul Nicklen, Amos Nachoum, David Doubilet, and Michael Aw photograph above and below polar waters.

1995-2000: Radarsat images used for first comprehensive maps of changing Arctic sea ice.

2004: Brian Skerry's *National Geographic* cover story exposes plight of Arctic harp seals.

2005: NOAA *Global Explorer* ROV photographs newly discovered finned octopods (nicknamed "Dumbos") in the Arctic's Canada basin.

2015: Biologist Ross Powell records snailfish using ROV lowered through the Antarctic ice shelf.

2015: Photographer Laurent Ballesta makes deepest Antarctic scuba dive, to 70 meters.

2017: Antarctic coral filmed from submersibles in 1,000 meters depth by BBC *Blue Planet II*.

2018: Viktor Lyagushkin and other photographers continue uncovering wonders of the deep.

I OUR LIVES

PART THREE

AND THE OCEAN

3

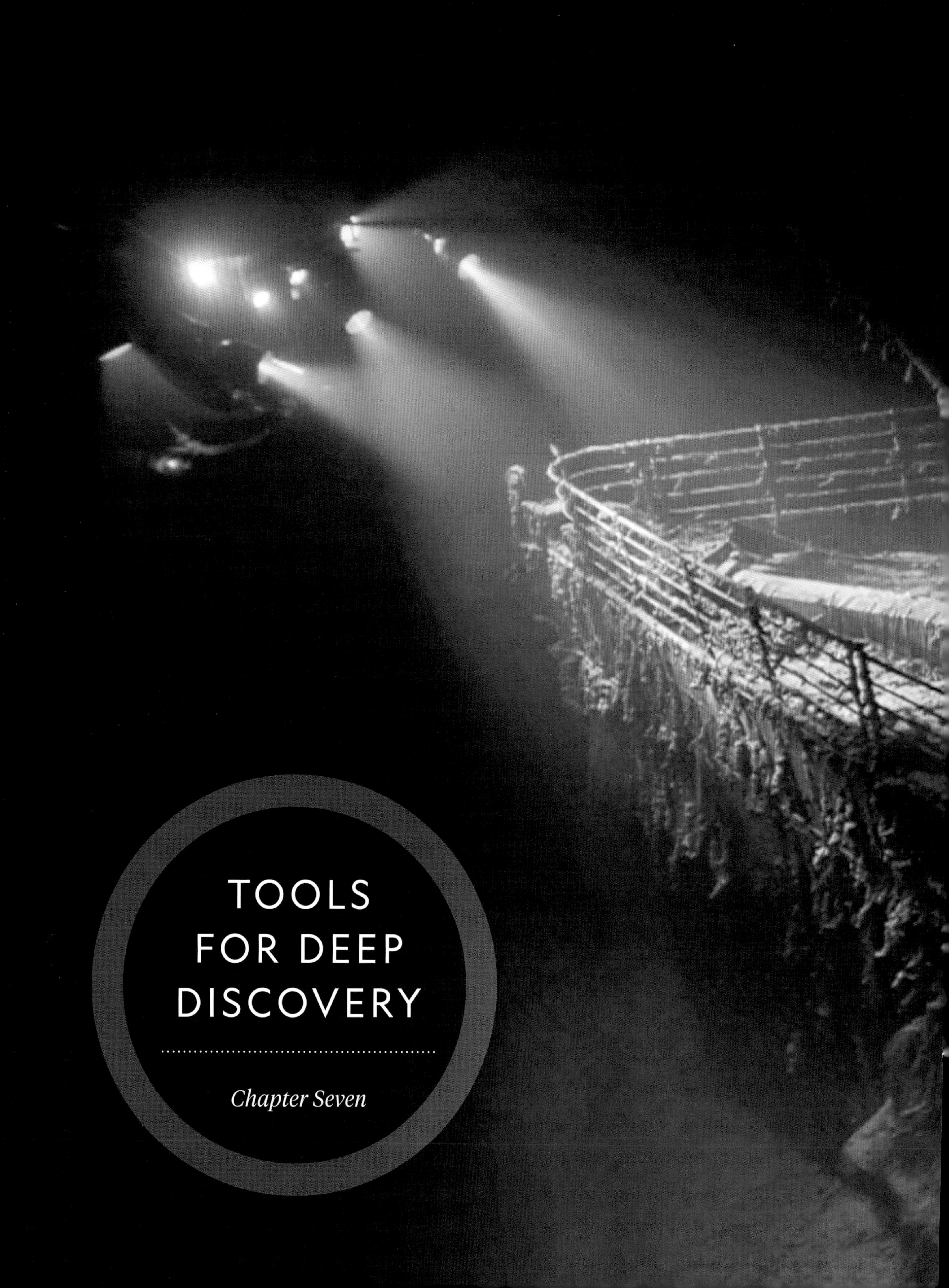

TOOLS
FOR DEEP
DISCOVERY

························

Chapter Seven

> "COULD IT BE MORE DIFFICULT TO SOUND OUT THE SEA THAN TO GAUGE THE BLUE ETHER AND FATHOM THE VAULT OF THE SKY?"
>
> —MATTHEW FONTAINE MAURY, *THE PHYSICAL GEOGRAPHY OF THE SEA*

Recent advances in technology have made possible the greatest era of ocean exploration in history, with each discovery leading to questions that no one previously knew enough to ask. It has not been for human lack of intelligence, curiosity, or longing to go deep that the ocean has been able to keep its secrets for so long. Rather, access to the depths of the sea, as well as to the skies above and space beyond, required a critical level of knowledge, will, funding, and the right sources of energy and materials before humans could venture high above the land and deep beneath the sea. But, despite remarkable advances, only a fraction of the ocean has been seen, let alone explored or fully mapped, and only a fraction of the enormous diversity of organisms has been identified, their habits known and their relevance to human existence fully understood.

Throughout time, people have been inspired to go beneath the surface of the ocean, often for food, sometimes for pleasure, for treasure, for salvage of sunken ships, for war, and for the most powerful reason of all: curiosity. But it should not be surprising that most of the ocean has been off-limits to humans for most of the time humans have existed. After all, the average depth of the ocean is nearly 4,000 meters, and the average human can dive less than four meters with a single breath of air. How long can you hold your breath? For most of us, in less than a minute there is an urgent need to breathe. With practice, some can go breathless for three, four, even 10 minutes or so—a skill used by some for round-trip dives to more than 100 meters depth. But that is still just the top of the largest reservoir of water—and life—on Earth. Depths below 500 meters have been reached by divers using specialized equipment and gas mixes for breathing. Remotely operated and human-occupied submersibles are just beginning to access the deepest parts of the sea.

Currently, the ocean is being surveyed with sophisticated research ships, spacecraft, new instruments, and a growing array of autonomous undersea technologies, buoys, and other equipment that did not—and could not—exist before recent decades. Knowledge is being gathered, compiled, linked, and communicated in ways that confirm an increasingly obvious reality: The greatest era of ocean exploration is actually just beginning.

RIGHT: Scientists submerge for days on a Florida reef in the undersea laboratory *Aquarius*, now operated by Florida International University. PAGES 268-269: Divers navigate a sea cave in Cocos Island, Costa Rica, a haven for schooling grunts. PAGES 270-271: In 3.8 kilometers depth in the North Atlantic, the HOV *Alvin* illuminates R.M.S. *Titanic* in 1985.

Taking the Plunge

Archaeological research in southern Africa shows humans living along the coast during the Middle Stone Age 160,000 years ago. This is the cultural period of the earliest *Homo sapiens,* according to a March 2020 report in *Science* by researcher Manuel Will. Shells of various marine mollusks found at the sites are evidence that our long-ago ancestors at least got their feet wet. For thousands of years, women divers—the ama in Japan, and the *haenyeo* in Korea—have used their skills to gather edible seaweeds and other marine life. For centuries, divers have pushed the limits of breath-hold excursions to obtain sponges, fish, precious coral, and pearl oysters from the Mediterranean and Red Seas.

In Tahiti and other South Pacific Islands there is a tradition of diving for pearls and fish and the joy of exploring the depths that continues today. There, some have learned to capture a bubble of air around each eye to make clear vision possible underwater. For at least a thousand years, the Bajau, a culture of expert breath-hold divers known as the Sea Nomads, have lived on houseboats and open canoes along the shores of Malaysia, Indonesia, and the Philippines, deriving sustenance by diving as deep as 70 meters.

Temperature, pressure, carbon dioxide buildup, and available oxygen are factors that limit how deep and how long humans can dive, and other animals are limited by these factors, too. In the sea, most have aquatic lung-equivalents called gills, thin membranes where oxygen dissolved in the surrounding water is transferred to their blood and internal organs. Many animals with lungs can dive and stay submerged for much longer than humans, and some seem more at home underwater than above. Most of the 60 or so species of sea snakes are lifelong seafarers, starting when they hatch from internal eggs and are released, fully formed, from their mothers. A few sea snakes related to Asian cobras are mostly aquatic but lay eggs on land—as do other marine reptiles, including several kinds of crocodilians, the Galápagos marine iguana, and all sea turtles.

ABOVE: From dugout canoes, the Bajau, known as sea nomads, fish among the coral reefs of Malaysia's Bodgaya Island. OPPOSITE: The centuries-old practice of breath holding allows Japanese ama divers to seek food during two-minute dives.

Numerous seabirds dive, most starting in the air from high above, including thick-billed murres that can descend to 200 meters depth in pursuit of small fish. Underwater, flightless penguins and Galápagos cormorants are as graceful and swift as their prey. Among mammals, dugongs and manatees tend to stay within 10 meters or so of the surface, but sea otters and sea lions can venture to about 100 meters. Harbor seals dive to at least 500 meters, while their enormous cousins, elephant seals, descend to 600 meters. Sperm whales go deeper than most other cetaceans, more than 1,000 meters, and stay longer, more than an hour. But the champion deep-sea diver is the sleek Cuvier's beaked whale *(Ziphius cavirostris),* documented by Duke University researchers to reach 3,000 meters. The only air-breathing animals able to go deeper are humans, thanks to engineering ingenuity.

Diving Deeper

Long before people devised ways of taking an air supply underwater, certain European freshwater spiders (*Argyroneta aquatica*) were spinning silken capsules shaped like an overturned vase, then filling them with air to create dry spaces underwater, where they retreated to dine on captured prey. The Greek philosopher Aristotle designed a much larger but similarly shaped structure in the fourth century B.C., and his student Alexander the Great is reputed to have explored parts of the Mediterranean Sea using a system based on that design. English astronomer Edmund Halley designed a wooden bell used in 1691 that, when lowered underwater, compressed air into the top, where he could breathe. By transporting air in casks, the bell's entire space could be filled. This allowed several men to sit inside, tending a tethered diver who could walk a short distance from the bell with another "personal bell" on his shoulders. In a report to the Royal Society, Halley said: "By this means I have kept three men 1¾ hours under water in 10 fathoms deep without any the least inconvenience and in as perfect freedom to act as if they had been above."

The concept of a diving helmet—basically a bell placed over a diver's head that is connected by a hose supplying compressed air from a surface pump—was developed by French inventor Sieur Fréminet in 1771. British engineer John Smeaton, called the father of civil engineering, is celebrated for his impact on the design and construction of bridges and canals, and also for devising the first widely used diving bell. Adopted in 1788 for use in salvage, construction, and repair operations, Smeaton's bells were made of cast iron and received air compressed with a hand pump, a hose, and non-return valves to keep the air from being sucked back up toward the surface when the pump stopped.

Valuable horses threatened by a barn fire sparked the next breakthrough in diving technology. Borrowing a helmet from a medieval suit of armor, a bystander named John Deane persuaded the firemen on the scene to pump air through one of their hoses into the helmet that he wore to enter the smoke-filled stable and safely rescue the animals. He patented the concept in 1823 for fire-fighting use, and soon thereafter produced a modified helmet for diving that rested on a diver's shoulders with compressed air pumped from the surface. With his brother Charles, John Deane operated a successful enterprise for many years, training divers and salvaging sunken ships. A German instrument maker, Augustus Siebe, took the concept further by sealing the helmet to a waterproof suit equipped with a weighted belt and lead-bottom boots. "Heavy gear" and "hard hat diving" are terms currently used to describe the many variations based on Siebe's diving dress, including sophisticated 21st-century derivations.

Siebe likely would be astonished to see adaptations of diving helmets in use today by recreational divers at tropical resorts around the world. Anyone willing to shoulder a colorful, lightweight helmet can be guided on a walking tour among corals, sponges, and clouds of curious fish.

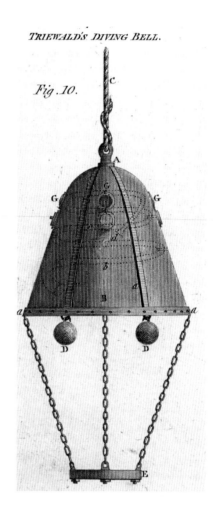

TRIEWALD'S DIVING BELL.

Fig. 10.

ABOVE: Circa 1730, Swedish merchant Marten Triewald saw a future in the deep and sketched this diving bell. OPPOSITE: Astronomer Edmund Halley's 1690 concept for an air-filled bell supporting a diver with a "personal bell" was successfully used into the 1800s.

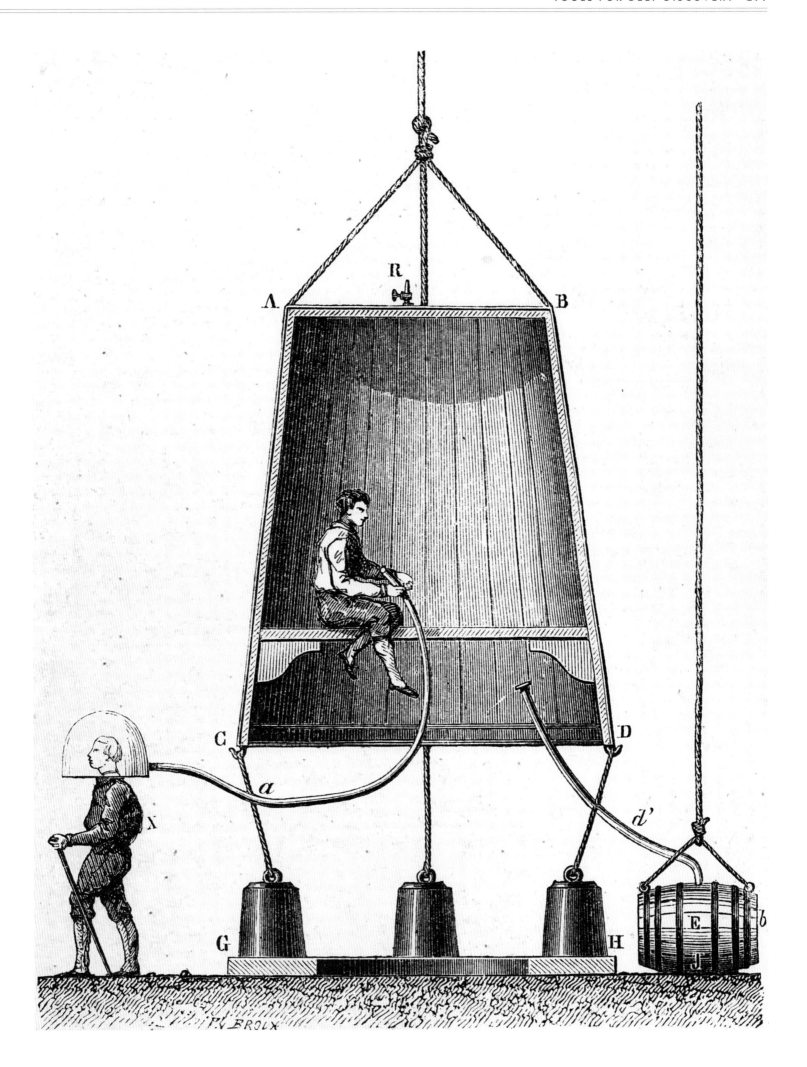

F. EROLX

Feeling the Pressure

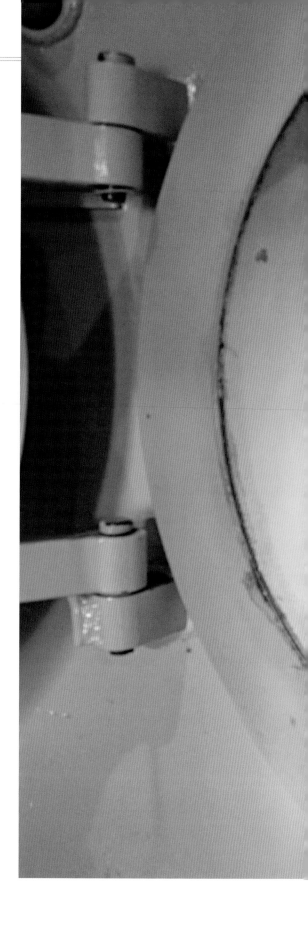

A freak accident in 1783 set in motion a series of breakthroughs in diving technology. The 100-gun British warship H.M.S. *Royal George,* while heeled over for maintenance in Portsmouth Harbor, flooded and sank with her full crew and more than 250 visiting women and children on board. The water was only 20 meters deep, but hundreds perished after being trapped. Following a salvage mission that commenced in 1834, a diving bell and Deane's hard-hat systems were used in 1839 to deploy underwater explosives to break up the ship. For the first time, diving in pairs—the now standard "buddy diving" system—was mandated. A severed air hose of one of the divers, Pvt. John Williams, caused the compressed air to rush out of his suit and helmet, sweeping his soft tissues toward the top of the helmet and causing his head to swell and blood to flow from his eyes and ears. He survived this classic case of "the squeeze" but divers operating in deeper water often did not. As a result, a simple but vital non-return valve was added on future diving helmets to solve this gruesome problem.

"Monsters of the deep"—exaggerated visions of giant octopuses, squid, sharks, and other specters of disaster—have haunted those who ventured underwater, but the greatest risks in diving have largely been caused by not understanding or respecting the laws of gas behavior. The composition of air—oxygen 20.9 percent, nitrogen 78.09 percent, and traces of other gases—was not known until the 1770s, and it took years of trial and error to discover the hazards of breathing air under pressure. Many commercial and military divers and bridge construction workers who operated for long hours in dry but pressurized chambers called caissons suffered from a mysterious malady when they returned to surface pressure. Now known as caisson disease, or decompression sickness, its afflictions can include a tingling or itchiness of the skin, shortness of breath, dizziness ("the staggers"), and acute feelings "like knives" in the joints, muscles, or abdomen. A common reaction to the excruciating pain is bending over—and thus the term "the bends."

In 1878, a French physiologist, Dr. Paul Bert, discovered that breathing air under pressure increases the amount of nitrogen in tissues. The nitrogen gas stays in solution in the blood under pressure but emerges as pain-inducing bubbles if the pressure is released too quickly. He deduced that nitrogen bubbles were the cause of the divers' misery. He also observed that the gradual reduction of pressure could allow the nitrogen to escape safely and that putting afflicted divers in a pressurized chamber could relieve the symptoms. Without treatment, wayward nitrogen bubbles could block the blood's access to vital organs and cause paralysis or death. J. S. Haldane, a physician conducting experiments with the British Navy in 1907, made diving even safer when he introduced the concept of staged, rather than just gradual, decompression. Recognizing how gases behave under pressure, he developed the first set of diving decompression tables, a formulation for dives as deep as 60 meters. The time needed to

In Italy's Gulf of Taranto, divers breathe oxygen in a decompression chamber to help eliminate nitrogen from their tissue.

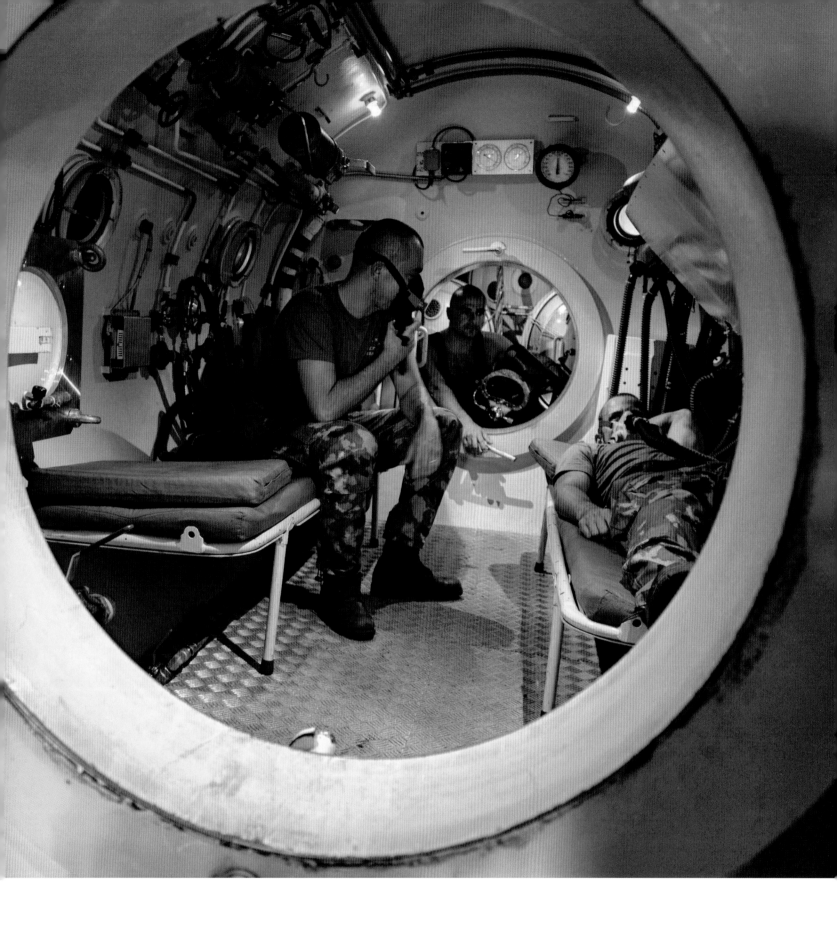

safely return to the surface depends on how long and how deep a diver goes.

Unknown at the time, compressed air delivers other unexpected effects on divers. Nitrogen in air induces a dangerous, mind-altering euphoria when breathed at depths greater than about 30 meters, and pure oxygen can cause convulsions and death below about 10 meters. The combined effects of nitrogen narcosis and oxygen toxicity limit diving with compressed air to about 50 meters. But, until someone tried it, who knew?

Our Business Is Going Under

Air supplied from hoses on the surface makes it possible to work underwater with diving bells and gear using helmets. But as early as 1772, the self-contained system developed by Frenchman Sieur Fréminet enabled a diver to be free of a surface connection by dragging along a tank of compressed air. In 1925, a less cumbersome approach was designed by Commandant Yves Le Prieur: a cylinder of compressed air slung across a diver's chest that released a continuous flow of air into a full face mask.

Master diver Henry Fleuss devised a different approach in 1876 using oxygen, not air, in a closed, self-contained rebreathing system that recycled gas using soda lime to remove carbon dioxide. It was successfully employed to rescue trapped miners and for work in underwater tunnels where hauling tanks or hoses was difficult and dangerous. Austrian underwater explorer and zoologist Hans Hass used updated oxygen rebreathers in the 1940s to silently approach and film sharks in the Red Sea. Similar systems were used for covert operations by the U.S. Navy during World War II.

After years of breath-hold diving in the Mediterranean Sea, diving pioneer Jacques Cousteau and his companions Frédéric Dumas and Philippe Tailliez were eager to go deeper and stay longer. They tried the Le Prieur system, but it proved cumbersome, and only short dives were possible. Like Hass, Cousteau tried a rebreather and reported in his book with Dumas, *The Silent World*, that "Swimming 25 feet down with the oxygen apparatus was the most serene thrill I have had in the water. Silent and alone in a trancelike land, one was accepted by the sea." It was reportedly safe to breathe oxygen 14 meters down, but when Cousteau reached that depth, he said his lips began trembling uncontrollably and his spine bent back like a bow before he lost consciousness. A few months later he tried again with a new oxygen rebreather that he helped design, but this time at 14 meters he "convulsed so suddenly that I do not remember jettisoning my weight belt."

Cousteau abandoned the use of "treacherous" oxygen rebreathers but was more determined than ever to find a safe, reliable approach to going deep in the sea. While serving in the French Navy, he continued experimenting with diving equipment—with occasional near-fatal failures. He justified the risks: "In testing devices in which one's life is at stake, such accidents induce zeal for improvement." His dream of a self-contained compressed-air diving lung came to fruition in 1942. Cousteau collaborated with engineer Émile Gagnan to adapt Gagnan's design of a gas-demand valve for automobiles into what became the first widely adopted "self-contained underwater breathing apparatus": scuba. They named it the Aqua-Lung. In *The Silent World*, Cousteau described his reaction to the system: "To halt and hang attached to nothing, no lines or air pipe to the surface, was a dream. At night I often had visions of flying by extending my arms as wings. Now I flew without wings."

WHY IT MATTERS

Technology Turnaround

While scuba technology opened the ocean to us, other technologies have broken it down—choking it with plastic, polluting it with industrial chemicals, acidifying it with carbon emissions. Today new technologies are helping restore ocean health. While NASA and NOAA satellites, drones, ROVs, and other instruments are monitoring ocean changes to implement worldwide solutions,

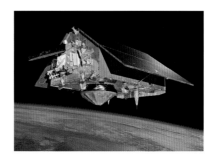

NGOs and private industries are making focused impacts. An alliance between Dow Chemical and the Ocean Conservancy helps developing nations close plastic "leakage" pathways from land to sea. Awarded the XPRIZE for design, Sunburst Systems is testing its affordable, easy-use pH sensor to measure coastal acidification. And the mFish Initiative engages governments, NGOs, and private companies in uploading real-time data onto mobile devices to help fishermen regulate their catches. More such solutions are coming, fast. The ocean's health—and ours—depends on them.

ABOVE: The Sentinel-6 satellite collects sea-level data worldwide. OPPOSITE: Diver Jacques Cousteau tests the revolutionary breathing apparatus he developed with engineer Émile Gagnan.

Flying Under the Sea

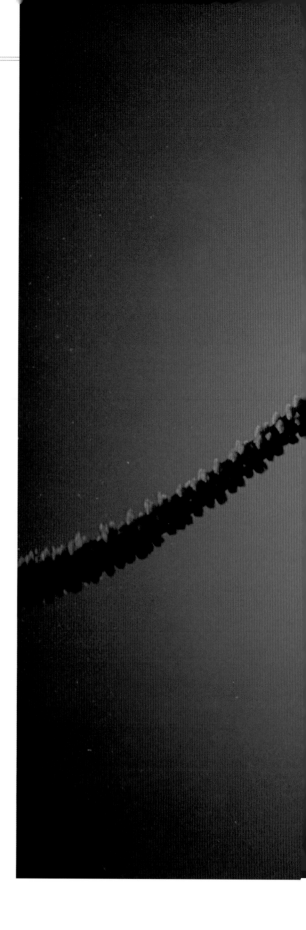

Millions of divers now "fly without wings," confidently and safely venturing into the ocean to explore places no one has ever seen before. Jacques Cousteau's vision of marketing the Aqua-Lung to recreational divers sparked a new industry and a new constituency for ocean exploration. His film *The Silent World* showcased the nature of life in the sea in ways few had seen before. The adventures he shared with his team caught the imagination of people worldwide. The film won an Academy Award for Best Documentary in 1957, and it brought unprecedented attention to the ocean as a new frontier.

Just as television was gaining momentum as the go-to place for entertainment and information, the action-adventure series *Sea Hunt*, featuring Lloyd Bridges as the heroic scuba diver Mike Nelson, inspired a new generation of hopeful undersea explorers. Then came the prime-time, 36-part documentary series *The Undersea World of Jacques Cousteau*, which ran from 1968 to 1976. Before then, it was popular for children to dream of becoming astronauts, but now many imagined being aquanauts, bravely exploring Earth's blue frontier.

While thousands of people began taking to the ocean with scuba, there was still much to learn about how deep and how long people could safely stay underwater. Many bought equipment and, with no instruction, just dived in. Concerned that training needed to be codified, several individuals pioneered the concept of certification for civilian divers. In 1955, Al Tillman, director of sports for Los Angeles County, and Bev Morgan, at the time a lifeguard, started the first civilian training program for divers. Tillman and a colleague, Neal Hess, incorporated The National Association of Underwater Instructors (NAUI) in 1960 with Jacques Cousteau, diving pioneers Drs. George Bond and Albert Behnke, and a noted scientist from Scripps Institution of Oceanography, Andreas Rechnitzer, as advisers. The Professional Association of Diving Instructors (PADI) was initiated far from the ocean, in Niles, Illinois, in 1966 by John Cronin and Ralph Erickson over a large bottle of a favorite libation and began operations in the basement of Cronin's home.

A third organization began when medical doctor—and diver—Col. Jefferson Davis set up a telephonic hotline so divers could get medical help 24 hours a day. Building on Davis's concept, in 1980, under the guidance of physiologist Peter Bennett, the Divers Alert Network (DAN) began operating at the Hyperbaric Medical Center at Duke University. It has grown into an international network of doctors and other professionals who provide diving research, data gathering, advice, and medical assistance to divers. PADI, NAUI, and DAN now provide services to millions of members globally, and dozens of other diving organizations have helped what was once a novelty sport become a profession for some and a coveted skill that enriches the lives of a wide range of men, women, and children drawn together by a shared love of diving.

Commercial divers have long recognized the importance of setting and

Framed by a spiral whip coral, a diver soars through the deep using a self-contained underwater breathing apparatus (scuba).

abiding by standards for safety and have adopted rigorous training and certifications that are in keeping with demanding and often dangerous underwater operations. When oil and gas industries moved offshore in the 1950s, a new level of diving sophistication and discipline was needed to service wellheads, platforms, pipelines, and other installations. This resulted in significant investment in ways to safely venture deeper and, importantly, to stay longer than had been possible before.

Living & Working Underwater

ntrepid divers known as "frogmen," so-called because of their green full-body suits and fins, performed special covert operations during World War II. They were an elite group in naval warfare operations that has since morphed into the legendary Navy SEALs. During the war, while developing underwater rescue techniques for Navy divers and submariners, American physician Dr. Albert Behnke made a discovery that led to the realization of what had long seemed an impossible dream: not just diving but living and working under the sea.

In Jules Verne's classic 1870 story *Twenty Thousand Leagues Under the Sea*, Captain Nemo tells the hero, Professor Pierre Aronnax, "You know as well as I do, Professor, that a human being can live underwater if he but carry with him a sufficient quantity of breathable air . . ." Behnke knew it wasn't that simple. His discovery that there is a point, depending on depth, when the tissues of a diver become saturated with the air or a mix of gases being breathed proved to be a breakthrough in making the vision of people living underwater a reality.

U.S. Navy physician Capt. George F. Bond began experimenting with the concept of "saturation diving" while serving as officer-in-charge at the Naval Medical Research Laboratory in Groton, Connecticut. He confirmed that once tissues are saturated, decompression time is the same, whether a dive lasts for 24 hours, 24 days, or even longer. Staged reduction of pressure over time would enable the gases to escape safely, with the amount of time dependent on the depth and the nature of the gases breathed. A series of trials called Project Genesis marked the beginning of the U.S. Navy's Man-in-the-Sea Program and earned Bond the title "Father of Saturation Diving" and, among divers, "Papa Topside."

Bond did not always remain topside. He often took part in the experimental dives that, in 1964, led to the U.S. Navy's *Sealab I* project, conducted off the coast of Bermuda. Four Navy divers stayed 11 days at 59 meters depth, breathing a mixture of helium and oxygen. Much was learned that was applied to the later *Sealab II*, including how to cope with the squeaky Donald Duck voice caused by breathing helium. Deployed in 1965 at 62 meters depth in the La Jolla Canyon near the Scripps Institution of Oceanography, the new lab was twice as big as the first and had amenities including humidity control, hot showers, refrigeration, and 11 viewing ports. Located on a slope, *Sealab II* became known as the "Tilton Hilton," with features that included delivery service of supplies by Tuffy, a trained bottlenose dolphin. *Sealab III*, placed near San Clemente Island, California, in 1969, in water three times as deep as *Sealab II*, ended in tragedy with the death of one of the aquanauts during the first mission, owing to failure of his rebreather system. The *Sealab* program ended, but the U.S. Navy, and other military forces globally, have continued to apply saturation research and "technical diving" to their operations—with significant spinoff benefits to those with commercial, scientific, and recreational interests.

OPPOSITE ABOVE: *Aquarius* undersea laboratory, operated by Florida International University, near Key Largo, Florida. OPPOSITE BELOW: Sylvia Earle passes a barrel sponge as she explores outside *Aquarius* laboratory.

In an Octopus's Garden

viator and ocean engineer Edwin Link adapted Capt. George Bond's work with the U.S. Navy to extend time underwater for research and exploration, especially to help satisfy his passion for underwater archaeology. Link was the first person to fully saturate underwater with a breathing mixture of helium and oxygen. He did this by spending eight hours at a depth of 18 meters in a submersible decompression chamber at Villefranche-sur-Mer in 1962. Soon thereafter, his colleague, Robert Sténuit, spent 24 hours in a submersible chamber at 61 meters depth. Then, in 1964, as *Sealab I* was getting under way, Link conducted an experimental dive in the Bahamas with two men staying at a depth of 132 meters breathing helium and oxygen for 49 hours.

Concurrently, ocean explorer Jacques Cousteau developed the concept for Conshelf (Continental Shelf Station), with the goal of submerging stations at a maximum depth of 300 meters. It was largely funded by the French petrochemical industry, and the expectation was that manned colonies underwater would facilitate exploration and exploitation of the deep sea. The first of three stations, *Conshelf I*—at 10 meters depth in the Mediterranean Sea near Marseilles—was the undersea home for Albert Falco and Claude Wesly for a week in 1962. A year later, six men spent a month during *Conshelf II* in the Red Sea off the coast of Sudan. During the third version, six divers spent three weeks 102 meters deep in the Mediterranean near Monaco, with a nearby mock oil rig where the "oceanauts" could simulate the work expected on a commercial platform.

The Beatles were singing about living in an "Octopus's Garden" and the first footprints were being planted on the moon when the U.S. Navy partnered with NASA, the Department of the Interior, and the General Electric Company to conduct an ambitious saturation diving project called *Tektite I.* NASA scientists wanted to observe the behavior of aquanauts in anticipation of having astronauts occupy stations beyond Earth's atmosphere.

Four men were selected to live for two months at 15 meters depth in a four-room habitat in Lameshur Bay, in the U.S. Virgin Islands. Cameras monitored them day and night, psychologists took notes, and medical examinations before, during, and after helped assess the effects of long-term isolation—and saturation. After satisfactory medical and psychological results, 50 scientists and engineers were selected for an expanded program, *Tektite II,* and given on-site training in 1970 prior to missions lasting 10 to 20 days. No one expected women to apply, but some did, with credentials and proposed projects that were impressive enough that five were selected for an all-woman team. Their success physically, psychologically, and as scientific and engineering professionals helped advance the acceptance of women in places where they were previously excluded, including as astronauts in the U.S. space program. But it did not keep them from being referred to as "Aquababes," "Aquabelles," and even "Aquanaughties."

Thirty meters deep in the Red Sea in 1963, a crew member of Jacques Cousteau's *Conshelf II* laboratory introduces his pet parrot to parrotfish of Shaab Rumi reef.

DEEPER DIVE

Excavation Grid

While Edwin Link and other visionaries developed ways to reach and explore the deep ocean, there was still a need to find and preserve archaeological sites and artifacts. That's where George Bass, described as "burly and dynamic" while "characteristically thorough," stepped in, turning underwater archaeology into a rigorous scientific discipline. Bass's secret: Approach every wreck as a traditional land-based archaeology site, using time-tested techniques to excavate and record artifacts. Bass was a Ph.D. student at the University of Pennsylvania in 1960 when American archaeologist Peter Throckmorton discovered a 3,000-year-old Bronze Age

George Bass, Bronze Age wreck

cargo vessel off the coast of Turkey. Bass, not yet a diver, was tapped to help. He took a YMCA dive course, then headed to the site. His team covered the area with a metal grid to break it into two-meter squares, giving each diver a specific workplace. They precisely measured, photographed, and made drawings. Dozens of shipwreck excavations followed. Bass founded the Institute of Nautical Archaeology and has mentored archaeologists and explorers, including Robert Ballard, who located the *Titanic*. Today, ROVs, high-definition photography, and advanced diving equipment aid underwater excavation, but the rules of demarcation are still Bass's rules.

Astronauts & Aquanauts

The Tektite saturation diving projects of 1969–1970 yielded significant insights about not only ocean ecology and the behavior of sea creatures but also the ecology and behavior of humans.

The parallels between humans adapting to prolonged missions under the sea and in space have resulted in mutually beneficial discoveries ranging from development of life-support systems and sophisticated rebreather concepts used for spacewalks and deep dives to provisions of appropriately tasty meals welcomed above and below the sea.

By 1975 there were about 50 underwater habitats in operation around the world, ranging from simple structures for use as underwater "camps" in shallow depths to sophisticated systems such as West Germany's *Helgoland* and Russia's *Chernomor*. The U.S.-built, four-person system, *Hydrolab*, was used by more than 300 researchers in the Bahamas and, for a while, in the U.S. Virgin Islands until it was retired in 1985. Years before the first space shuttle transported astronauts to a space station, *Hydrolab*, in 1975, served as home base for scientist-aquanauts, who were shuttled to the edge of a nearby deep-water drop-off in the *Johnson-Sea-Link I* submersible. The sub, one of a pair, was designed by Edwin Link and featured a one-atmosphere, clear acrylic sphere for a pilot and an observer and a separate chamber that could be pressurized to match the outside ambient pressure. A hatch at the bottom of the chamber opened to allow a diver to "lock out" for an excursion, then return to the sub and be transported safely under pressure back to the *Hydrolab*. The concept of moving saturated divers vertically in pressurized chambers is used today by commercial divers and in the mid-1980s was incorporated into the design of the 28-meter-long submarine *Saga*. Built and developed by the French Research Institute for Exploration of the Sea (IFREMER) and Compagnie Maritime d'Expertises (COMEX), *Saga* succeeded in deploying human and robotic divers from a submarine at 317 meters.

Aquarius Reef Base has been located since 1992 in 18 meters depth nine kilometers offshore from Key Largo, Florida. Originally owned and operated by NOAA, and since 2013 by Florida International University, the six-person system has hosted hundreds of researchers who live in dry quarters but swim into their laboratory—the living reef—day and night. The years of research provide unique long-term data and insights about reef ecology, behavior, and physiology, including detailed studies of individual corals, sponges, and fish in their natural homes.

In 2013, Fabien Cousteau led a 31-day mission in *Aquarius*, a tribute to his grandfather's 30-day saturation dive 50 years prior. Since 1991, NASA has used *Aquarius* for its NEEMO (NASA Extreme Environment Mission Operations) analog missions, sending groups of astronauts to simulate human spaceflight missions. The weightlessness and isolation in both environments are comparable, but surrounding *Aquarius* there are numerous distractions not encountered in space, from curious barracuda and squid to galaxies of luminous plankton.

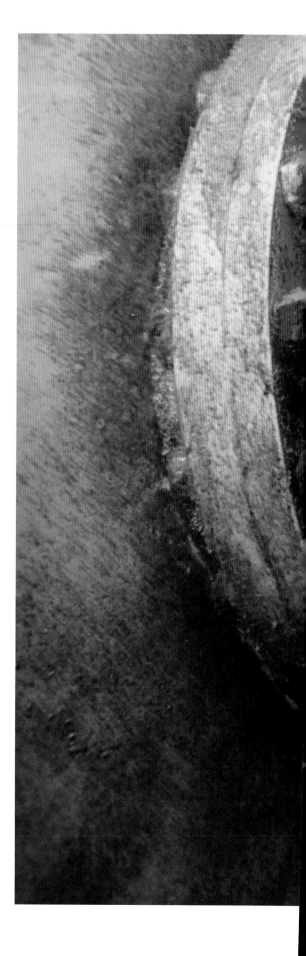

Tektite II aquanauts Margaret Lucas (left) and team leader Sylvia Earle meet during a two-week saturation dive in the U.S. Virgin Islands.

Diving for Science

The first recorded scientific diving was in 1844, when biologist Henri Milne-Edwards and a naturalist friend submerged beneath the Strait of Messina off the shores of Italy. To "pursue animals in their most hidden retreats," they took turns using breathing equipment designed by the Paris Fire Brigade for use in flooded cellars. During the century that followed, diving helmets were used by various scientists, including zoologist William Beebe, who was so enamored with them that he wrote in *Beneath Tropic Seas,* "Don't die without having borrowed, stolen, purchased, or made a helmet of sorts, to glimpse for yourself this new world."

Marine scientist Eugenie Clark relied on a face mask and fins to explore reefs in the South Pacific and Red Sea in the 1940s, but in the 1950s, she and other scientists discovered the advantages of using scuba to extend the time and depth of their observations. John E. "Jack" Randall, Gerald Allen, and John "Charlie" Veron are among the pioneers who have spent thousands of hours underwater studying the ecology, systematics, and behavior of fishes and corals, but they had to obey rules not required of their terrestrial counterparts. Spending the day in a forest or desert? No problem! Doing the same thing underwater? Not so easy! It is possible for them to work submerged 20 meters from the surface for about an hour, but to go just 50 meters down reduces observation time to about five minutes, with stops while returning to the surface to allow accumulated nitrogen to safely escape their tissues. Over the years, thousands of researchers have learned to work within the time and depth constraints of diving and now routinely use the ocean as their laboratory.

Mindful of the risks and concerned about the special nature of scientific diving in contrast to recreational and commercial diving, the American Academy of Underwater Sciences (AAUS) was formed in 1977 to set guidelines for training and certification that have become standard requirements in academic institutions internationally. This includes a special category for rebreather certification using air and various combinations of gases to go deeper—and stay longer—for those willing and able to master the physical and technical challenges.

The rewards of exploring far into the twilight zone, where light fades to darkness, are worth the risks to Richard Pyle, an ichthyologist at Hawaii's Bishop Museum. In the 1990s he teamed up with Bill Stone, president and CEO of Austin-based Stone Aerospace, to fine-tune Stone's rebreather technologies that are used in space and also for exploring some of the world's longest and deepest underwater caves. Pyle's focus is on reefs that are 50 to 200 meters down. He says, "The twilight zone is the most neglected part of the ocean. I have found 13 new species of fish in an hour there. What else are we missing because we are limited by how deep we can go?"

VISIONARIES

Eugenie Clark
The Shark Lady

Revered as a scientist-explorer who had a way with fish and people, Eugenie Clark spent her life enlightening a public whose knowledge of sharks was laced with myth and fear. Among the earliest women in the 1950s male-dominated field of ichthyology, Clark credited her love of the water to her sea-centric Japanese heritage. Dedicated to meeting fish in their home environment, she undertook unprecedented fieldwork by scuba and submarines in little-explored waters. In the Red Sea she discovered the Red Sea Moses sole (*Pardachirus marmoratus*), which emits a natural repellent that can halt sharks and set their heads thrashing from side to side. "Sleeping sharks" were another landmark find: Unlike most other species, whitetip reef sharks off the coast of Mexico do not have to swim continuously to breathe. Founder of Florida's Mote Marine Laboratory, Eugenie Clark continues to inspire love and care for her other laboratory—the sea.

OPPOSITE ABOVE: Shark expert Eugenie Clark examines a bull shark released by a fisherman. OPPOSITE BELOW: Scientists study Mediterranean seagrass.

HOPE SPOT

Gulf of California

Celebrated by Jacques Cousteau as "the world's aquarium," the Gulf of California is globally recognized for its exceptional diversity of marine life. Nearly a thousand kinds of fish and more than 5,000 species of invertebrates and at least 170 kinds of seabirds prosper in this magnificent steep, deep chasm that extends from the Colorado River Delta to the tip of the Baja California Peninsula. The Gulf is an original Pristine Seas site and a Hope Spot since 2009.

Panamanian porkfishes and Mexican goatfishes gild a Cabo Pulmo reef.

Onward & Downward

ushing the limits of human physiology, the French engineering
company COMEX conducted experimental open-water saturation
dives in 1988 to a depth of 534 meters, and to 701 meters in 1992,
with returns to the surface requiring many days of gradual reduction in pressure. In reality, safe working access for specialized commercial
and military saturation divers currently extends to about 300 meters. But
it is possible to go deeper, even to the greatest depths of the sea, if tucked
inside a submarine that is strong enough to withstand the ocean's pressure
on the outside while inside maintaining sea-level pressure: one
atmosphere.

Leonardo da Vinci is credited with designing the first submarine early in
the 1500s. He believed that as a "ship to sink another ship" it would be so
effectively destructive in war that he kept the concept secret. British mathematician William Bourne, in 1578, proposed a cylindrical wooden vessel
covered in waxed leather and outfitted with holes for oars that could be
rowed by the occupants inside. The versatile Dutch engineer Cornelis Dreb-
bel, known for developing a microscope with compound lenses as well as an
incubator for chicken eggs, designed and built at least three multipassenger
submarines made of greased leather stretched over a wooden frame.

The first working "diving machine" was developed in 1715 and used for
salvage work by the English inventor and entrepreneur John Lethbridge.
He constructed a wooden barrel fitted with a small glass window and holes
with waxed leather sleeves attached so a man could be lowered on a cable
to about 18 meters depth.

Numerous military systems evolved starting in the 1800s, mostly to operate undetected at a depth of a few hundred meters. Diesel-electric-powered
systems dominated during both 20th-century world wars, but the advent
of nuclear power in 1951 marked a new era. The first U.S. nuclear sub, the
U.S.S. *Nautilus,* set records for distance and time underwater, crossing from
the Pacific to the Atlantic under the Arctic ice pack in 1954. One-atmosphere
systems, from one-person suits to massive submarines, advanced significantly during the latter half of the 20th century. U.S. nuclear systems and
their equally sleek and fast counterparts, the Soviet Alfa-class attack submarines, traveled throughout much of the world. Some engaged in exploration, notably the U.S.S. *Queenfish,* which charted unknown terrain under
the Arctic ice in 1970. About the same time, the U.S. Navy commissioned
one special small, fast, and deep-diving nuclear sub, *Nuclear Research
(NR-1),* deliberately designed for research. Since 1969, scientists have
explored many parts of the ocean world with *NR-1,* including the Mediterranean Sea, where geophysicist and explorer Robert Ballard used the sub's
unique capabilities to find and document numerous ancient shipwrecks.

Technologies developed since the 1950s greatly advanced knowledge of
the ocean from high above. Now, at last, technologies are making possible
exploration of the ocean from deep within.

Dawn outlines a Trident nuclear
submarine's conning tower housing
the helm, periscopes, radar, and
torpedo-firing controls.

The Hunley

Deep-sea artifacts remain more intact than those on land, explorer Robert Ballard often reminds us, because the ocean preserves them. When the Confederate submarine *Hunley* was found in Charleston Harbor in 1995, its story filled a lost page in Civil War history. The cigar-shaped sub, the first ever to destroy a wartime vessel (the U.S.S. *Housatonic*), was a marvel of modern engineering, financed by Horace L. Hunley. At 12 meters long, the *Hunley* was small and stealthy, tapered at both ends to move smoothly and undetected; one Housatonic crew member thought it was a dolphin—until it was too late. With a torpedo attached to a spar on its

H. L. Hunley sinks U.S.S. *Housatonic*

bow, the *Hunley* rammed the Union ship in 1864, sinking it in five minutes. With a lantern, the *Hunley* signaled its own safety to shore—but it never arrived. A century later, archaeologists found it on its side in nine meters of water. A silent crypt, it held the remains of its eight crewmen. Holes indicate it may have been damaged as the *Housatonic* exploded. Archaeologists have since found a pipe that may have pulled away from *Hunley*'s hull during the blast, letting water flood in. Cutting-edge technology brought the sub up and conserves it, including scanning electron microscopes that reconstruct the story of shoes, a wallet, and the lantern that signaled "safety."

Alone in the Deep

The lure of the unknown and the search for treasure have provided powerful incentives to develop various submersible systems. The German manufacturing firm Neufeldt and Kuhnke created a one-man armored diving suit used in 1930 to recover five tons of gold bullion from the vessel *Egypt,* sunk in 130 meters of water off the coast of France. The same year, zoologist William Beebe and engineer Otis Barton descended in a hollowed-out steel ball, the *Bathysphere,* off the south shore of Bermuda, and later set a record for human exploration of the sea at a depth of 923 meters. Meanwhile, English engineer Joseph Peress developed a refined version of a one-atmosphere diving system, *Jim,* named after his colleague Jim Jarret, who tested the suit during various perilous trials. Modern versions were developed for commercial use, and derivations of the one-person, one-atmosphere systems include *Wasp* (a yellow and black machine that "flies" with thrusters) and *Mantis* (armed with a pair of powerful manipulators), both crafted by engineer Graham Hawkes.

All one-atmosphere, one-person diving suits were operated on a cable attached to the surface until 1979, when Oceaneering International teamed up with the University of Hawaii and National Geographic to explore deep reefs offshore from Oahu. The team transported *Jim 9,* strapped to the front of the two-person submersible *Star II,* then released the suit—with its human occupant—at 400 meters to walk freely on the seafloor. There it was connected to *Star II* by a slender communication line. The advantages of being free of a surface connection were clear. A new generation of hardmetal dive suits built by the Canadian company Nuytco were designed with the option of being tethered or free. In addition, one-person microsubmersibles, including *Deep Rover I* and a fleet of Nuytco subs called *DeepWorker,* were designed to operate without a tether.

During the Sustainable Seas Expeditions (SSE) from 1998 to 2003, more than 50 scientists and technicians operated *DeepWorker* subs with a 600-meter depth range, as well as the 1,000-meter *Deep Rover I,* both known to be "so simple to drive even a scientist can do it." Sponsored by the Goldman Foundation, the National Geographic Society, and NOAA, the SSE included more than 50 partner institutions and universities. Most operations took place in 12 U.S. national marine sanctuaries, Mexico, and Belize. Michael Guardino, a California high school science teacher, helped some of his students to get scuba-certified and conduct underwater studies in depths to 25 meters near Monterey, where kelp, starfish, sea lions, and sea otters abound. Just offshore, Guardino recorded what he witnessed from the SSE sub, descending 10 times as deep as his students into a sunless realm populated by bright red shrimp and legions of sea cucumbers, brittle stars, and basket stars. The SSE project was the first time scientists and teachers were encouraged to become submersible pilots and engage with surface-based researchers in a comprehensive public-private-industry project to explore and gather baseline data, engage students, and regularly communicate to the public.

Why Small Research Subs?

For the same reasons that it makes sense to have astronauts, not just instruments, go to the moon, it is vital to deploy humans, not just robots, into the depths of the sea. Ever since Woods Hole's manned research sub *Alvin* debuted in 1964, the concept has proved true. In the 1970s, Project FAMOUS used *Alvin* and France's *Cyana* to make unprecedented discoveries of new geologic and life forms. In the 1980s

and '90s *Alvin* and Russia's *Mir I* and *II* documented hydrothermal vents in the Atlantic and explored the *Titanic;* high-definition footage from a *Mir* sub reached millions via James Cameron's 1997 film *Titanic.* Today, with even higher definition cameras and flexible arms that coddle the animals and artifacts they collect, human-guided submersibles document new ecosystems and clues to our past. "Everyone who goes underwater becomes an amateur scientist," wrote legendary author Arthur C. Clarke. If he is right, small subs will lead citizen scientists to understand the ocean and protect it.

ABOVE: A *DeepWorker* HOV: "So simple to drive even a scientist can do it." OPPOSITE: Sylvia Earle set the untethered solo deep-dive record in the *Jim* suit in 1979.

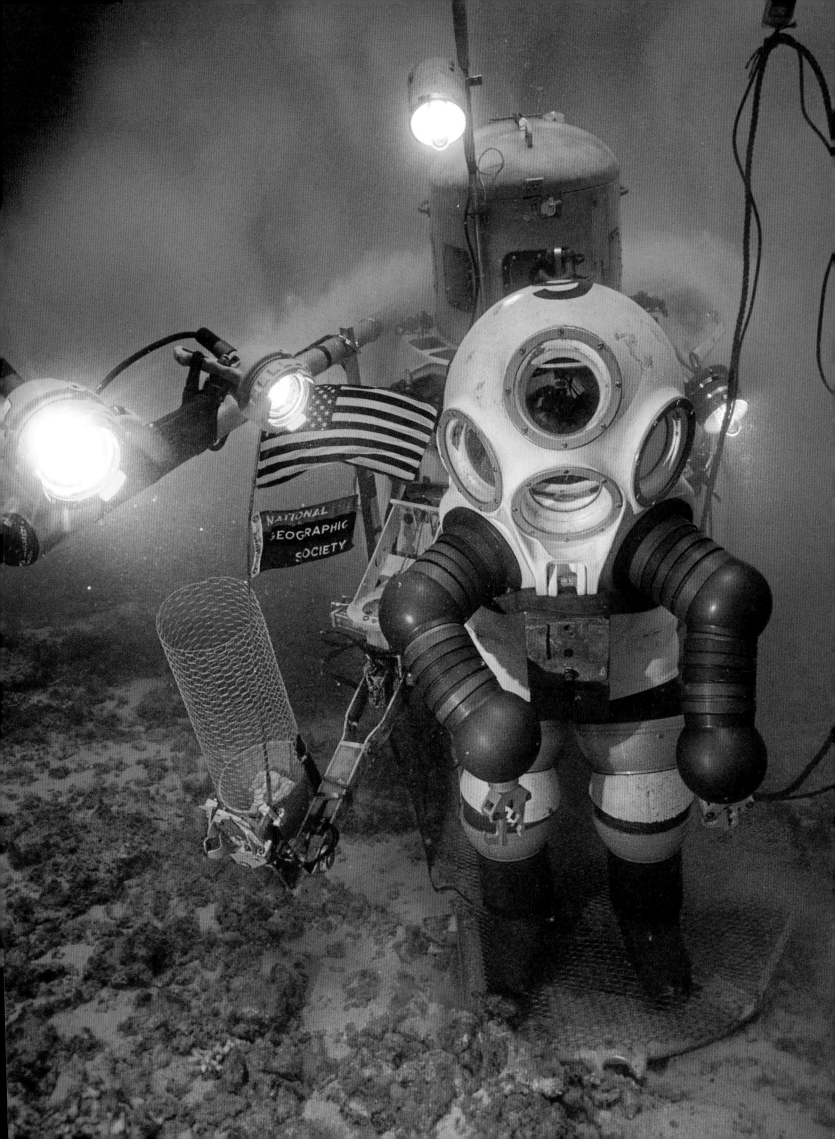

Deep, Deeper, Deepest

By the mid-1970s, 12 countries had deployed more than a hundred small manned submersibles. The three-person *Alvin,* operated by Woods Hole Oceanographic Institution, launched in 1964. It has continued with significant upgrades as the undisputed workhorse of research submersibles, now capable of reaching 6,500 meters depth. Several Canadian-built *Pisces* subs have supported ocean exploration to 2,000 meters for more than five decades, and in the same time frame, two four-person *Johnson-Sea-Link* research subs have taken hundreds of scientists to depths as great as 1,000 meters.

In 1948 Swiss physicist Auguste Piccard and Belgian physicist Max Cosyns tested a deep-diving system they called a bathyscaphe. In 1954 they used it for dives to a record 4,050 meters depth near Dakar, in West Africa. Six years later, the bathyscaphe *Trieste,* with Piccard's son, Jacques, and Lt. Don Walsh of the U.S. Navy piloting, dived to the deepest place in the world's ocean, the bottom of the Mariana Trench.

Fifty-two years later, filmmaker and National Geographic Explorer James Cameron emerged from the bright green submarine *Deepsea Challenger,* the first person to return to the Mariana Trench since the *Trieste* dive in 1960—and the first one to do so solo. Getting to the bottom of the Mariana Trench was, for Cameron, a boyhood fantasy come true, a dream motivated, he said, "not to set records but by the same force that drives all science and exploration . . . curiosity." The sub took seven years to design and construct, but, he said, "The only way to make my dream a reality was to build a new vehicle unlike any in current existence."

Chinese scientist Weicheng Cui led the development of China's 7,000-meter *Jiaolong (Dragon),* launched in 2012, and expects to have the full-ocean-depth *Rainbow Fish* operational soon. Meanwhile, explorer and Texas-based businessman Victor Vescovo commissioned the construction of a full-ocean-depth, two-person HOV, *Limiting Factor,* and set out to make solo dives in the ocean's five deepest places. He reached the first, the Puerto Rico Trench, in 2018. In 2019 he visited the South Sandwich Trench, the Java Trench, the Molloy Deep, and the Mariana Trench. Vescovo—who has summited the seven highest mountains and skied both poles—remarks, "Who says there's nothing left to explore on this planet? There is plenty to explore, and learn, in the oceans."

So, how deep is the deepest ocean? Unlike the peak of Mount Everest, a singular point visible from miles away and measurable to within centimeters, the bottom of the Mariana Trench is out of sight, and it is generally flat with gentle slopes and mounds extending over 1,000 kilometers. Vescovo measured a maximum of 10,927 meters—11 meters deeper than the 10,916-meter depth recorded by Walsh and Piccard in 1960 and 19 meters deeper than the 10,908 meters reported by James Cameron in 2012. Other scientific measures place it at 10,984 meters. The deepest place on Earth, therefore, is something close to 11,000 meters, give or take a few.

VISIONARIES

James Cameron
Filming the Deep

Bringing the thrill of deep-ocean exploration to a global audience is the mission of diver, engineer, and celebrated filmmaker James Cameron. Before filming the 1997 box office hit *Titanic,* he made 12 submersible dives to document the wreck with high-definition cameras. Later, he and engineer Vince Pace developed a revolutionary 3D camera system to produce the IMAX documentary *Ghosts of the Abyss* in 2003. Making the world's first solo dive to the floor of the Mariana Trench in 2012, Cameron dropped nearly 11 kilometers over two and a half hours in *Deepsea Challenger,* co-designed with engineer Ron Allum to endure pressure 1,000 times greater than at the surface; he then documented the trench for several hours with 3D cameras and LEDs. His quest to bring audiences the mysteries of the sea continues with his film *Avatar 2.* It is fitting that the sequel to his 2009 Academy Award–winning *Avatar* takes place in the ocean depths.

Deepsea Challenger takes a test dive before James Cameron's record descent to the bottom of the Mariana Trench.

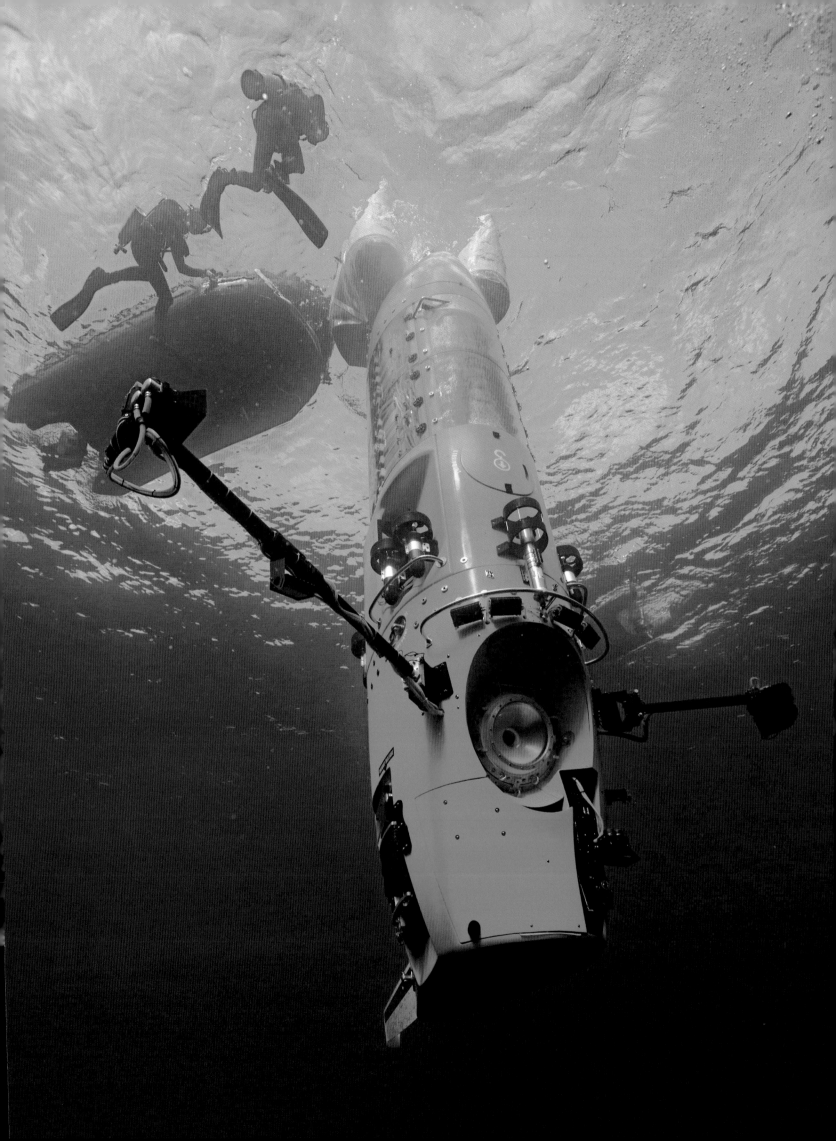

Send in the Robots

"I am sitting in the darkened control room of a mother ship, while an ROV flies through the water column 3,000 meters below us. High-definition video fills a large screen in front of me. Microwave links connect us in real time to experts around the world." So says Bruce Robison, deep-sea biologist at the Monterey Bay Aquarium Research Institute, describing his experience as an "armchair explorer," witnessing the ocean through the camera "eyes" of a remotely operated vehicle (ROV) and sharing the view with colleagues thousands of kilometers away.

ROVs, some the size of trucks, others small enough to fit in a briefcase, were initially developed for military, scientific, and industrial uses in the 1960s. The first discovery of hydrothermal vents near the Galápagos Islands was made using towed ROV systems from Scripps Institution of Oceanography in 1976 and Woods Hole Oceanographic Institution in 1977.

Growing demand for ROVs for commercial uses provided incentive to develop powered vehicles with manipulator arms and specialized tools. By the late 1980s, remotely operated systems were regarded as essential for underwater inspection, construction, maintenance, and repair in the offshore oil and gas industry. Navies around the world developed or adapted their own versions for harbor monitoring and locating and removing mines. ROVs were used to document and eventually cap the blowout in 2010 of the *Deepwater Horizon* well 1,500 meters deep in the Gulf of Mexico.

Scientists began using powered ROVs in the 1980s and, working with engineers, ultimately developed systems to explore below polar ice and even the deepest parts of the ocean. The Japanese ROV *Kaikō* reached the bottom of the Mariana Trench in 1995 and the Woods Hole Oceanographic Institution's hybrid ROV, *Nereus,* did so in 2009, both returning with images that helped define the existence of diverse forms of life prospering at that stupendous depth. Sadly, both of these robotic explorers were subsequently lost at sea.

ROVs, such as the University of Hawaii's 6,500-meter *Lu'ukai,* are proving to be essential tools for exploring and monitoring deep-sea locations where mining is planned to extract minerals from manganese nodules and deep-sea crusts that form around hydrothermal vents, mostly in depths between 2,000 and 6,000 meters. *Lu'ukai*'s cameras are gathering evidence of robust ecosystems that counter claims of life being sparse.

Vehicles with cables provide direct control, real-time images, data, sampling, and, usually, the means to recover the system at the end of a dive. Autonomous underwater vehicles (AUVs), driven by onboard, preprogrammed computers, can cover long distances and do not have the cable entanglement issues that plague ROVs. But they are vulnerable to loss if their homing or tracking devices fail. In the 21st century, hundreds of AUVs and ROVs and dozens of manned submersibles are being used, sometimes independently, sometimes together—"the right tool for the right job."

A research ship launches the ROV *Global Explorer* to photograph and collect data as deep as 2,743 meters. Its manipulator arm delicately captures fragile specimens.

Droning for Science

The saildrones are here. Resembling oversize windsurfers, these bright orange sailboats are digitally savvy, sturdy (carrying up to 90 kilograms of equipment), and flexible enough to explore hard-to-reach ocean surface areas. They report data fast—a requirement in the rapidly changing ocean environment. Since 2015, NOAA and its international partners have sent these remotely controlled boats into the Arctic Ocean to track melting ice and endangered right whales; from California to the Equator to support the multinational

NOAA saildrone

weather forecaster TPOS (Tropical Pacific Observing System); and, in an "audacious" venture by the lone *Saildrone 1020,* to complete the first circumnavigation of Antarctica by an autonomous vehicle. In a rain-blotched "selfie," the little drone plows bravely through a churning Southern Ocean, gathering carbon dioxide measurements. Its data hit home: Once thought to be a "sink" for carbon from the atmosphere, the Southern Ocean now appears to be emitting carbon during the winter—a wake-up call for the planet.

The Power of Images

A t a 1967 Sea Rovers meeting in Boston, underwater photographer Stanton Waterman lifted the audience out of their seats with the first filmed images of half-ton manta rays dining on a shimmering blizzard of plankton, each black-and-white giant pirouetting, arching, and gliding with gargantuan grace. Being in the midst of a dizzying throng of mantas was a rare experience, but the ability to record it and share it with thousands was unprecedented.

Before the 1960s, underwater photographs of any sort were hard to come by. A French biologist with a passion for mollusks, Louis Boutan, is credited as the pioneer who first crafted underwater camera systems in the 1890s. Prior to that time, scientists such as those aboard the oceanographic research vessel H.M.S. *Challenger* relied on artists to document sea creatures brought to the surface by hooks or in nets. Lacking suitable imaging systems in the 1930s, zoologist William Beebe engaged artist Else Bostelmann to render paintings of animals that he described when he returned from dives off Bermuda in the *Bathysphere.*

The first commercial use of underwater cinematography began in 1916, when J. Ernest Williamson filmed dramatically staged scenes for Jules Verne's *Twenty Thousand Leagues Under the Sea* for Universal Studios. His father, a sea captain, had devised a flexible tube made of interlocking iron rings that enabled a person to descend underwater to conduct salvage and ship repair in more than 50 meters depth. Williamson adapted the design by adding a viewing chamber with a glass window that housed not only cameras but the photographer as well.

Underwater color photography began in 1926 with an encounter with a cooperative hogfish in the Florida Keys. Scientist William Longley and National Geographic photographer Charles Martin encased a camera in a waterproof housing and, for lighting, tripped a battery that set off on an explosion of magnesium flash powder on a raft that they dragged along. The resulting image of the miraculously unperturbed fish graced a *National Geographic* article later that year. Austrian scientist Hans Hass was a teenager in the 1930s when he made his first underwater camera system; and in the 1940s he pioneered underwater filmmaking.

A turning point for National Geographic photography began in 1934 when an adventurous young man, Luis Marden, arrived equipped not with the usual 90 kilograms or so of heavy gear used by most serious photographers, but rather, with a lightweight Leica camera. Some 64 years and 55 *National Geographic* articles later, Marden could look back as a witness and key participant in a transformative time for underwater photography. Harold "Doc" Edgerton, the distinguished Massachusetts Institution of Technology engineer and creator of the electronic stroboscope, joined Jacques Cousteau

ABOVE: A hogfish made history as the first published underwater color image, by Charles Martin in 1926, in *National Geographic* magazine. OPPOSITE: Shells of planktonic foraminifera were sketched by 1870s H.M.S. *Challenger* artist A. T. Hollick.

and Marden in making significant strides in underwater photography. Affectionately known as "Papa Flash," Edgerton made possible some of the first images of sea life thousands of meters deep taken by a safely housed "drop camera" and flash lowered from the side of a ship. Modern versions are used regularly, sometimes remaining for hours on the seafloor to document wary creatures lured into view by strategically positioned bait.

The underwater films Cousteau and his team created reached millions of people globally. He also promoted a small underwater still camera with O-ring seals that did not require a special housing. Called the Calypso-Phot, it was designed by Jean de Wouters and first released in 1961. A modified version was produced by Nikon and marketed as the Nikonos in 1963. The camera became a widely used 35mm-film system through versions up to a sophisticated single lens reflex Nikonos 6. The Nikonos, and many other cameras designed to use film, were largely displaced by 1996 with the introduction of increasingly effective digital systems.

Two AT&T Bell Laboratory scientists, George E. Smith and Willard Sterling Boyle, developed the basis for digital photography when they invented the charge-coupled device, or CCD. Light strikes a tiny grid of photosensitive silicon cells, and each builds a charge proportional to the light that hits it. The charge can be measured precisely, and that determines exactly how bright that portion of the image should be. With filters, color can be discerned as well.

The individual elements of the grid form tiny squares, pixels, that together create an image. Using the CCD technology, a young Kodak engineer, Steve Sasson, built the first digital camera—the size of a breadbox—in 1975. The first consumer models arrived in the 1990s and exploded in popularity in 1997 when engineer Philippe Kahn created a prototype cell phone camera and shared a picture of his newborn daughter over his wireless network.

It took time and significant improvement of image quality for expert photographers including David Doubilet—renowned for his masterful use of light, exquisite portraits of small sea creatures, and split above-and-below underwater images—to shift from film to digital. An early adopter was Emory Kristof, celebrated for creatively deploying cameras on ROVs and documenting deep-sea creatures and shipwrecks, including iconic images of the wreck of the *Titanic*. The advantages of digital systems are seductive. When Luis Marden returned after weeks on his first expedition with Jacques Cousteau in 1956, he had amassed the largest collection on film of underwater color images then in existence: 1,200. Digitally, on a fingernail-size card, that many photographs can be acquired in an hour underwater.

The quality of an image relates to the quality of available light, and an indication of how sensitive a film or digital system is to light is measured as its ISO (the acronym for International Standardization Organization). Typically ranging from 100 to 1600, the higher the ISO number, the greater the sensitivity to light. Canon, whose ME20F-SH system was used by the British Broadcasting Corporation (BBC) to film glowing bioluminescent organisms 1,000 meters underwater in Antarctica, won an Emmy in 2020 for its camera's unparalleled performance—achieving an unprecedented four million ISO.

"Seeing with sound" is a phenomenon well known among bats, moths, and many marine mammals, but digital systems that are sensitive to sound

"HOW INAPPROPRIATE TO CALL THIS PLANET EARTH WHEN IT IS CLEARLY OCEAN."

—ARTHUR C. CLARKE

have evolved from sonar technologies to increasingly effective levels of resolution that make for clear images. Sound waves are more than 2,000 times longer than light waves and do not get blocked or easily scattered in dark or murky water. That means sound waves can image minute plankton, schools of fish, and the nature of the seafloor—even in zero visibility.

Creatures in the sea perceive light and color in ways that are now being made visible to humans through the use of fluorescence-detecting cameras. Marine biologist and National Geographic Explorer David Gruber and his colleagues have discovered scores of biofluorescent compounds in marine animals, as well as the means to visualize for humans what the creatures likely see that we normally cannot.

Many concur that the most meaningful, powerful, and influential image ever recorded was taken by astronaut William Anders on December 24, 1968: Earth viewed from the far side of the moon. Decades later, as exploration of the dominant blueness of Earth continues, images from the depths confirm not only the wondrous nature of this planet but also how much more there is to discover.

Lighting the deep in neon blue, bioluminescent firefly squid migrate from offshore Pacific waters to the coast of Japan to breed.

The Science of Where

Flying high above Earth in the space shuttle *Challenger,* astronaut-oceanographer Kathryn Sullivan observed "a stunning tableau of blue and white . . . In the sun's glint, I saw major circulation features, like the wall of the Gulf Stream. I traced fine filaments of sediment far away from major rivers and marveled at the milky white swirls of huge plankton blooms. My spaceship was moving almost 1,300 times as fast as a typical oceanographic ship, letting me see almost all of the Atlantic Ocean in just a few hours."

From space, the entire surface of the world has now been seen and mapped in exquisite detail, but most of the land under the sea, and therefore most of the actual surface of the world's terrain, has yet to be seen or charted with accuracy equivalent to the land above.

Limited visibility through seawater is the key. Dolphins and whales have overcome the problem by using sophisticated acoustic techniques, and, starting during World War I, so have humans. Methods for "seeing" with sound underwater were further developed during World War II as sonar (sound navigation and ranging), by which a beam of sound is sent into the water and the length of time of its return echo is used to find the seafloor as well as objects such as enemy submarines. When multiple sonars—multibeams—are arrayed across the hull of a ship, a swath of images is generated that shows the seafloor more accurately than do single beams. Later advances identified the various sound frequencies best for determining the depth of the bottom, as well as what was beyond, probing into several kilometers of sediments and rock structures underlying the seafloor. Seismic surveys by the oil and gas industry followed, enabling geologists to identify promising formations before undertaking expensive drilling operations.

In 1995, altimetry data from the Navy's GEOSAT satellite provided a generalized idea of the configuration of much of the ocean floor by measuring the sea surface height and gravity field through microwave pulses. In 1999, the U.S. government twice convened a panel of ocean explorers, scientists, and educators to consider developing a strategy for "discovering Earth's final frontier." After much deliberation, the panel called for "mapping of the physical, geological, biological, chemical, and archaeological aspects of the ocean . . . exploring ocean dynamics and interactions at new scales . . . developing new sensors and systems for ocean exploration." These are mandates much like those given to the scientists aboard H.M.S. *Challenger* a century and a half ago—"to explore all aspects of the ocean." In 2008 a new NOAA vessel, *Okeanos Explorer,* was launched with similar lofty goals. Globally, a network of observing stations—the Integrated Ocean Observing System (IOOS)—has gained the support of many nations.

In 2006, Google assembled a group of 30 international experts to determine how its publicly available portrayal of three-dimensional images of Earth—called Google Earth—could be extended from the land into the depths of the sea. In collaboration with the U.S. Navy, the most up-to-date,

Dawn Wright
Mapping the Ocean Floor

Taking the lead on applying geographic information systems (GIS) technology to ocean sciences could be an intimidating role. Dawn Wright (affectionately known by colleagues as "Deepsea Dawn"), chief scientist at Environmental Systems Research Institute (ESRI), takes it in stride: "I make maps of the ocean floor to unlock the buried treasure of scientific insight, to understand how the ocean works, and how to better protect it." Behind her modest explanation is her development of cutting-edge technology intent on securing the ocean's future. Inspired as a graduate student by GIS data playing a role in the discovery of the R.M.S. *Titanic* in 1985, she became a key player in developing the first GIS data model for the ocean, in both two and three dimensions. Today advancing UN Sustainable Development Goal 14 (Ocean and Coast) with the highest quality data is now a key interest. This is vital to making maps that tell a full story and inspire us to "sustain and protect" the blue part of the planet.

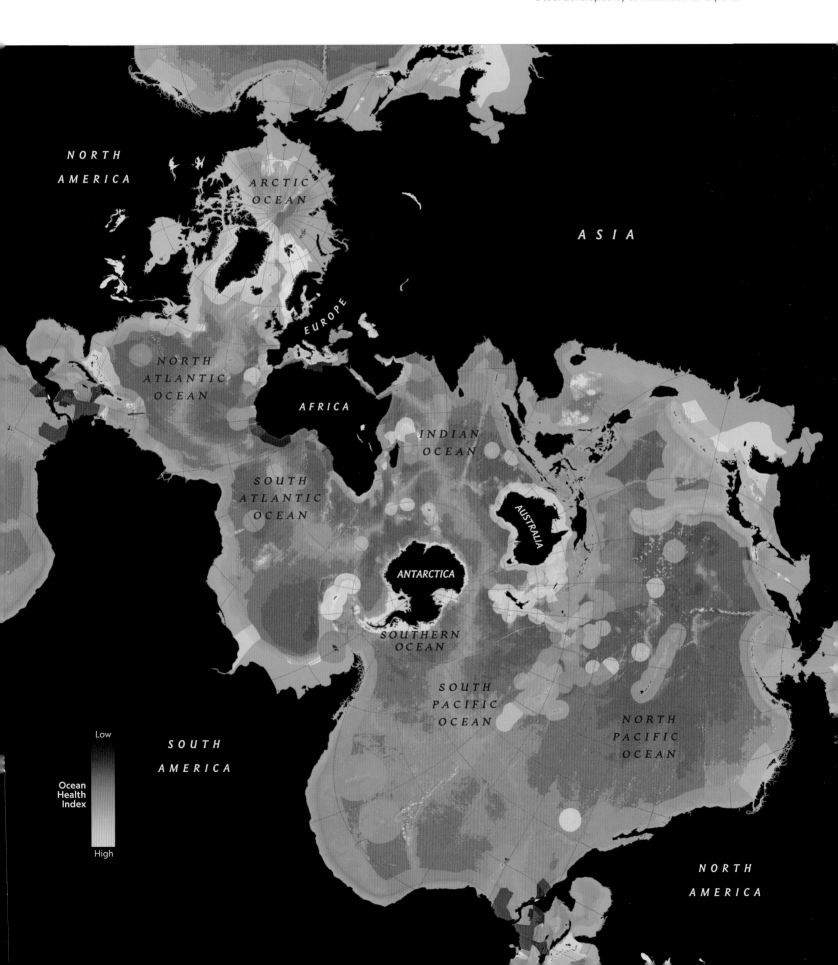

A view above Antarctica shows world ocean health, assessed annually by the Ocean Health Index, a tool developed by 65 multinational experts.

publicly available ocean bathymetry made possible the launch of "Google Ocean" in digital form in 2009. Google Ocean also provided an opportunity for National Geographic to update all of its seafloor maps in its World Atlas series for the first time in 30 years, published in a book as well, *Ocean: An Illustrated Atlas.*

Broad outlines now exist of major undersea mountains, valleys, and plains, and the bathymetry of some coastal regions and targeted areas in the deep sea is increasingly well known. Nonetheless, an article in *Smithsonian* magazine quoting oceanographer Robert Ballard said, "Our knowledge of what's beneath the ocean is about as detailed as a set dinner table with a wet blanket thrown over it. You can see the outlines, but how do you tell the candelabra from the turkey?"

To better understand not just the seafloor but the entire ocean, top to bottom, the California-based company Environmental Systems Research Institute (ESRI), a pioneer in developing global information systems (GIS), is now applying its sophisticated data-layering techniques to compile maps of the sea. A crucial first step was launched in 2011 with the release of the first Ocean Basemap on the geospatial cloud ArcGIS Online. Compilations of thousands of datasets from dozens of sources now come closer to representing the ocean's bathymetry, with grids ranging in resolution from one kilometer to 50 meters in some areas.

In 2016, the intergovernmental partnership Group on Earth Observations (GEO) commissioned a global map of ecological marine units (EMUs) to provide a framework for understanding ocean processes and to detect and monitor changes. A public-private partnership led by ESRI and the U.S. Geological Survey (USGS) is working in collaboration with NOAA, NASA, the National Geographic Society, Duke University, the Woods Hole Oceanographic Institution, the National Institute of Water and Atmospheric Research (NIWA), NatureServe, the Marine Conservation Institute, the University of Auckland, and Norway's GRID-Arendal. Data are being integrated from satellites, buoys, sonar, submersibles, underwater cameras, and other devices, forming a multilayered reconstruction of underwater geographic reality.

In 2019 an international team of oceanographers launched an ambitious effort to create a comprehensive map of the world's ocean with greatly enhanced resolution by the year 2030. One ship equipped with current multibeam bathymetry systems would take approximately 200 years to map, in detail, the targeted 363 million square kilometers of ocean. To condense the time frame, the nonprofit group General Bathymetric Chart of the Oceans (GEBCO), with support from the Nippon Foundation, is undertaking the Seabed 2030 project, recruiting about 100 ships to engage in the initiative. These will be complemented by free-swimming autonomous underwater vehicles that can gather precise sonar images, as well as data on water chemistry and temperature along predetermined transects.

Much remains to be done, but for the first time, mapping the ocean means embracing the reality that the ocean is three-dimensional, and it is alive.

VISIONARIES

Bob Ballard
Ever Exploring

....

G eophysicist Bob Ballard is one of the ocean's most prolific explorers, his name synonymous with that of Earth's most iconic shipwreck, the *Titanic,* lost on its maiden voyage in 1912. Not only did Ballard lead the 1985 expedition that found it but he has championed the preservation of its site as a sanctuary not to be salvaged. His quest to fill in the blanks in human history through the stories of ships lost at sea has led him to Phoenician ships sunk in a Mediterranean storm around 750 B.C.; the World War II German battleship *Bismarck,* torpedoed in 1941 off France; and dozens more underwater discoveries, from the Black Sea to the Pacific's Guadalcanal. Cutting-edge technologies he develops with cartographers, ROV and submersible innovators, and videographers are his keys to these treasure troves. His most recent expedition? Looking for pilot Amelia Earhart's plane, lost in 1935. Evidence led him in August 2019 to the remote Pacific island Nikumaroro, far off her flight path—but his search was inconclusive. Still, he's said, "That plane exists. That means . . . I'm going to find it."

....

OPPOSITE ABOVE: Researchers finish an Atlantic dive in the three-person HOV *Alvin.* OPPOSITE BELOW: The ROV *Jason* recovers fourth-century amphora from the Mediterranean.

GREAT EXPLORATIONS

Diving Milestones

From Greek sponge divers breathing air trapped in kettles lowered into the water to today's cutting-edge ROVs scouring depths where humans may never venture, we have sought through time to perfect means of exploring the ocean's deep frontier. Whether divers go solo or in multipassenger descents, to find food, explore, appreciate, combat, or make scientific discoveries, every extraordinary advancement in technology—from the simple breath-hold technique used by Japanese ama divers to gather shellfish to the air held inside a pressurized capsule descending to the floor of the Mariana Trench, the ocean's deepest place—is cause for celebration. Each date in this time line has brought humans a step closer to understanding and appreciating the ocean, and to finding ways to protect the future of its complex systems and varied inhabitants.

In his sub *Limiting Factor,* Victor Vescovo surfaces after reaching the ocean's deepest point, Challenger Deep.

TOP: In 1898 engineer John Holland peers from his submarine, soon christened the U.S. Navy's *Holland*. ABOVE: Jacques Piccard's *Bathyscaphe* prepares for an Atlantic dive in 1963, after reaching 1,000 meters depth in 1960.

Deep Discovery

4th century B.C.: Aristotle describes Greek sponge divers breathing air trapped in kettle.

1st century B.C.: Japanese ama divers free dive.

1663: Salvagers in diving bell in Stockholm harbor recover cannon from *Vasa*.

1690: Edmund Halley's diving bell uses weighted barrels.

1715: John Lethbridge's early diving dress resembles armor.

1776: David Bushnell's *Turtle* submarine attacks H.M.S. *Eagle,* New York Harbor.

1797: K. H. Klingert invents diving dress with helmet using surface-pumped air.

1800: Robert Fulton builds the *Nautilus* submarine.

1819–1837: Augustus Siebe fashions a diving suit with surface-pumped air.

1839: British Navy founds first diving school.

1863: Siméon Bourgeois, Charles Marie Brun invent the compressed-air submarine *Le Plongeur*.

1864: Confederate submarine *Hunley* sinks after sinking the U.S.S. *Housatonic*.

1865: Benoît Rouquayrol, Auguste Denayrouze develop the back-mounted breathing apparatus.

1872: H.M.S. *Challenger* starts global research voyage.

1876: Henry Fleuss builds the first self-contained oxygen-rebreather lung.

1892–99: Louis Boutan takes the first underwater photographs.

1897: John Philip Holland launches an electric-motor submarine, soon approved by U.S. Navy.

1903: Sir Robert H. Davis designs a submarine escape lung.

1906: John Haldane makes decompression tables to 61 meters.

1908: British Navy launches the first diesel-electric submarine, *D-1*.

1912: Sir Robert H. Davis designs the first pressurized submersible decompression chamber.

1913: German company Neufeldt and Kuhnke patents an armored diving dress with jointed legs and arms.

1927: Helium tested for diving by U.S. Navy.

1930–34: William Beebe and Otis Barton dive in a bathysphere off Bermuda to 923 meters.

1943: Jacques Cousteau and Émile Gagnan perfect the Aqua-Lung.

1948: Auguste Piccard invents the bathyscaphe, *FNRS-2*.

1955: Nuclear submarine *Nautilus* launched by U.S. Navy.

1960: Jacques Piccard, Lt. Don Walsh descend Mariana Trench in *Trieste;* U.S. nuclear submarine *Triton* circumnavigates globe.

1962: Cousteau's *Conshelf I* team lives seven days underwater in Mediterranean.

1964: U.S. Navy *Sealab* team live off Bermuda and California. Navy launches HOV *Alvin* for research.

1969–70: Tektite Project: 11 science teams live underwater in U.S. Virgin Islands.

1972: *Hydrolab* begins two decades of scientific saturation dives.

1979: Sylvia Earle makes the first solo untethered dive in *Jim* suit, to 381 meters off Oahu.

1985: First back-to-back solo dives to 1,000 meters in *Deep Rover I* by Graham Hawkes, Sylvia Earle, and Phil Nuytten, off California.

1990–present: NOAA's Argo project deploys robotic probes to monitor the ocean.

1992–present: *Aquarius* undersea lab, Florida, hosts coral reef researchers.

1998–2003: National Geographic, NOAA, Goldman Foundation sponsor Sustainable Seas Expeditions.

1999–2001: Southampton Oceanography Centre's *Autosub* reaches remote places.

2009: Pristine Seas and Mission Blue Hope Spot programs begin.

2012: James Cameron makes first solo dive to Mariana Trench floor in *Deepsea Challenger*.

2016: Deep Ocean Exploration and Research designs 1,000-meter subs to explore Hope Spots.

2019–2020: Victor Vescovo reaches the deepest spot in every ocean basin in submersible *Limiting Factor*.

HUMANS AND THE SEA

Chapter Eight

> "THE FUTURE CAN STILL BECOME WHAT
> WE REASONABLY AND REALISTICALLY WANT."
>
> —AURELIO PECCEI, FIRST PRESIDENT, CLUB OF ROME

NASA's "Black Marble Map" of Earth from space at night and the later, more detailed "Cities at Night" project portray what astronauts witness from high in the sky: a concentration of sparkling lights indicating the distribution of people globally, largely defined along the shores of inland lakes, lacing the banks and deltas of rivers, and especially in a brilliant border marking coastal regions where land and sea converge. About half of the global population lives within 100 kilometers or so of the sea. Access to water—fresh water—is the key to human survival and prosperity. But unbeknownst to many who live far inland and never see or touch the ocean, the ocean touches them with every drop of water they drink, every bath they take, every crop they harvest.

Water is restless, shifting endlessly in form and context. A molecule in the foggy exhalation of a whale may be in the next breath of a seagull, then return to the sea with the remains of the gull's meal. That same molecule may once have been part of a desert spring or in sweat glistening on the face of a farmer in China, or Kenya, or Canada. Vapor arises from the ocean—the reservoir of 97 percent of Earth's water. It leaves the salt behind and forms misty clouds of water that rain or snow on land and sea. Flowing through time and space, water sometimes pauses for millennia in a glacier, or travels through channels far beneath the ocean floor before eventually bursting forth in a mineral-laden geyser that returns vapor again to the sky.

In a way, all humans are sea creatures, as dependent on the existence of the ocean as any fish or whale. Poet W. H. Auden put it succinctly: "Thousands have lived without love; none without water." Mindful of that dependence or blissfully complacent, all people everywhere owe their past, present, and future to the ocean. The impact of human actions, especially in the most recent two centuries, means that the future of the ocean now is inextricably linked to us. Without us, the ocean and most of life on Earth can prosper. But changes will endure as a consequence of what we have already taken out of and what we have already put into the ocean. Armed with knowledge that did not and could not exist centuries or even decades ago, there is time to protect what remains of intact areas and to restore damaged ocean systems—and to do so as if our lives depend on it, because now we know they do.

RIGHT: A light-studded Europe imaged by satellites in NASA's Black Marble program shows concentrated human activity along the coastlines. PAGES 312-313: Off Mexico's San Benedicto Island, curious young snappers investigate a free diver.

Who Needs the Ocean?

Our absolute dependence on the ocean has not always been obvious. Even now, the significance of the ocean to everyone, everywhere, all the time, needs considerable explanation for many. Some uses are clear: Food habits of coastal dwellers are recorded in the shells and bones of animals dined upon thousands of years ago. In modern times, millions of tons of sea life are extracted every year to support the needs and desires of those who live near the ocean, as well as those who are far inland. Many of the world's oldest and largest cities are ports, strategically located to support transportation of goods across the ocean, to access ocean wildlife for food, and to ease the disposal of wastes into the sea.

Transportation is high on the list of the ways the ocean serves humankind. Over the ages, people probably have taken to the sea wherever people and the sea converge. Rafts and canoes crafted from logs, reeds, bamboo, and animal hides likely provided access to areas beyond native lands. Travel by sea made possible the arrival of people in Australia about 50,000 years ago. Over millennia, Polynesian voyagers reached widely separated islands of the Pacific. Following coastal routes, people also moved from Asia to North America at least 10,000 and maybe 20,000 years ago. The sea has imposed formidable barriers to human societies throughout history, but it has also provided a comprehensive liquid highway that connects us all.

Rose George observes in her account of modern shipping, *Ninety Percent of Everything,* that today's commercial trade mostly travels by sea. She explains, "On ship-tracking Web sites, the waters are black with dots. Each dot is a ship; each ship is laden with boxes; each box is laden with goods . . . Without all those dots, the world would not work . . . These ships and boxes belong to a business that feeds, clothes, warms, and supplies us. They have fueled if not created globalization."

In addition to shipping, extractive industries—the taking of wildlife, oil, gas, minerals, and even seawater itself from the ocean—account for some of the many ways the ocean benefits humankind. The ocean also serves as the ultimate place to dispose of trash, toxins, wastes, garbage, and whatever else people want to "throw away." These are among the usual measures of how the ocean is thought to be useful.

But far and away the most important thing that we take from the sea is our existence. Living on the moon, or Mars, or in some distant galaxy may someday be an option for humankind, but we are now anchored to the one place in the universe hospitable for life as we know it—life that is possible because this planet is blue. No ocean, no life. No ocean, no us. If the ocean is in trouble, so are we. It is not possible to retrieve all of the noxious things we have put into the sea that have already altered planetary chemistry. Nor can we expect full recovery of the whales or cod or seabirds—or any of the other now greatly depleted forms of ocean wildlife. But there is no question that we can heal much of the harm with benefits not only to the ocean, but to ourselves as well.

OPPOSITE ABOVE: Dusk settles over container ships at eastern Thailand's state-of-the-art Laem Chabang port. OPPOSITE BELOW: Outriggers of centuries-old design masterfully ply Papua New Guinea waters.

Living on the Edge

L iving at the edge of the sea is at once exhilarating and perilous, whether in a salt-seasoned cottage on Cape Cod, perched along a rocky cliff in Chile, or even nestled in a high-rise apartment in any of the world's great coastal cities. Exhilarating because, for humans, the attraction to coasts is primal, touching aesthetic senses that many find irresistible. Dennis Wilson of the Beach Boys said it simply: "On the beach, you can live in bliss."

Coasts are captivating to many because that's where much of the action happens that drives civilization forward. Port cities are centers of commerce, with their proximity to shipping, the world's most important mode for national and international trade. Most high-population areas are coastal, with a supporting infrastructure that includes financial centers, medical facilities, cultural institutions, schools, employment, and other attractors in their own right. Recreation and tourism are often the basis of economic prosperity in coastal regions and in island countries globally.

But the benefits of coastal living are countered by the perils. A single storm can rip thousands of tons of sand from a beach and relocate it to an offshore bar. A flooding river can sweep thousands of tons of debris from the land into the sea. Rocky shores erode as well, yielding to the relentless force of waves and the destructive action of boring mollusks, urchins, and other creatures that drill, chew, or dissolve places for themselves in the rock.

Great storms—hurricanes, cyclones, monsoons—are among the risks threatening coastal dwellers, with impacts measured in thousands of lives and billions of dollars every year, as well as enduring social, environmental, and economic consequences. In 2005, Hurricane Katrina devastated the port of New Orleans with a storm surge of three to eight meters, overwhelming critical levees, flooding 80 percent of the city, displacing millions, and taking hundreds of lives. The 2012 Hurricane Sandy careered across five Caribbean countries before striking the New York-New Jersey area, impacting 22 other states and more than 50 million people with enormous costs in terms of lives and property. The list of monster storms is long and has no end.

Tsunamis, too, though rare, devastate coastal communities with sudden swiftness that gives little time to retreat. Low-lying coasts are especially vulnerable, illustrated by the magnitude 9.1 earthquake off Indonesia's island of Sumatra in 2004 that caused a series of tsunami waves up to 30 meters high. The impact caused the death of more than 200,000 people in 14 countries surrounding the Indian Ocean, one of the greatest natural disasters in history. A 2011 earthquake offshore of Japan and the tsunami that followed killed more than 15,000 people and inflicted incalculable damage to Japan's coastal cities. High water invaded Japan's Fukushima Daiichi Nuclear Power Plant, causing a level-7 nuclear meltdown and releasing radioactive materials into the sea. An estimated five million tons of broken buildings, sections of docks and boats, along with thousands of household items, floated to shores on the far side of the Pacific. Effects of the tsunami were felt as far away as

Forever wrestling the elements, a community along New York's Fire Island recovers from Hurricane Sandy in 2012.

DEEPER DIVE

Coastal Living

A sleek high-rise condo, a low-slung bunga-low, a thatched hut. Any could house the more than 2.4 billion people who live by the sea. Besides residents, some 50 percent of all tourists vacation along a coastline. For locals, their lives may depend on the bounty below: About 97 percent of the world's fishermen live in developing nations, and many rely on marine life for food and income. Imagine a video of Earth's eclectic coastal environments: a dizzying rush of lush rainforest with quiet sand beaches and open-air huts in Bali; homey cottages rising on rocky cliffs in France; sprawling clap-board communities along stretches of Arctic ice; walls of glass-

Inuit village, Greenland

plated high-rises on the beaches of Dubai and Miami; and harbors galore, with massive tankers and fishing skiffs little bigger than a windsurfer. When National Geographic photographer David Alan Harvey documented North Carolina's Outer Banks in 2011, he captured the hard work and joy of locals and visitors in a quiet realm protected by mid-Atlantic barrier Islands—one he fre-quents. A decade later, it, and every other coastal community, is threatened by increasingly powerful storms and rising sea levels linked to the impacts of climate change. Still, most coastal residents would echo Harvey's words: "Nothing seems sweeter than the view from my front porch."

Norway, where water in fjords sloshed back and forth, and in Antarctica, where tsunami-generated waves tore huge chunks of ice off the Sulzberger Ice Shelf. Wildlife suffered, too. As the tsunami swept across the Pacific, waves inundated the low-lying Midway Atoll National Wildlife Refuge and nearby islands, killing thousands of nesting seabirds and upending shallow coral reef systems.

Despite concerns about vulnerability to storms, wind, waves, and the inexorable reality of modern sea-level rise, homes, hotels, and high-rises hug tightly to the edge of shorelines, as if daring the ocean to challenge human engineering. As more people crowd into coastal areas, pressures on land and sea increase. Natural systems are altered, overwhelmed, and destroyed. Marshes, wetlands, coastal forests, coral reefs, and kelp forests that have provided buffers against destructive forces of the ocean in the past have given way to human occupation and uses.

So desirable is coastal real estate that marshes, lagoons, and entire bays are transformed into land by draining and filling them with rubbish or soil that's been dredged from the ocean or mined from inland sources. Mumbai, India, is largely built on land that 300 years ago was underwater. Once a collection of coastal islands, it was made one contiguous landmass after British colonizers arrived in 1661 and later filled in the gaps to hold buildings. About 25 percent of the island nation of Singapore was once marshland or submerged, and about 70 percent of Hong Kong's commercial real estate is on land that was once underwater. From Honolulu to Tokyo Bay, and Monaco to Miami, land has been taken from the sea to gain additional waterfront property, often for homes and marinas for pleasure boats, but also for offices, shops, and parking lots. Coastal forests have given way to human structures from the tropical coastal forests of Brazil to New York's once wooded Manhattan. In San Francisco, under the tall buildings that mark the financial center, hotels, and high-end shops, dozens of old ships from the gold rush era lie buried. In the late 1800s, some shipowners deliberately sank their boats at the dock and dumped rocks and other debris on top of them to create and claim highly coveted coastal real estate. Most pedestrians are not aware they are walking over what once was water, bristling with the masts of sailing vessels.

People simply want to live along the coast. Not only are vistas beautiful and soothing, but sunlight penetrating to the seafloor and nutrients flowing from the land have combined to make coastal systems among the most productive and biologically diverse places on the planet. Wild food from the sea enabled people to prosper on islands and coasts for thousands of years. Now, however, not only has land reclamation and commercial building blighted the natural flow of coastal life, but modern industrial fishing technologies, global markets, and changing dietary habits have promoted widespread depletion of once common animals now considered luxuries too valuable to be consumed by the fishermen who catch them. Knowledge exists about how to restore the natural systems that have made possible the prosperity of people of the coasts—and beyond. What is needed is the will to do it.

VISIONARIES

Tommy Remengesau, Jr.
Protector of Paradise

....

"Preserve the best and improve the rest" is the policy Tommy Remengesau, Jr., introduced to boost Palau's tourist industry and establish environmental stability as an island nation in Micronesia. Elected the nation's president two separate times since 2001, his definition of the "best" is Palau's extraordinary underwater world encompassing more than 200 islands in the Pacific. A diver's paradise, it is one of the world's most diverse, stunning, and productive ecosystems. Remengesau has become a global leader in environmental initiatives, including the Micronesia Challenge, in which Palau and other South Pacific nations make the public aware of how global warming is changing their world—and how it will change ours. After receiving the UN Champion of the Earth award in 2014, he led the campaign for 80 percent of Palau's waters to become a no-take sanctuary, with 20 percent reserved for domestic fishing. Today the Palau National Marine Sanctuary is the world's sixth largest fully protected marine area. Still, says Remengesau, "We're a paradise in peril."

....

OPPOSITE: Fifty percent of all tourists annually seek coastal getaways; many flock to beaches like Brazil's Recife.

Earth's Blue Realm

Most of the creatures living on Earth are adapted for life above, on, or in the open ocean. Humans are not. The anatomy, physiology, food habits, and dependence on air and fresh water foster the human inclination to be terrestrial. But inexorably, people have been drawn to the sea. It is innately human to explore, to yield to the insistent need to know what is over the horizon, to discover the nature of the blue expanses beyond the comfort of coastal waters.

Early ventures into the open sea may not have been deliberate. Even in modern times, people sometimes become unwilling ocean drifters, as happened to the Mexican fisherman José Salvador Alvarenga in 2012. Caught in a violent storm in an open eight-meter boat with a failed engine and dead radio, he was swept by ocean currents from Mexico's Costa Azul across the Pacific for nearly 10,000 kilometers before making landfall, exhausted but alive, in the Marshall Islands 438 days later. Floating trash enabled him to build shelter and provided the means to capture and store rainwater. Unwary seabirds, fish, and an occasional sea turtle provided sustenance.

Maritime history resounds with harrowing stories of castaways who have survived perilous journeys, but most seafaring has not been by chance but by deliberate preparation. Nearly every part of the ocean's surface has now been reached by vessels of one sort or another, initially achieved by "dead reckoning"—a way of knowing where you are by using a previously determined position, or fix, and estimating speed, course, and direction from that point over time. Even with modern instrumentation, some do it the hard way. Rosalind "Roz" Savage is among an elite few who have rowed a boat solo across the Pacific, Atlantic, and Indian Oceans. Savage—aided by modern navigation technologies, a satellite phone, a desalination system, and thoughtfully selected food provisions—had the audacity to rely on "fair winds and following seas," muscle power, and favorable currents, with the willingness to face the risks of capsizing, injury, freak waves, and encounters with storms and large ships oblivious to her presence.

While the ocean serves as a formidable barrier, it has also connected people and cultures throughout recorded history. Ships that captured the wind with sails and swept their way across vast ocean spaces were vital for fishing, trade, travel, and warfare until well into the 19th century. Steamboats, powered by coal, came into favor in the mid-1800s; and it is said that the "Golden Age of Sail" ended when the British vessel H.M.S. *Devastation* was launched in 1871—the first of a class of battleships that did not carry sails. Civilization changed early in the 20th century when oil-powered engines revolutionized access to the sea. Thousands of sailboats still propel millions of people, but in the 21st century most seagoing vessels—from three-meter skiffs to cruise ships, freighters, and oil tankers more than 200 meters long—are powered by oil or gas. Some, such as the 173-meter Russian icebreaker *Ural*, and ships and submarines from five other countries, are powered by nuclear energy.

Aboard craft from racing boats to freighters, navigators brave storm-driven seas for exploration, adventure, and opportunity.

During the Age of Sail, from the 16th through the 18th centuries, warships like this British frigate escorted a nation's merchant ships.

Despite prudent preparation, ships sink. From the remains of canoes from 5,000 years ago to somewhere in the sea a few days ago, millions of seagoing craft have ended their journeys permanently submerged. Thousands have been lost during wars and thousands more during storms. Even in recent times, with modern communication and satellite tracking, large vessels sometimes disappear without a trace. According to Lloyd's/Allianz insurance data reported by Jennifer Lang in 2014, nearly 100 large ships sink every year.

To avoid the dreaded fate of being lost at sea, skills in celestial navigation and ingeniously engineered devices have been developed over the ages to guide safe travel on the ocean's unmarked highways. China is thought to be the first civilization to create a magnetic compass to indicate direction, as early as the 11th or 12th century, and it is likely that the great Chinese ocean-voyaging pioneer Admiral Zheng He used them during his ocean travels in the 1400s. A translation of He's deeds, inscribed in granite, notes traversing over 500,000 kilometers of vast ocean, "beholding great ocean waves rising as high as the sky . . . like mountains. We spread our cloud-like sails aloft and sailed by the stars . . ."

The Portuguese nobleman Ferdinand Magellan had the best tools available to European seafarers in 1519 when he set off in command of five Spanish ships to find a western route across a mostly unknown ocean. For navigation, he had a compass and an astrolabe or "star taker," an elaborate, flat, multi-disk instrument, usually made of brass or wood, that Smithsonian Institution writer Laura Poppick refers to as the "original smartphone." Used for centuries in Islamic and European countries starting around 200 B.C., the astrolabe was critically important for calculating latitude in the open sea and helped guide Christopher Columbus, Vasco da Gama, and other notable explorers.

By the mid-1700s, the sextant, a double-mirrored device used to measure the altitudes of the sun and stars relative to the horizon, largely displaced the astrolabe in determining latitude. But it took British carpenter and clockmaker John Harrison to resolve the critical problem of determining longitude. He invented a reliable marine chronometer, first tested in 1761, that could determine the time while at sea and compare it to the time at a known longitude on land, designated as the prime meridian.

The gyrocompass, invented early in the 19th century, uses a spinning gyroscope to follow Earth's axis of rotation, rather than magnetism, to point to true north. Modern compasses can calculate and adjust for motion, variation, and deviation between magnetic and true north. The position of the sun can also determine direction, but 21st-century humans tend to rely instead on their cell phones, cameras, or vehicles equipped with global positioning system (GPS) receivers connected to Earth-orbiting satellites.

Oceania, the vast region of the Pacific peppered with myriad small islands and the large island nations of Australia, New Zealand, Papua New Guinea, and part of Indonesia, has been populated for many thousands of years by people whose home is mostly aquatic and who have close relatives living

A replica of the seventh-century Viking Oseberg ship, based on the richly decorated ship that became a grave, excavated in Norway in 1904

DEEPER DIVE

What Shipwrecks Tell Us

If, as explorer Robert Ballard says, "The deep sea has more history in it than all the museums of the world," what do its shipwrecks tell us about humanity? The iconic 20th-century luxury liner R.M.S. *Titanic* tells of a pleasure voyage turned fatal, and of hubris. "The boat is unsinkable," claimed its owners in 1912, weeks before it struck an iceberg and sank. The World War II battleship *Arizona*, attacked in Pearl Harbor, tells of wartime savagery, but also of bravery and sacrifice. The oil tanker *Exxon Valdez,* spewing its contents when grounded in Prince William Sound in 1989, tells of environmental assault. Earth's oldest known shipwreck—a Greek trading vessel from 400 B.C.—

Shipwreck in the Red Sea

evokes the mythical hero Odysseus and the sirens who lured ships to shatter along treacherous coasts. Centuries later, in 1628, the Swedish *Vasa* toppled in Stockholm harbor because King Gustav added an extra row of showy cannons. Pirate Blackbeard's 18th-century *Queen Anne's Revenge,* wrecked off North Carolina, ended a bloody treasure pursuit. Ravaged lifeboats racing to rescue sinking ships off stormy North Atlantic coasts relate their own dark stories. None is as dark, perhaps, as that of *São José-Paquete de Africa,* barely 40 meters long. In 1794 it left Mozambique for Brazil with 543 chained captives packed in her hold only to smash into reefs off Cape Town, killing 212.

on specks of land separated by hundreds or thousands of kilometers of water. In Polynesian cultures, navigators are held in especially high esteem for their uncanny ability to steer voyaging canoes over the open sea with guidance only from knowledge they have acquired and carry in their minds. They begin as children to learn the language of the winds, clouds, and stars; to understand waves by "becoming a wave"; to observe and absorb the habits of whales and fish and seabirds; to hear and smell and see what others do not. Only then can they master the highly sophisticated art and science of navigation known as wayfinding.

In the collision of cultures when aggressive Europeans arrived in the 1700s, native societies were crushed and knowledge acquired over many millennia was lost, including how to construct voyaging canoes and acquire the ineffable skills of wayfinding. By the middle of the 20th century, the last of the master navigators were aging without having conveyed their wisdom to a new generation. Their almost mythical insights and accomplishments might have been recorded merely as legends and historical footnotes. But rather than ending, a resurgence of pride in the history and heritage of the people of the Pacific ensued. In 1971, island nations convened the Pacific Islands Forum to recapture their shared ocean identity as the Blue Pacific Continent.

"I AM GOING TO CROSS THE PACIFIC ON A WOODEN RAFT . . . WILL YOU COME?"

—THOR HEYERDAHL, *KON-TIKI*

Racers in a traditional Polynesian outrigger off Kaui, Hawaii, begin as children to learn the language of the sea.

VISIONARIES

Nainoa Thompson
Wayfinder

....

Raised to revere his ancestors, Hawaiian wayfinder Nainoa Thompson realized as a young man that he actually knew little about his heritage. Ancient practices, such as wayfinding—navigating the sea by reading the environment—had been subsumed into Western ways. Hawaii's cultural renaissance in the 1970s inspired Thompson and friends to construct a canoe to navigate from Hawaii to Tahiti by traditional methods. But no one knew the traditional methods. In Micronesia they found the youngest of six surviving wayfinders, Pius "Mau" Piailug, in his 40s, and he agreed to teach them. It was all about memorizing direction, Mau said—knowing patterns for the rising and setting stars and sun, the swell of the waves, the flight paths of birds. A trial journey with Mau aboard *Hōkūle'a* ("Star of Gladness") readied the crew for a global voyage, and from 2013 to 2017 they engaged communities in 18 nations to practice sustainable living. Now a wayfinding mentor himself, Thompson continues his voyages of tradition and teaching. "If you can read the ocean," he quotes Mau, "you will never be lost."

....

In Hawaii, a spiritual revival was manifest in the Polynesian Voyaging Society and its construction of an authentic, seaworthy double-hulled canoe, *Hōkūle'a*—a movement inspired by Herb Kāne, a Hawaiian artist. One of the last of the master navigators, Mau Piailug, from Satawal Island in the Caroline Islands of Micronesia, agreed to lead a voyage in *Hōkūle'a* from Oahu to Tahiti in 1976. It is possible that only one person, an earnest young Hawaiian, Nainoa Thompson, could have persuaded Mau Piailug to take him on as an apprentice, not only to learn but to share and keep alive the ancient wisdom of wayfinding.

Working with the Polynesian Voyaging Society, Thompson initiated and led an ambitious three-year, 65,000-kilometer, around-the-world voyage in *Hōkūle'a*, from 2013 to 2017. Largely crewed by young Pacific Islanders intent on becoming wayfinders in their own right, the vessel visited 13 marine World Heritage areas and 150 ports in 20 countries. The expedition was named Mālama Honua, meaning "to care for our Earth." Thompson observed, "*Hōkūle'a* has sailed using wayfinding for many years, and her navigators find their destinations using nature as a guide. Now *Hōkūle'a* and her sister canoe, *Hikianalia,* and their navigators are taking that knowledge to find a new destination—a healthy ocean and island Earth. We are a blue planet and an ocean world."

HOPE SPOT

Florida Gulf Coast

Hugging the Florida coast from Apalachicola Bay on the northern panhandle to Ten Thousand Islands near the southern tip of the peninsula, the Florida Gulf Coast Hope Spot is a swath of bays, inlets, salt marshes, creeks, mangrove swamps, and barrier islands sparkling with white sand beaches. Local conservationists, business owners, and residents work together to maintain what remains of the pristine systems displaced by rapid population growth and comprehensive conversion of the coastline for human purposes.

. . . .

Deep reefs offshore from Florida's Gulf Coast abound with life.

The Green Economy:
It's Basically Blue

The "green economy" is a concept arising from concerns in the 20th century about how to maintain economic growth, health, and social justice for a rapidly expanding human population at the same time as the natural systems underpinning civilization—air, water, land, and wildlife—are sharply declining. Conservation biologist Carl Safina sums up the problem: "How can we use the planet without using it up?" Most of the focus has been on the land and terrestrial systems. As creatures of the land, we have largely taken the ocean for granted and treated it as free, whether for transport, extraction, or disposal of wastes. Until now. Now, the concept of a "blue economy" is emerging.

Not until the middle of the 20th century did policymakers begin to take seriously the concept of limits to growth, even though Earth is obviously finite. The limits relate to the number of people and what energy and materials are needed to keep civilization functioning, in conjunction with how much can be taken from the Earth and ocean without doing unacceptable harm.

The goal of having your planet and consuming it too led to the seemingly contradictory term "sustainable development," words first used in 1969 in a document signed by 33 African countries under the auspices of the International Union for Conservation of Nature (IUCN). It was the year that astronauts first set foot on the moon, and through their eyes and images, people saw with stunning clarity that no matter who we are or where we live, we are all connected by one atmosphere, one ocean, and one shared planet. That year, in the United States, the Environmental Protection Agency (EPA) was formed as part of the National Environmental Policy Act. The act defines sustainable development as "economic development that may have benefits for current and future generations without harming the planet's resources or biological organisms." That is a tall order, given the level of what is being taken from the land and sea that cannot be replaced—whether it is the drain on minerals laid down billions of years ago, fossil fuels that originated hundreds of millions of years ago, intricate coral reef systems that developed half a million years ago, redwood trees that started as seedlings a thousand years ago, or a Greenland shark that emerged from its mother up to four centuries ago.

Earth hosts numerous animals with admirable intelligence, from elephants, dolphins, and parrots to our fellow primates, but only humans have the capacity to draw on knowledge acquired over many millennia, share information globally, and, armed with knowledge, anticipate future consequences on a global scale. The Club of Rome—an international group of government officials, scientists, economists, and business leaders—commissioned a 1972 report, *The Limits to Growth*, based on a computer simulation of exponential economic and population growth globally using a finite supply of resources. Throughout history, families, cities, and even

ABOVE: Colorful subtidal dweller *Janolus cristatus,* a nudibranch mollusk, takes its name from the two-headed Roman god Janus. OPPOSITE: Before 1970, barges plied Dubai's Jebel Ali, today the world's ninth busiest port.

CALCULATING OCEAN HEALTH

Data from 220 countries and territories are analyzed yearly to measure ocean health. This map shows the impact of 19 stressors, including ecological, economic, and political ones, with tourism and food production at the top.

Human Ocean Impact

High ——————— Low

No data

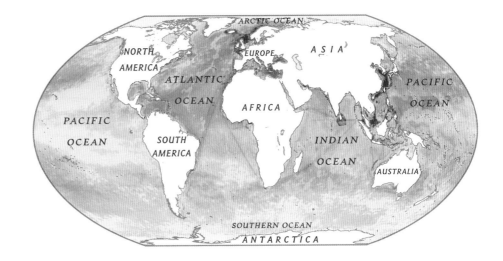

countries have looked at their assets, considered past trends, and made plans for the future accordingly. But for the first time, the report introduced the concept of a global system in which everyone and everything connects to everyone and everything else.

The United Nations commissioned a three-year assessment of the way forward, chaired by Norway's prime minister Gro Harlem Brundtland, with results published in a 1987 report entitled *Our Common Future*. Brundtland said, " 'The environment' is where we all live; and 'development' is what we all do to improve our lot within that abode. The two are inseparable." Sustainable development, she said, "is development that meets the needs of the present without compromising the ability of future generations to meet their own needs.

So what is the blue economy, and how does it relate to what is regarded as "green"? The answer depends on whom you ask. During the 2012 Rio+20 summit in Rio de Janeiro, Pacific small-island developing states said that, for them, a green economy was in fact a blue economy, since their jurisdictions embraced far more ocean than land. As of 2020, for the European Commission, it means, "all economic activities related to oceans, seas, and coasts." The Commonwealth of Nations considers it "an emerging concept which encourages better stewardship of our ocean or blue resources." The Middlebury Institute of International Studies' Center for the Blue Economy says, "It is now a widely used term around the world with three related but distinct meanings—the overall contribution of the oceans to economies, the need to address the environmental and ecological sustainability of the oceans, and the ocean economy as a growth opportunity for both developed and developing countries."

It is the latter meaning—economic growth opportunity—that is attracting the greatest attention and the largest investments. NOAA has a term for marine life with economic potential: "underutilized species." The term "undeveloped waterfront property" implies greater value if a natural area is converted to a housing development or parking lot than conserved as a marsh or natural beach. Space in the ocean is now being targeted for aquaculture pens and offshore wind farms, thereby making "unused ocean" useful.

The World Bank defines the blue economy as "the sustainable use of

> "NATURE HAS NEITHER SENTIMENT NOR MERCY. WHAT IT DOES HAVE IS LIFE, TRUTH, AND LOGIC."
>
> — CARL SAFINA, *THE VIEW FROM LAZY POINT*

OPPOSITE: The protected waters of Costa Rica's Bat Islands host myriad species, including these schooling bull jacks.

ocean resources for economic growth, improved livelihoods, and jobs while preserving the health of ocean ecosystems." Some would argue that the value of the ocean as the engine that makes the existence of all life possible is, essentially, priceless. However, using traditional measures, the World Wildlife Fund in 2015 put the worth of key ocean assets at more than $24 trillion (U.S.), noting that fisheries are now overexploited but suggesting there is room for the economic growth of aquaculture and offshore wind power. That same year, in September, the United Nations General Assembly adopted 17 2030 Sustainable Development Goals (SDGs), based on the theme of "leaving no one behind." Some of the goals are: No Poverty, Zero Hunger, Good Health and Well-being, Affordable and Clean Energy, and Climate Action. Goal 15 considers Life on Land; goal 14 is Life Below Water.

For SDG 15, much attention is being given to the connection between saving and restoring forests and addressing atmospheric carbon absorption, climate stability, poverty alleviation, and biodiversity loss. A 2019 analysis, the *Global Assessment Report on Biodiversity and Ecosystem Services* showed that of the 1.8 million known plant and animal species currently identified, about one million, mostly terrestrial, are threatened with extinction within a few decades. Every day, the economic potential, the individual genetic treasury, and the light species bring to the universe are being snuffed out forever. Since only about 10 percent of the ocean has been explored, the biodiversity loss in the ocean is likely as great as or greater than on the land.

Human activity has altered about 75 percent of the surface of the land, eliminating natural systems millions of years in the making and squeezing wildlife into fragments of their former ranges. This might appear to be progress: replacing forests with farms, homes, cities, highways, and industrial facilities that maintain the increasing demands of the growing population. But by current accounting methods, a nation's gross national product, or GNP, focuses on benefits, neglecting to fully account for the costs. Air, water, and wildlife are regarded as free for the taking. The value of a barrel of oil is based on what it can be sold for, not the millions of years invested in creating

DEEPER DIVE

Healing Estuaries

Great Barrier Reef Marine Park

Estuaries—places where rivers and the sea converge—are notoriously productive, attractive to large numbers of people and wildlife. They are notorious, too, for the challenges they present: how to protect them from waste and fertilizer overload, major storms due to climate change, and overfishing. Earth's largest estuaries—Canada's Gulf of Saint Lawrence and the U.S. Chesapeake Bay—together cover about a million square kilometers. With thousands of others around the world, they are life's blood for millions and a priority for protection. The Ocean Conservancy works to diminish acidification in American estuaries; New Zealand's NIWA (National Institute of Water and Atmospheric Research) helps restore areas of economic and cultural significance to the native Maori; and more. Virginia Tippie, former director of the U.S. EPA's Chesapeake Bay Program, is founder of the Coastal Ocean Initiative of the Coastal America Foundation. Strategic partnerships have helped her group remove old railway beds and dams to restore estuary flow, open migration routes and spawning grounds for seabirds and sea turtles, and restore coastal reefs. Her oyster deployment program invites citizens to sponsor an oyster. Each oyster filters 113 liters of water a day, nature's cleaning service.

Barbara Block
Dining With Sharks

Stanford marine biologist Barbara Block wanted to find out why great white sharks migrate for the winter to a Colorado-size area between Hawaii and California, which satellite images had shown to be a nutrient desert. In the spring of 2018, she and a team of researchers attached tracking devices to the fins of 37 sharks and followed their movements. It turned out that in deep water, this area is actually so rich with tuna, squid, and other marine life that Block gave it the name "White Shark Café." Each day, like clockwork, sharks descended over 400 meters to revel in a buffet of deep mid-water life. The females returned to shallower water only at night, but the males moved up and down about 120 times a day. Why? For feeding only? Is mating involved? Why do males and females travel differently? Block looks to tagging to answer these and other questions about how sharks, tuna, and other large animals move, behave, and use the open ocean ecosystem.

the shales it comes from. Swimming in the ocean, fish have an accounting base of zero; they are considered no-cost. The value is based on the market price of the animals after they have been taken from the sea.

SDG 14, Life Below Water, articulates aims to sustainably manage and protect marine and coastal ecosystems from pollution, addresses the impacts of ocean acidification, and acknowledges the role of the ocean in driving climate, weather, planetary temperature, and chemistry. While recognizing the sharp decline in ocean wildlife, there is a long tradition of taking ocean life as commodities, not just to support the needs of people who have limited food options (food security), but to support sales to upscale markets, where dining on wild fish and other sea life is an option (food choice). Peter Thompson, the United Nations Special Envoy for the Ocean with special oversight of SDG 14, is cautiously optimistic. "There is still time to protect the ocean," he said in 2019, "and in so doing, we have a recipe for humanity's survival."

ABOVE LEFT: A great white shark courses its popular feeding grounds, the waters of Neptune Islands Conservation Park, South Australia.

Not Too Big to Fail

Throughout most of human history, the ocean has appeared to be so vast, so resilient, that it is not surprising that people have imagined it to be infinite in its capacity to yield whatever we wanted to take out of it and to accept whatever we wanted to put into it. What is surprising is that even with 21st-century evidence of ocean decline brought about by human impacts, the myth that the ocean is "too big to fail" persists.

Offshore oil and gas development has impacted the ocean both ways: Infrastructure goes into the ocean and then takes out, transports, and processes petrochemicals. Accidental spills and sometimes deliberate application of toxic chemical dispersants also go into the sea. Among the most damaging substances released into the air and ocean is excessive carbon dioxide generated by burning oil, gas, and coal. In the atmosphere, CO_2 contributes to increasing the temperature of the planet overall, especially the ocean. In the sea, it forms carbonic acid, causing the ocean to become more acidic. Increased temperature reduces the ability of the ocean to hold oxygen, alters the distribution range of species and ecosystems, and is a major factor in causing the loss of about half of the highly temperature-sensitive, shallow-water coral reefs and kelp forests globally. Coal is the source of soot that dusts the surface of the sea, diminishes the reflectivity of polar ice, and transports mercury into the ocean. There, bacteria help convert mercury to the neurotoxin methylmercury, which makes its way through the food chain to marine animals and the people who consume them.

Motorized shipping, military testing, seismic surveys, and even scientific mapping are putting into the ocean something new: high levels of noise. Communication among marine mammals has long been known, but now the importance of communication and the impact of high levels of sound have been demonstrated on fish, crustacea, cephalopods, and other marine life. Human interference with normal animal communication and outright killing of creatures with sounds produced by shipping, drilling, sonar, and seismic surveys have been unintentional but costly to ocean life.

Dredging and mining sand and mud from underwater sources to replenish eroded beaches and obtain construction materials, and mining the deep-sea bed for minerals: All involve "taking out" but also "putting in" noise, plumes of fine, smothering sediment, and destruction of sensitive benthic life. Coastal areas have been disturbed by mining and so-called reclamation activities for centuries, but mining metals and minerals taken from fields of manganese nodules, crusts, and chimneys around hydrothermal vents in the deep sea has become a new threat to ancient communities immune until now from human interference since their origin, millions of years ago. Before converting them into products for current markets, it would be prudent to consider their value—and that of other wild ocean species and systems—to planetary health and security.

This loggerhead sea turtle is one of thousands snared in derelict fishing gear each year. Most die.

Ocean Subtractions

For most of human history, the great majority of ocean life has been protected by being inaccessible. That has changed dramatically since the 1950s as new technologies make possible unprecedented abilities to find, capture, and—most significantly—market marine mammals, fish, and other marine life to increasingly large numbers of consumers.

Even before the 20th century, fishermen had already made deep inroads into populations of coastal ocean wildlife and were forced to venture increasingly far to find and capture their prey. Herman Melville's classic, *Moby-Dick,* is based on Yankee whalers who historically worked from shore, but by the mid-1800s, after having depleted Atlantic whales, were operating half a world away near the Galápagos Islands. In 1871 the U.S. government established the Office of Commissioner of Fish and Fisheries to address complaints over depletion of fish stocks. In his insightful book *The Ocean of Life,* biologist Callum Roberts documents similar concerns in England. "Even late in the 19th century," he writes, "fishermen had begun to grumble of declining catches . . ." Nonetheless, the vision of unlimited ocean resources was hard to shake. In a famous 1883 speech, the highly esteemed British scientist Thomas Huxley wrote, "The cod fishery, the herring fishery, the pilchard fishery, the mackerel fishery, and probably all the great sea-fisheries, are inexhaustible; that is to say that nothing we do seriously affects the number of fish. Any attempt to regulate these fisheries seems consequently . . . to be useless."

There is a deeply rooted perception that ocean life exists to be monetized. Unmarketable species are known ignominiously as "trash fish." Bottom trawling, a widely used method of fishing, uses nets dragged across the seafloor, indiscriminately taking whatever is in their path. The amount of fish, invertebrates, and seaweed taken as troublesome "bycatch," whether by long lines, traps, or nets, often exceeds the amount of life retained for sale. Even if people agree that the best use for the ocean is to extract wildlife for food and products, the approaches used currently have not achieved and likely cannot attain the desired "sustainable yields." The complex realities of ocean dynamics, life history information, and causes of the natural ups and downs of species and systems are yet to be unraveled.

In a 2006 review, "The Sustainability Myth," by Alan Longhurst, fishery science is described as a discipline that "produced a corpus of theory that was taught in universities and applied at sea, but which has since proved to be wrong." Since the peak year of global catch in 1989 of 90 million tons, the level of take has consistently declined. Nonetheless, sea creatures continue to be the subjects of the largest global trade in wildlife, most of it legal. The evidence is growing that no form of life—not whales nor cod nor krill—can for long accommodate predation on the scale humans now impose.

VISIONARIES

Heather Koldewey
Saving the Seahorse

She calls them "beautiful and fragile," and she should know. The world's leading expert on seahorses, marine biologist Heather Koldewey is also founder of Project Seahorse, which advances marine conservation by protecting these small, complex creatures. A 2018 National Geographic Fellow and scientific co-leader for the National Geographic Society program to address the global impact of plastics, Koldewey engages local communities to solve conservation challenges. Through her award-winning Net-Works program, coastal residents collect discarded fishing nets to recycle into nylon yarn for clothing and carpet tiles—fighting both debris and poverty. Head of global programs at the Zoological Society of London (ZSL), Koldewey also runs the #OneLess campaign, which aims to make London the first capital city to ban single-use plastic water bottles. Through ZSL's Bertarelli Programme in Marine Science, she works to protect coral reefs and sentinel species like sharks, vital to a thriving food web.

Bycatch—an unintended catch—crowded with rays and guitarfish is tossed from a shrimp boat.

FISHING ACTIVITY

While the global fishing industry burgeons, with Asian countries taking about 70 percent of the catch, since 2015 the UN's Sustainable Development Goal (SDG) 14 has catalyzed new fisheries and aquaculture policies.

 Heavier

Lighter

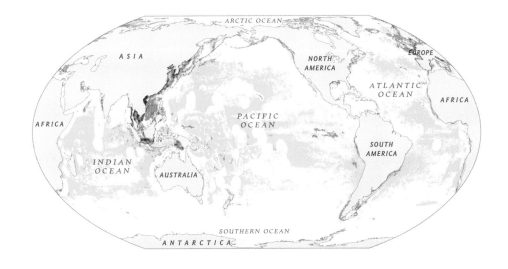

Ocean Additions

Damage to ocean health by what we have taken out is more than matched by what we have put in. Changing the chemistry of the liquid medium that is home to sea life is as disruptive to them as it would be to us if our atmosphere became suffused with toxic chemicals, the oxygen level reduced, production of basic food sources diminished, and our senses of smell, hearing, and taste corrupted. Fish and other sea dwellers have senses—including chemoreceptors vital for finding home, food, and one another—that we can barely imagine.

Significant changes to ocean chemistry through human actions have tracked with our population growth. Our impact was probably negligible for most of our history, but that changed two centuries ago when our numbers exceeded one billion. Even prior to 1900 there was evidence of coastal pollution in places such as New York City, where raw sewage was channeled into the sea, smothering once vast oyster beds. Despite serious depletion, the mollusks were still being mined for fine dining. In U.S. waters overall, a 1968 National Academy of Sciences report estimated an annual dumping into the ocean of 100 million tons of petroleum products, up to four million tons of acid wastes from pulp mills, more than a million tons of heavy metals in industrial wastes, and more than 100,000 tons of organic chemical waste.

At three sites offshore from the West Coast, more than 55,000 containers of radioactive wastes were dumped between 1946 and 1970; another 34,000 containers went into three offshore locations along the Eastern Seaboard. And that was just in the United States. Globally, the ocean has been and still is widely regarded as the best place to dispose of chemical and industrial wastes, trash, munitions, sewage sludge, and contaminated dredged material, with the assumption that either it does not matter or that the ocean has an unlimited capacity to mix and harmlessly disperse wastes. "The solution to pollution is dilution" is a commonly used rationale for ocean dumping, despite evidence that decades of uncontrolled dumping have led to severe depletion of oxygen levels in some areas and high levels of harmful pollutants—heavy metals, chlorinated hydrocarbons, and other toxic chemicals—in others.

A turning point was reached in 1972, when mounting evidence of harm caused by ocean dumping led to the international Convention on the Prevention of Marine Pollution by Dumping of Wastes and Other Matter, known as the London Dumping Convention. While it now holds in check some deliberate use of the ocean for waste disposal, offshore dumping persists; in addition, the flow of contaminants from inland sources has created toxic rivers and downstream pollution.

Under U.S. law, it is still legal to dispose of vessels, fish wastes, and dredged material at sea, but no permission is ever obtained from the legions of creatures smothered under the avalanche that descends upon them, nor

VISIONARIES

Jenna Jambeck
Tracking Trash

When engineer Jenna Jambeck sailed across the Atlantic Ocean in 2014, she and the crew of 13 other women were on a mission: to preserve the future by encouraging women to enter STEM (science, technology, engineering, and math) disciplines. The odyssey, called eXXpedition, showcased destructive impacts of plastic on women's health, inspired followers worldwide, and launched yearly eXXpeditions to explore how ocean waste harms us. The 2018–2021 National Geographic Fellow is scientific co-lead for National Geographic's Sea to Source Expedition: Ganges, a plastics-tracking initiative. A distinguished professor at the University of Georgia, Jambeck is "fascinated by waste's very intense and complex relationship with people." The Marine Debris Tracker mobile app she co-developed guides citizen scientists to document litter and marine debris for removal—with over three million "trashings" to date.

Plastic trash, like these samples from a New York beach, is projected to reach 600 million metric tons by 2040.

has an accounting been made of their loss as part of the cost of waste disposal.

Greatly increased application of chemical fertilizers and pesticides for agriculture, lawns, and golf courses from the 1970s onward, which wash down rivers to the coast, has led to creation of notorious "dead zones" in the sea. High levels of nitrates and phosphates stimulate rapid growth of a few kinds of phytoplankton. As they decompose, oxygen is depleted and other organisms die. The number of resulting toxic areas has nearly doubled every decade since the 1960s. More than 500 such places now exist globally. One, offshore from Los Angeles, was recently reported to hold perhaps a million barrels of DDT amid other toxic waste pumped from a local plant between 1947 and 1982. A 2016 report by UNESCO's Intergovernmental Oceanographic Commission (IOC) indicates that ocean deoxygenation is taking place all over the world as a result of the human footprint. Professor Robert Diaz at the Virginia Institute of Marine Science said the speed of ocean suffocation is literally "breathtaking."

The synthetic materials collectively known as plastics were hailed as miracles of modern chemistry when they began to appear early in the 20th century. Plastics are now "fully integrated" into manufactured goods from buttons to buses and are widely embraced for single-use containers for medicines, hardware, drinks, and food—even including such products as bananas and coconuts, which come naturally packaged in their own skins. Plastics are also fully integrated into the ocean. Escaped party balloons, discarded beverage bottles, and puffy plastic bags drift like gatherings of jellyfish and are consumed by unwary sea turtles, fish, and whales.

A University of Georgia engineering professor, Jenna Jambeck, calculated that between 5.3 million and 14 million tons of plastic reach the sea

SEAS OF PLASTIC

Increasing human impact on the ocean includes dumping plastic, a product integrated into every aspect of life. Since the 1950s plastics produced, consumed, and discarded have filled the sea with an estimated 450 million metric tons in 2015, devastating the ocean and its life.

Management Problems
Metric tons of mismanaged plastic per year
More than 100,000 · 100–999 · 10,000–100,000 · 10–99 · 1,000–9,999 · Less than 10

Plastic Accumulation Zones
Probability of plastic waste build-up
Most likely — Least likely

Cleaning Up Plastics

....

I n 2019 Canada, Peru, and the European Union pledged to ban single-use plastic; the District of Columbia banned plastic straws; and Great Britain's Royal Statistical Society named 90.5 percent "the stat of the year." That's the amount of plastic waste that has never been

recycled. Awareness is rising fast, and our lives depend on it. Studies show that as much as 12.7 million tons of plastic waste fill the ocean each year, never breaking down but building to a current 51 trillion microplastic particles, per United Nations Environment Programme estimates, and killing more than a million animals every year. A 2018 UN report says we may be on the list: Plastic-marinated shrimp, clams, and oysters land on our dinner plates, and plastic toxins may trigger cancers, immune system deficiencies, and birth defects. In late 2018, 250 major plastic-producing organizations made a pact with the UN to do their part by 2025 to reduce plastic production and clean up what's already in the water. Citizens can start helping by visiting the Ocean Conservancy website's "Fighting for Trash Free Seas."

....

ABOVE: Volunteers scan world beaches to remove plastic. LEFT: A brown pelican is rescued after the Gulf of Mexico's 2010 *Deepwater Horizon* oil spill.

every year, mostly from land-based sources. Since 1950, more than 6.3 billion tons of plastic generated have never entered a recycling bin, according to scientists who made the calculation in 2017. Most of it has eventually made its way into the ocean, and most of it will still be there as large chunks, and as degraded micro- and nano-particles, and as even smaller stable manufactured molecules, for centuries to come. Or forever.

By far the largest amount of obviously damaging plastic trash originates from fishing operations. Hundreds of thousands of tons of lost and abandoned "ghost gear"—nets, lines, pots, and traps—unintentionally kill millions of marine mammals, birds, sea turtles, fish, and invertebrates every year. Intentional fishing with such equipment takes more than 90 million tons of ocean wildlife annually, a figure that does not include unintended bycatch and illegal, unreported, and unregulated fishing.

War & Other Chaos

Although justified for security reasons, some of the most devastating harm to the ocean has come about from wartime activities. In addition to at-sea turmoil, explosions, noise, debris, and sunken vessels, massive arsenals of unused chemical weapons from World War I and World War II were "disposed of" in the ocean—relocated, but not gone. Between 1946 and 1958, 23 nuclear weapons were detonated by the United States at Bikini Atoll in the Marshall Islands, vaporizing pristine reefs, and contaminating the surrounding ocean. Johnston Atoll, about 1,500 kilometers southwest of Hawaii, was once a thriving haven for seabirds and coral reefs and was designated a Wildlife Sanctuary in 1926. But in 1950 it was used as a test range for thermonuclear warheads. Dredging and filling increased the land area by more than 10 times for its use as a chemical weapons disposal and incineration site for mustard gas, sarin nerve gas, Agent Orange, and other toxins from 1964 to 2004. Russia has dumped nuclear reactors and obsolete nuclear submarines into the ocean as well. Before 1993, no international laws prevented countries from using the sea for nuclear waste disposal.

Another way that humans have unwittingly created chaos in the ocean is by taking creatures from one part of it to another. Two species of lionfish (*Pterois volitans* and *Pterois miles*), native to the Indo-Pacific, have wreaked havoc as voracious predators on small fish in places in the Atlantic Ocean where they were unintentionally introduced, perhaps as stowaways in the ballast water of ships or released from home aquariums. Populations of sharks, groupers, and other large fish that might have kept them in check have declined sharply in recent years, making it easier for the newcomers to become established.

Port cities are especially vulnerable to the introduction of nonnative species, in part because the local ecosystems are usually disturbed, and new arrivals are more likely to find places to fit in when they arrive. A 2018 report by the Nature Conservancy states that 84 percent of the world's coasts are being affected by nonnative species. More than half of San Francisco Bay's subsea residents are recent arrivals, from Chinese mitten crabs to New Zealand sea slugs. It is unrealistic to imagine that the ocean can be restored to its state of a century ago. But as we've become armed with new knowledge, new actions are beginning to reverse the damaging trends.

Dozens of international conventions, agreements, protocols, and treaties addressing environmental concerns have come about since the 1950s. Among those relevant to the ocean are the 1973 and 1978 International Convention for the Prevention of Marine Pollution from Ships (MARPOL), the 1972 London Dumping Convention, the 1973 Convention on International Trade in Endangered Species (CITES), the 1980 Convention for the Conservation of Antarctic Marine Living Resources (CCAMLR), the 1992 Convention on Biological Diversity, and, most significantly, the 1982 UN Convention on the Law of the Sea (UNCLOS) that came into force in 1994.

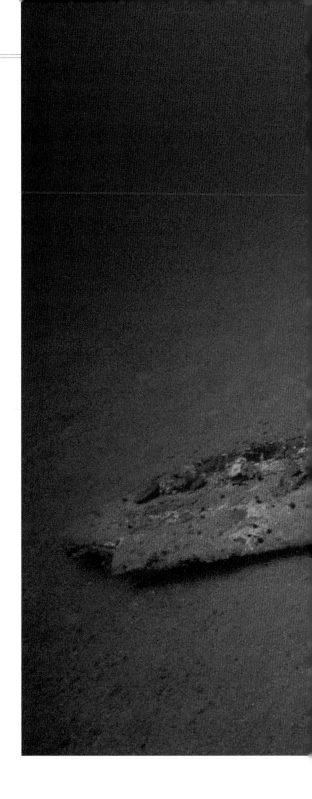

Off Papua New Guinea's Bismarck Archipelago, a diver explores the wreckage of a Japanese fighter, downed during World War II.

WHO CONTROLS THE OCEAN?

Beyond a nation's exclusive economic zone (EEZ), the High Seas—nearly half of the world—are open to all with policies guided by the UN Convention on the Law of the Sea.

Maritime Divisions
— Maritime boundary
Median line
— Joint Development
Area boundary
▦ Territorial waters
EEZ
High Seas

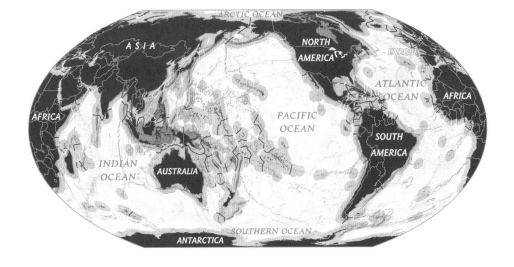

Making a Comeback

Measures to save and restore wildlife and natural landscapes began early in the 20th century with the establishment of a system of United States national parks, often known as the "best idea America ever had." A system of protected areas in the ocean is an idea late in arriving, but it may be the best hope for ocean restoration.

The International Union for Conservation of Nature (IUCN) defines a marine protected area (MPA) as "a clearly defined geographical space, recognized, dedicated, and managed through legal or other effective means, to achieve the long-term conservation with associated services and cultural values." In 2020, the Marine Conservation Institute's digital Atlas of Marine Protection reported that 5.3 percent of the ocean is protected and managed, but only 2.5 percent is safeguarded from fishing. Fully protected areas are proving to be the most effective strategy by far in maintaining intact systems and restoring degraded ones. Numerous studies demonstrate an increase in size, number, and diversity of species within a few years.

The National Geographic Society's Sustainable Seas Expeditions from 1998 to 2003 focused on exploring and documenting the young but promising system of 13 U.S. National Marine Sanctuaries, as well as places in Mexico and Belize, engaging with more than 50 government agencies, universities, and organizations. For the first time, one-person submersibles and remotely operated vehicles were piloted by scientists in depths to 1,000 meters. Students were enlisted to help. The organization Mission Blue partners with the IUCN to review proposed protected areas called Hope Spots and works with government leaders, some 200 partner organizations, local champions, and communities toward achieving maximum protection for more than 140 places, ranging from intact wilderness areas such as deep seamounts and the High Seas to urbanized regions such as San Francisco Bay and Florida's Gulf Coast. National Geographic Explorer Enric Sala launched the Pristine Seas project in 2009 to identify and help save the last wild places in the ocean. He leads teams of scientists to explore and document ocean wilderness areas and works with government leaders to achieve protection for millions of square kilometers of ocean. Several large organizations—the World Wildlife Fund, Conservation International, the Nature Conservancy, the Pew Charitable Trusts, and the Wildlife Conservation Society—as well as hundreds of smaller NGOs are working to expand ocean protection.

In 2016, the IUCN called for at least 30 percent of the ocean to be highly protected by 2030. Harvard Professor E. O. Wilson initiated the Half Earth project, aiming for half of the land and half of the ocean to be fully protected for Earth to have some hope for recovery. Safeguarding the High Seas global commons—half the world—would achieve a large part of that goal. Currently, humans are occupying most of the planet, one way or another. Finding a place for billions of people within the natural systems that sustain us means treating all the natural world with dignity and respect.

They may look alike, but diagonal-banded sweetlips in Australia's marine protected Challenger Bay have distinctive individual markings and behaviors.

Citizen Scientists for the Ocean

Satellites circle the planet, ROVs glide below the ocean surface, and scientists and governments monitor data and set policy—all for ocean health. Then there are the citizen scientists. Along rocky and sandy coasts, in bays and estuaries, on remote islands, and around plastic gyres, they're counting species, recording behaviors, disposing waste, documenting tides and temperature, and more—doing their part to help the ocean revive and survive. The best scientists have the attributes of kids—they ask questions, report honestly what

A volunteer observes a crab.

they see, and have a sense of wonder. Anyone can join efforts to track whale population trends, fight harmful algal blooms, record trash accumulation, collect GPS data to monitor sea-level rise, or help out at a marine sanctuary. Apps, like the Marine Debris Tracker and EMU Explorer, guide and connect volunteers to others around the globe. A citizen science journey can start with a government agency such as NOAA, or a local aquarium or private consortium. Or it can start with an individual. No one can do everything, but everyone can do something.

Marine Protected Areas

The concept of protecting the ocean has ancient origins. For millennia, residents of Oceania closed fishing grounds when they became depleted and protected vital spawning and nursery areas. A ruler might proclaim an area untouchable while it gradually recovered. However, by the 1950s, catching, distributing, and consuming ocean wildlife had become globalized and increasingly industrialized. In response to concerns about overfishing and pollution, between 1958 and 1972, several international conventions laid the framework for protecting ocean resources, including legislation establishing the U.S. National Marine Sanctuaries.

In 1975 an international group of scientists, policy makers, and activists came together in Tokyo to acknowledge the plight of the ocean—and to do something about it. Sponsored by the IUCN, the group aimed to establish protected areas across the globe to help restore ocean health. Each area was to be "a defined geographical space" effectively managed to achieve long-term conservation, but the definition gives latitude for interpretation. While some MPAs allow limited human activity such as regulated fishing and boating, others do not. Data accumulated over the decades following that meeting show that the most successful MPAs meet five criteria: fully protected, with no fishing allowed; strongly enforced; older than 10 years; having an area greater than 100 square kilometers; isolated.

The first large-scale MPA, designated in 1975, was the 344,000-square-kilometer Great Barrier Reef in Australia, but initially just 4 percent of it was fully protected. Some four decades after the IUCN began its efforts, the *Protected Planet Report* in late 2016 announced "just under 15,000 Marine Protected Areas (MPAs) spread across 18.5 million square kilometres of ocean and sea. Up to 13 percent of territorial waters are now protected." Research on no-take reserves worldwide has documented an average increase of 446 percent in total marine life across MPAs and an increase in number of species by an average of 21 percent.

Still, despite international efforts, the Marine Conservation Institute's digital Atlas of Marine Protection reports in 2020 that just 5.3 percent of the ocean is now protected and managed, with only 2.5 percent safeguarded from fishing.

The good news is that highly protected areas are seeing results that can resonate worldwide. In 2019, the state of California alone reported 124 MPAs covering 3,550 square kilometers, 16 percent of waters off the state. In a study published in the journal *Ocean & Coastal Management,* Samantha Murray at Scripps Institution of Oceanography and environmental lawyer Tyler Hee documented rebounding fish populations and ecosystems in the state's carefully monitored waters, including a 52 percent increase of biomass (total marine life) in the Channel Islands reserves. In addition, there was a 23 percent increase just outside those waters—a spillover effect that demonstrates how strong protection is a gift that keeps on giving.

WHY IT MATTERS

Hope Spots: Protecting the Blue

The ocean embraces 97 percent of Earth's biosphere, yet two decades into the 21st century, less than 3 percent of the ocean is fully protected. About 15 percent of the land is maintained in wild parks and reserves, but neither land nor sea can prosper without the ocean's regulating system, now diminishing as

human impacts magnify. To restore planetary health and safeguard civilization, there is widespread support for strongly protecting at least 30 percent of the land and sea by 2030. Working with the National Geographic Society and more than 200 partner organizations, the Mission Blue NGO has designated more than 140 community-supported Hope Spots. In partnership with the IUCN, Mission Blue asks for nominations from the public for places that shelter diverse or endangered species, have potential to rebound from human impact, are migration corridors, or have other ecological and cultural values. The Hope Spot initiative matters because it empowers people to protect the ocean. And it's working.

ABOVE: A Hope Spot since 2018, Indonesia's Bunaken Marine Park thrives. OPPOSITE ABOVE: A fur seal frolics in False Bay Hope Spot off Cape Town. OPPOSITE BELOW: As if bejeweled, a giant clam reigns in a protected reef off the Pacific's Nikumaroro Island.

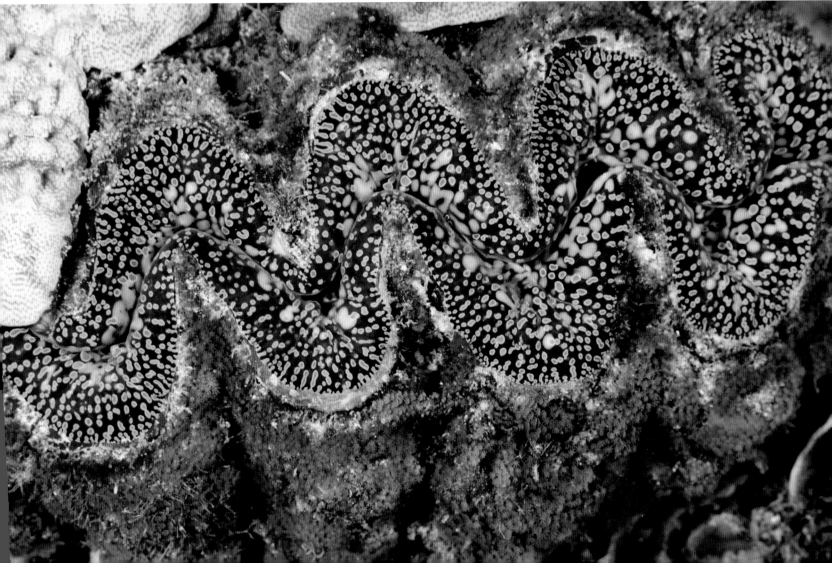

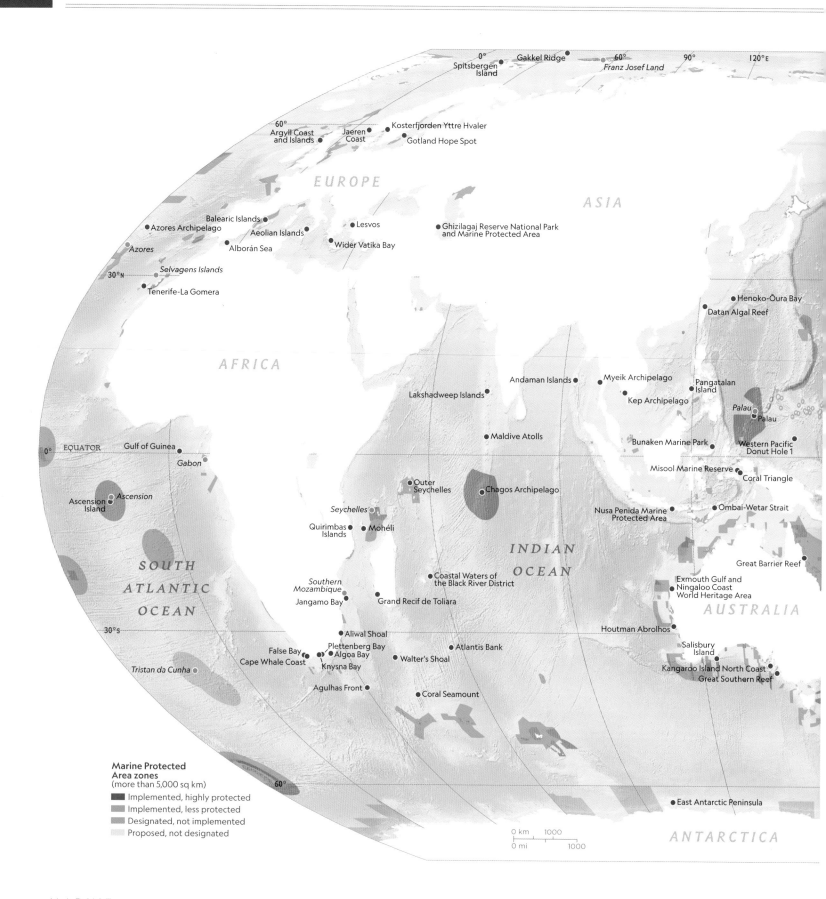

**Marine Protected
Area zones**
(more than 5,000 sq km)

- ▮ Implemented, highly protected
- ▮ Implemented, less protected
- ▮ Designated, not implemented
- ▯ Proposed, not designated

MARINE
PROTECTED AREAS

Encircling the globe, some 15,000 marine protected areas (MPAs) across 18.5 million square kilometers include more than 140 Mission Blue Hope Spots and 22 National Geographic Pristine Seas expedition sites. MPA definitions, as defined by the Marine Protection Atlas, are as follows:

Implemented, highly protected: An MPA has regulations in force on the water that prohibit or highly restrict fishing.
Implemented, less protected: An MPA has regulations in force on the water, but they are minimally restrictive or do not regulate fishing.
Designated, not implemented: An MPA is specifically codified or legally recognized by an

authoritative rule. The MPA exists "on paper" and in law or other formal process, but regulations are not yet in force on the water.

Proposed, not designated: The intent to create an MPA is made public; however, it is not yet legally recognized (designated) or in force on the water (implemented).

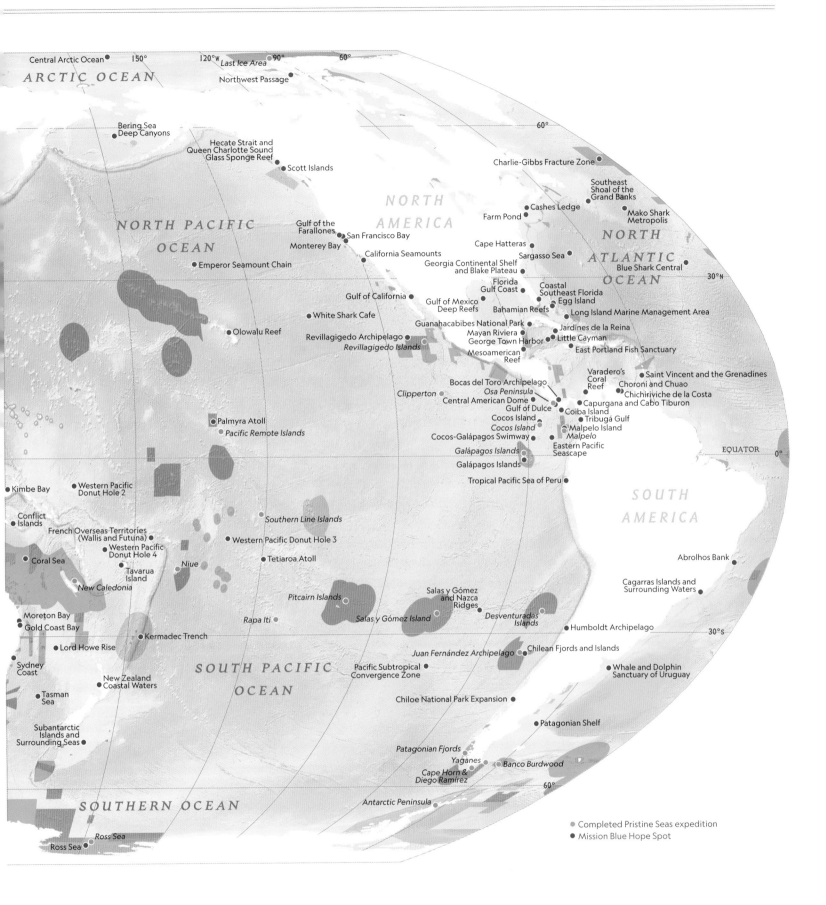

ARCTIC OCEAN

Central Arctic Ocean
150°
120°W
Last Ice Area
90°
60°
Northwest Passage

Bering Sea
Deep Canyons
60°
Charlie-Gibbs Fracture Zone

Hecate Strait and
Queen Charlotte Sound
Glass Sponge Reef
Scott Islands

Southeast
Shoal of the
Grand Banks

NORTH
AMERICA
Farm Pond
Cashes Ledge
Mako Shark
Metropolis

NORTH PACIFIC
OCEAN
Gulf of the
Farallones
San Francisco Bay
Monterey Bay
Cape Hatteras
Sargasso Sea
NORTH
ATLANTIC
OCEAN

Emperor Seamount Chain
California Seamounts
Georgia Continental Shelf
and Blake Plateau
Blue Shark Central
30°N

Gulf of California
Florida
Gulf Coast
Coastal
Southeast Florida
Egg Island

White Shark Cafe
Gulf of Mexico
Deep Reefs
Bahamian Reefs
Long Island Marine Management Area

Olowalu Reef
Guanahacabibes National Park
Mayan Riviera
Jardines de la Reina

Revillagigedo Archipelago
Revillagigedo Islands
George Town Harbor
Little Cayman
East Portland Fish Sanctuary
Mesoamerican
Reef

Varadero's
Coral
Reef
Saint Vincent and the Grenadines
Bocas del Toro Archipelago
Clipperton
Osa Peninsula
Choroni and Chuao
Chichiriviche de la Costa
Central American Dome
Capurgana and Cabo Tiburon
Gulf of Dulce
Coiba Island
Palmyra Atoll
Cocos Island
Tribugá Gulf
Pacific Remote Islands
Cocos Island
Malpelo Island
Cocos-Galápagos Swimway
Malpelo
Eastern Pacific
Seascape
EQUATOR
0°
Galápagos Islands
Galápagos Islands

Kimbe Bay
Western Pacific
Donut Hole 2
Tropical Pacific Sea of Peru
SOUTH
AMERICA

Conflict
Islands
French Overseas Territories
(Wallis and Futuna)
Southern Line Islands
Western Pacific
Donut Hole 4
Western Pacific Donut Hole 3
Abrolhos Bank

Coral Sea
Tetiaroa Atoll
Cagarras Islands and
Surrounding Waters
Tavarua
Island
Niue

New Caledonia
Pitcairn Islands
Salas y Gómez
and Nazca
Ridges

Moreton Bay
Rapa Iti
Salas y Gómez Island
*Desventuradas
Islands*
Gold Coast Bay
Humboldt Archipelago
30°S
Kermadec Trench
Lord Howe Rise
Juan Fernández Archipelago
Chilean Fjords and Islands

Sydney
Coast
SOUTH PACIFIC
OCEAN
Pacific Subtropical
Convergence Zone
Whale and Dolphin
Sanctuary of Uruguay
Tasman
Sea
New Zealand
Coastal Waters
Chiloe National Park Expansion

Subantarctic
Islands and
Surrounding Seas
Patagonian Shelf

Patagonian Fjords
Yaganes
Banco Burdwood
Cape Horn &
Diego Ramirez
60°

SOUTHERN OCEAN
Antarctic Peninsula

● Completed Pristine Seas expedition
● Mission Blue Hope Spot

Ross Sea
Ross Sea

"WE HAVE CREATED CONDITIONS NOT
AMICABLE TO OUR OWN FLOURISHING."

—SIR KEN ROBINSON, BRITISH AUTHOR

National Geographic Society Ocean Initiatives

Marine parks are now being designated by nations around the world as a proven means of saving and restoring declining ocean wildlife and ecosystems that in turn impact the global climate, the economy, and human health, well-being, and security. Long a champion for national parks and reserves on the land, the National Geographic Society co-sponsored the Sustainable Seas Expeditions with the Richard and Rhoda Goldman Fund, NOAA, and more than 50 partners. This private-and-public-industry program supported exploring and expanding the U.S. National Marine Sanctuary system, launching scientist-piloted submersible expeditions to research and document deep-water regions and hosting student summits and public meetings. A multiyear collaboration followed with NOAA, the U.S. Navy, and Google to improve public access to knowledge of the ocean's underwater terrain and showcase marine protected areas on Google's widely used digital globe, Google Earth. Since 2008 the Society has fostered two global ocean initiatives, Mission Blue and Pristine Seas, led by National Geographic Explorers-in-Residence Sylvia Earle and Enric Sala, respectively. In 2020 National Geographic partnered with the Wyss Campaign for Nature in a 10-year program to help communities, Indigenous peoples, and nations conserve 30 percent of the planet in its natural state by 2030, by creating and expanding protected areas, investing in science, and inspiring conservation. It is a goal consistent with the IUCN's vision of protecting 30 percent of the ocean by 2030, and is wholeheartedly endorsed both by Pristine Seas and Mission Blue.

Mission Blue/Hope Spots

Given a chance at the 2009 Technology Entertainment Design (TED) conference to make a wish "big enough to change the world," along with help to make it happen, Sylvia Earle said, "I wish you would use all means at your disposal—films, expeditions, the web, new submarines, campaigns—to ignite public support for a global network of marine protected areas, Hope Spots, large enough to save and restore the ocean, the blue heart of the planet." In the same week, Google featured the first 12 Hope Spots on its inaugural portrayal of the ocean on its digital globe, backed by short National Geographic films that enabled users to "dive in" to places from polar seas and the Galápagos Islands to the Patagonian Shelf and the Sargasso Sea.

The following year, a TED-at-Sea expedition aboard the National Geographic-Lindblad ship M/V *Endeavor* engaged more than a hundred scientists and leaders in government, industry, conservation,

ABOVE: Sylvia Earle, here in a *DeepWorker* sub, champions ocean exploration and care. OPPOSITE: A young polka-dot ribbonfish (*Desmodema polystictum*) swims at night in Indonesian waters.

communication, and entertainment to consider ways to implement the wish and commit it to action. Hope Spots have become the signature program for Mission Blue, a nonprofit foundation now with more than 200 partner organizations to promote research, exploration, education, and a vigorous media program to inspire ocean care. Several expeditions to Hope Spots are conducted every year to explore, film, document, and work with local champions toward long-term, community-based protection. Working in partnership with the IUCN, a Hope Spots Council chaired by British scientist and conservationist Dan Laffoley reviews proposals for new places to be designated as Hope Spots and guides champions to achieve local participation, develop scientific monitoring, and enlist government support aimed at full protection. In partnership with the Environmental Systems Research Institute (ESRI), using its story maps and a consistent framework for data, images, and stories, Mission Blue is securing a global network of hope for enduring ocean protection.

Pristine Seas

As a university professor and marine biologist at Scripps Institution of Oceanography from 2000 to 2007, Enric Sala studied marine ecosystems and helped create the Scripps Center for Marine Biodiversity and Conservation. He was also a researcher at the Spanish National Research Council (2007–08) and held the first position in marine conservation ecology there. After years as a research scientist, Sala realized that as an academic, "I was writing the ocean's obituary."

> "KNOWLEDGE COMES, BUT WISDOM LINGERS."
> — ALFRED, LORD TENNYSON

He says, "I thought saving the ocean was a lost battle, but then I decided I wanted to be part of the solution." In 2008 he joined the National Geographic Society as a fellow before becoming an explorer-in-residence. He soon launched the Pristine Seas program to fulfill the mission of exploring and protecting the last wild places in the ocean. "Now I go to these [untouched] places and see what it used to be like," he says, "to see what the future could be elsewhere with regeneration. To know what marine ecosystems are truly capable of, we must look at the few pristine places that remain."

To identify pristine areas and determine how to manage them effectively for years to come, Sala and a team of scientists and filmmakers spend weeks at sea. They dive thousands of hours in remote regions to explore and document areas that are the least disturbed by humans. Key to the mission is collecting data on how such places function as healthy marine ecosystems. Understanding the engine of such an ecosystem helps the team establish a baseline that guides future monitoring. The baseline can also be applied to damaged ecosystems, which helps researchers understand how to restore them to better health. As a Young Global Leader at the World Economic Forum in 2008, Sala began to combine the scientific evaluation of ocean ecosystems with economic and social analyses that clearly demonstrate that people, along with fish and whales, benefit from a healthy ocean.

Opportunities for success in achieving protection require that governments with jurisdiction over the areas also have the willingness to take action. One such area is the British Overseas Territory, the Pitcairn Islands. This group of small islands, with about 50 people living on 47 square kilo-

meters of land, lies in an exclusive economic zone (EEZ) three times the size of the United Kingdom. In partnership with the Pew Charitable Trusts and numerous British scientific and conservation organizations, Pristine Seas supported the 2015 designation of the 834,334-square-kilometer Pitcairn Islands Marine Reserve. During a Pristine Seas expedition in 2012, Sala reported: "My first dive in the Pitcairn Islands revealed the most incredible and pristine living reef, hundreds of feet down in gin-clear water, with thousands of species and fish, including top predators and sharks, and a mammoth bed of leaf coral, like a bed of blue roses, stretching out into infinity."

Pristine Seas has become an effective program of exploration, scientific research, economic analysis, policy change, and global media. Over 5.8 million square kilometers—including more than 22 large regions of ocean around the coasts of the Pitcairn Islands, Chile, the Seychelles, French Polynesia, Gabon, Kiribati, the high Arctic, and elsewhere—have been championed by Pristine Seas as no-take marine reserves: safe havens for marine life and their underlying ecosystems. Alerting the public to these reserves is a vital part of the Pristine Seas program. Documentaries, television, social media, articles, and books are all part of a global awareness campaign. Five yearly expeditions can be joined in real time through Instagram and Twitter.

Director of National Geographic's Pristine Seas initiative Enric Sala explores a "jellyfish lake" in Palau.

GREAT EXPLORATIONS

Keepers of the Coasts

Mariners from Satawal Island, a mecca for wayfinding, ply Polynesian waters in outriggers.

Almost as long as there have been humans, there have been stories of coastal settlers discovering, appreciating, exploiting, and protecting the ocean.

In obscure Liang Bua Cave on the tiny Indonesian island of Flores, archaeologists in 2016 uncovered stone tools dating to 840,000 years ago, as well as the lower jaw and teeth of humans some 700,000 years old. Separated from the Asian continent by 24 kilometers of water, these Floresians may have been the first to cross the ocean to settle a new coast.

More recent journeys have been tracked by combining carbon-dated artifacts, computer simulations of early populations, ancient sea levels, and seafloor topography. In one case archaeologists estimate that about 50,000 years ago seafaring migrants may have braved the waters from Southeast

ABOVE: Centuries-old *makaus,* or fishhooks, crafted from turtle shell symbolize safe ocean passage. RIGHT: Free-diving fishermen descend off the coast of Polynesia's Tuamotu Archipelago.

Asia to Australia. In 2016, excavations of 23,000-year-old fishhooks crafted from snail shells and 35,000-year-old fish and human bones led researchers to determine that humans lived on today's Okinawa and fished its waters.

California's Channel Islands, a National Marine Sanctuary, likely welcomed migrants who had crossed the Bering Land Bridge from Asia 15,000 years ago. But the roasted bones of dwarf mammoths found on one island tell us that others may have come 25,000 years earlier—perhaps by boat.

Some 6,000 years ago along Australia's Northern Territory coast, Aboriginal residents developed sustainable fishing practices still followed today. Deft management has kept their fishing grounds, called maritime estates, healthy through modern climate change and sea-level rise.

New Zealand's Maori community Ngāi Tahu since 850 has followed a philosophy of sustainable resource management. While their livelihood now includes whale-watching tourism, they continually patrol the Kaikōura coast, home to young male sperm whales, to stop potentially harmful human activity and resolve environmental stressors.

More recent are the human stories recorded off Atlantic Ocean coasts—largely of commerce and warfare. In the North Sea, salvagers in 2019 discovered a trading vessel from 1540, filled with copper plate probably destined to become the Netherlands' first copper coins. The Stellwagen Bank National Marine Sanctuary, off the coast of Massachusetts, harbors historic shipwrecks that tell of deadly ocean conditions, even while their hulls live on today as artificial reefs that shelter humpbacks, minke, orca, and the critically endangered North Atlantic right whale.

Among 21st-century ocean sanctuaries, the imprints of ancient coastal cultures demonstrate lives long intertwined with the sea—and respecting it. Along Washington's Olympic coast in an area now home to a national park and a National Marine Sanctuary, a rock carving by early Ozette people depicts a breaching whale, a symbol of the enduring connection of people to the sea.

TIME LINE

Coastal Connections

840,000 years ago: Early humans may have started fishing on island of Flores, Indonesia.

60,000 years ago: Early Africans may have crossed Red Sea to Asia.

50,000 years ago: First humans arrived on Australian coast.

40,000 years ago: Dwarf mammoths likely eaten by residents of Channel Islands, California.

20,000 years ago: Ice sheets lock up Earth's water, lowering sea level, allowing coastal settlement in areas now underwater.

17,000 years ago: Ice sheets retreat, sea levels rise; barbed fishing darts are used on Japan's coast, similar to those later found off California, Peru, and Chile.

15,000 years ago: Channel Islanders become Chumash and Tongva nations, who develop sophisticated maritime trading systems.

6,000 years ago: Australian Aboriginals establish fishing havens and sacred sites along the north coast, some on reefs and islands 80 kilometers offshore.

1,500 years ago: Polynesian settlers arrive at Rapa Nui (Easter Island).

A.D. 850: New Zealand Maoris develop sustainable fishing practices.

1540: Merchant ship carrying copper plate for Dutch coinage sinks off the Netherlands; later found in 2016, the oldest shipwreck in Dutch waters.

1862: Civil War battleship U.S.S. *Monitor* sinks in treacherous coastal waters off Cape Hatteras, North Carolina.

1975: IUCN plans worldwide marine protected areas (MPAs). First U.S. National Marine Sanctuary protects U.S.S. *Monitor.* Australia's Great Barrier Reef Marine Park established.

2016: Twenty-four countries and European Union designate world's largest MPA, Antarctica's Ross Sea. UN aims for 30 percent of the ocean fully protected by 2030.

2020: British island territory Tristan da Cunha designates 90 percent of its EEZ for full protection.

CLIMATE— IT'S ABOUT THE OCEAN

Chapter Nine

"CLIMATE CHANGE KNOWS NO BORDERS."

—ANGELA MERKEL, CHANCELLOR OF GERMANY

As terrestrial beings surrounded by air, we might be forgiven for focusing on the atmosphere as the primary driver of Earth's climate—a hundred years ago. But today, the reality is clear: The ocean is where the action is. While the nature of the atmosphere is fundamental to shaping climate and the habitability of Earth, all things considered, it is most important to follow the water, and 97 percent of Earth's available water—that is, not bound in rocks deep within the Earth—is in the ocean. Energy from the sun reaching the Earth is mostly absorbed by the ocean, where the capacity to soak up heat is a thousand times that of the atmosphere.

Eighty-six percent of global evaporation and 78 percent of precipitation occur over the ocean, according to Raymond Schmitt, a senior scientist at Woods Hole Oceanographic Institution. He notes, "Storing 23 times the water on land and a million times the water in the atmosphere, the ocean's air-sea fluxes are many times larger than the terrestrial equivalents." Movement of cold and warm water masses by ocean currents shapes the climate over land and sea, and the flow of water evaporating from the sea to the atmosphere and back again is largely governed by the ocean's temperature and salinity. Warmer temperature affects the water cycle that underpins climate and weather, impacting everything from storms to floods and droughts and the overall distribution of life on Earth.

Over billions of years, life in the ocean converted an atmosphere on Earth consisting mostly of carbon dioxide (CO_2) into what we breathe today, a remarkably stable mix of oxygen, nitrogen, and just enough carbon dioxide—with water, sunlight, and chlorophyll—to power photosynthesis and also to provide a sufficiently robust, heat-retaining blanket that makes Earth hospitable to diverse forms of life. Analysis of fossils, ice cores, deep-sea sediments, and other measures shows the nature of past climates.

Evidence gained in recent decades confirms that the current rise in global temperature is linked to increased levels of carbon dioxide and methane in the atmosphere, caused by mining, by raising industrial livestock, and mostly by burning coal, oil, and gas. Deforestation, land use, and overfishing destroy natural carbon capture processes, exacerbating the problem. We can look to the ocean as the foundation of Earth's climate, and we can look in the mirror to find the cause of current climate disruptions.

RIGHT: Alive, a single tablespoon of seawater hosts millions of microscopic bacteria and other plankton. PAGES 358-359: Adrift in winds and currents, an iceberg carries chinstrap penguin passengers in the Southern Ocean.

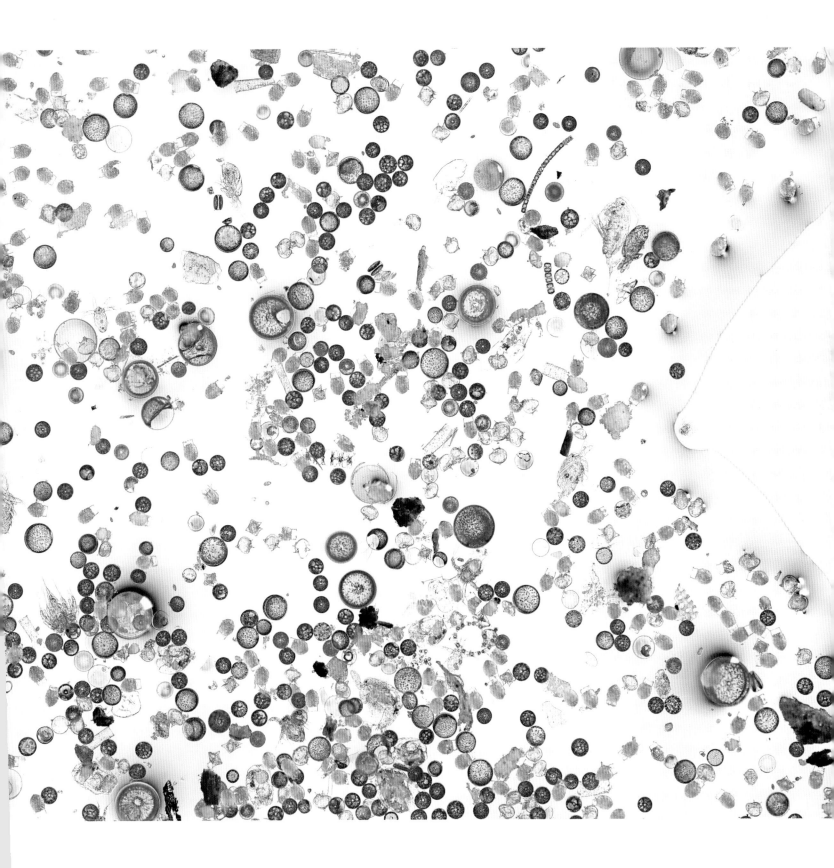

The Ocean Impacts Climate

Like master detectives linking clues that span centuries, climate scientists, oceanographers, geologists, biologists, historians, and anthropologists gradually pieced together the connections between the cyclical behavior of an ocean current along the coast of Peru and its dramatic consequences for human affairs globally. The Inca people who lived along the coast and in the high mountains of Peru for thousands of years were familiar with the current's two- to seven-year cycles, and conquistador Francisco Pizarro noted unusual rain in the Peruvian desert in 1525. But it took scientists of many disciplines and 20th-century technologies to see how oceanic and atmospheric conditions combine to form the El Niño phenomenon and its counterpart, La Niña, together called the El Niño–Southern Oscillation (ENSO), and their global connections. In 1982 scientists documented the impact of a powerful, warm El Niño event as it generated a fury of storms in California with mudslides and floods, an anomalous warm winter across much of North America and Eurasia, and droughts in Indonesia and Australia, as well as rains in the Peruvian coastal desert. The back-and-forth phases of ENSO events are now known to be the most influential natural climate pattern on Earth, impacting weather around the globe. Unraveling the mysteries of the ENSO phenomena brought focus to what now seems obvious: The ocean and atmosphere are one integrated system, with perturbations anywhere having impacts planetwide.

From the land, measurements of temperature, humidity, and other factors have been collected since the 1600s, but the advent of satellites gathering data from space made possible unprecedented overviews of the dynamic nature of Earth's atmosphere and ocean surface. Even before the Russian satellite Sputnik I began orbiting Earth in 1957, plans were under way to put cameras into space to observe clouds in ways that would improve weather forecasting. By 1959 the first U.S. weather satellite, Vanguard 2, was launched, paving the way for TIROS-1 in 1960, regarded as the first system capable of remotely sensing Earth's weather. Since then, polar-orbiting satellites have been launched that globally measure ocean surface temperature, atmospheric temperature and humidity, and ocean dynamics—data vital for weather analysis and forecasting and for climate research and prediction. Additionally, geostationary satellites—proposed by scientist and science fiction writer Arthur C. Clarke 20 years before they became a reality, in 1965—now hover over a fixed point on the Equator, moving at exactly the same rate that Earth rotates on its axis. Only a portion of Earth is visible from these satellites, but they allow a constant view of moving weather patterns, such as hurricanes.

The Nimbus program, from 1964 to 1978, marked the first time that instruments positioned in space could probe vertically through the Earth's atmo-

VISIONARIES

Roger Revelle
Pioneer of Climate Awareness

....

"Grandfather of the greenhouse effect," Roger Revelle used his career in oceanography and public policy to define how rising levels of carbon dioxide affect the atmosphere, and to educate the public. In the 1950s he joined the fledgling Scripps Institution of Oceanography and built it into an internationally renowned research complex. At the time, experts theorized that the increase of carbon dioxide from burning fossil fuels could not be harmful because the ocean would absorb it. Revelle thought otherwise. With radiocarbon dating expert Hans Suess, he determined that seawater can absorb only so much carbon dioxide. Too much makes the water acidic, or it remains in the atmosphere, warming surface waters, melting glaciers, and altering ocean circulation. To find solutions for carbon's impending effects, Revelle founded the International Geophysical Year in 1957–58. His students included John Knauss, founder of the University of Rhode Island's Graduate School of Oceanography, and global warming expert and former U.S. vice president Al Gore.

....

WHY IT MATTERS

Reading El Niño's Past

In Pacific coastal nations including the United States, Australia, Chile, and Peru, climate scientists and archaeologists are using ancient marine core samples, remains of mollusks and fish, and specimens from farmlands to find historic links between climate change and more intense El Niño events—and their effects on humans. The coast of Peru may be yielding the greatest insight, says a 2020 report by the National Academy of Sciences. Researchers have determined that in the warmer years of the Holocene epoch, sea surface temperatures rose along Peru and Chile, and El Niño distur-

bances roiled the waters, covering them with a warm cap and diminishing cold-water upwelling that brought bounties of fish to the surface. Fishermen eked out a living. Between those years, cooler La Niña brought back the cold upwelling, and fishermen prospered. As the globe warms in the 21st century, El Niño events will intensify, with increased global effects. By understanding the consequences, we can plan for the future.

ABOVE: Fishers thrive during La Niña, when coastal upwelling promotes cool water and vital nutrients. LEFT: Hurricane Linda, coincident with the 1997–98 El Niño, batters Pacific Mexico.

sphere to measure temperature and moisture. "It was like putting a thermometer at all different levels of the atmosphere and all over the globe," said Chris Barnet, a senior adviser for atmospheric sounding for the NOAA/ NASA Joint Polar Satellite System. "The polar orbit would see everywhere, twice a day." Data are now gathered by hundreds of Earth-observing satellites operated by six countries.

In the United States, measurement at sea of factors influencing climate and weather effectively began in 1939, when U.S. Coast Guard vessels were used as weather ships, and in the 1940s, when buoys called NOMADs (Navy Oceanographic Meteorological Automatic Devices) were deployed to monitor sea conditions. Currently, thousands of buoys deployed by many countries take daily measurements of ocean conditions, complemented by data gathered from ships and satellites. Carrying the first real-time instruments for detecting the initial advance of El Niño, the TOGA-TAO (Tropical Ocean Global Atmosphere-Tropical Atmosphere Ocean) buoy array, initiated in 1985, became fully operational in 1994, comprising 70 moorings across the Pacific Ocean near the Equator. By measuring sea surface temperature over this wide area, the mooring instruments were effective in predicting the onset of a powerful 1998 El Niño event. The TPOS (Tropical Pacific Observing System) 2020 in 2014 began delivering advanced ocean and atmospheric data over a wider area, an effort supported by 12 countries.

In 1999, Scripps Institution of Oceanography researcher Dean Roemmich chaired a meeting of scientists to address the urgent need for consistent data from within the ocean depths to better understand and predict weather and climate. A number of approaches have emerged since, including the thousands of instrument-bearing floats named Argo that descend to 2,000 meters depth while measuring temperature and salinity and determining current velocity. They rise briefly to the sea surface to transmit information via satellites to data-processing centers on land, then return to the depths to continue their journey. It is possible for acquired data to be openly shared with the public, available in near-real time, and accessible through data centers in the United States and Europe.

An ingenious way to fill gaps in data gathered by buoys and floats has been under way at Stanford University since 1994. Oceanographer Barbara Block and her team have engaged thousands of deep-diving seals, sea lions, tuna, sharks, and sea turtles that bear sensors to acquire and transmit via satellites refined physiological and oceanographic data taken as the animals travel horizontally and vertically over hundreds of kilometers of open ocean.

It took more than 20 years, but the idea of having an integrated ocean observing system became recognized as a U.S. national program in 2011. Currently, the Intergovernmental Oceanographic Commission (IOC), a coalition of 149 member states, oversees a growing network of ocean data collection and distribution called the Global Ocean Observing System (GOOS). Authors of a lead article in *Nature* in June 2020 say that it is time to "open up, share, and network information so that marine stewardship can mitigate climate change, overfishing, and pollution." Linking ocean and atmospheric data globally is yielding an unprecedented ability to analyze patterns and predict both short-term weather events and long-term climate trends. Building on evidence from decades of measurements, 195 nations gathered in Paris in 2015 for the United Nations Conference of the Parties (COP 21). There they signed a climate agreement that, for the first time, noted the vital role of the ocean.

VISIONARIES

James Hansen
Sounding the Early Warning

....

I t was a 38.3°C day that felt like 39°C when NASA climate expert James Hansen took his seat before a U.S. congressional committee. The year was 1988. "Global warming is here," he announced. He went on to explain that a buildup of greenhouse gases from burning fossil fuels and other carbon products pouring into the atmosphere was gradually smothering the Earth. And Earth was on its way to mirroring the nearby planet, Venus. For his Ph.D. dissertation in astronomy, and later as a Venus expert at NASA, Hansen had determined that Venus's atmosphere is mostly carbon dioxide, much like that of Earth's early atmosphere. Gradually, a greenhouse effect—gases in Venus's atmosphere trapping the nearby sun's heat—led to the planet's hot, barren state. For his testimony, he calculated that by 2017 the five-year mean temperature would rise 1.03°C above the average from 1950 to 1980. Actual warming turned out to be 0.95°C. Hansen's calculations, said Clara Deser, climate analysis chief at the National Center for Atmospheric Research, were "astounding" in their accuracy. "I wish I hadn't been right," said Hansen.

....

OPPOSITE ABOVE: Cyclone Aila leveled dikes in 2009, flooding Bangladesh.
OPPOSITE BELOW: Stressed by warming waters, Great Barrier Reef coral expels its life-giving algae.

Climate Impacts Ocean

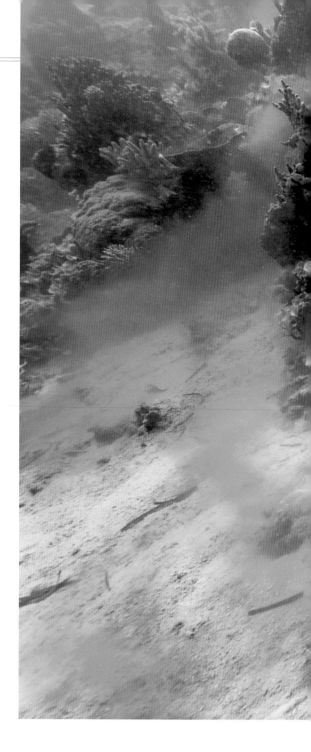

n the 1950s, it was thought that no matter how much carbon dioxide was lofted into the atmosphere by burning fossil fuels, it would not stay airborne long but would be taken up by the ocean and ultimately buried in the deep sea. Research by Roger Revelle, an oceanographer at Scripps Institution of Oceanography, and Hans Suess, a nuclear physicist and physical chemist, thought otherwise, based on their knowledge of ocean chemistry. In a joint report published in 1958, they demonstrated that much—but not all—of the carbon dioxide released by humans would be taken into the sea, and that the amount that stayed in the atmosphere "may become significant during future decades." They concluded that "Human beings are now carrying out a large-scale geophysical experiment of a kind that could not have happened in the past nor be reproduced in the future."

Charles Keeling, a geochemist who had developed an accurate method of measuring carbon dioxide in the atmosphere, was recruited to Scripps to undertake extensive measurements of the gas over the land and sea, including at a station on the Mauna Loa Observatory in Hawaii. Since his first recordings there in 1958, the station has continued reporting regular measurements, steadily gathering evidence. The results, when graphically portrayed, show a sharp upswing in carbon dioxide—the "Keeling Curve"—which tracks closely with the rise in global temperature. In 1988, NASA climate scientist James Hansen told a U.S. congressional hearing that he could declare "with 99 percent confidence" that a recent rise in global temperature was a result of human activity.

In addition to carbon dioxide, two other gases are notable for their heat-trapping nature: nitrous oxide (N_2O), famously known as "laughing gas," and methane (CH_4), popularly referred to as "natural gas," widely used as an energy source, and also known as a by-product of digestion by creatures from microbes to cows, sheep, and humans. Carbon dioxide persists in the atmosphere for hundreds of years; N_2O for 114 years; methane for about 12 years. But their impact varies. Methane is estimated to be 80 times more effective in trapping heat than CO_2. And, according to Caitlin Frame, a researcher at the Woods Hole Oceanographic Institution, "One molecule of N_2O has the same greenhouse warming power of 300 molecules of CO_2,"—but "air has about 1,000 times less N_2O than carbon dioxide."

About 75 percent of the increase in N_2O is attributable to the use of massive amounts of nitrogen-rich fertilizers by industrialized agriculture and wastes from intensive livestock cultivation. Runoff from feedlots is carried by groundwater and rivers to the ocean. There, phytoplankton and bacteria, like crops, respond with rapid growth. Decay follows, with emissions of CO_2, N_2O, and CH_4. Overfertilized lawns and manicured recreation areas add to the increased emissions, and so do various forms of transportation, from ships to automobiles.

Most of Earth's methane is largely the result of biological activity, in marshes, in Arctic soil, and in marine sediments several hundred meters thick,

TOP: Carpeting a once healthy reef, certain species of algae take over, favored by increased nutrients and decreased numbers of fish. ABOVE: Vital to the food web, phytoplankton feed krill, which sustain life as large as blue whales.

NOAA's CarbonTracker

Greenhouse gases—gases that trap heat in the atmosphere—include carbon dioxide from burning fossil fuels like coal; methane, from production of natural gas, coal, and oil; and nitrous oxide, emitted during agricultural or industrial practices such as wastewater treatment. In 1972 the U.S. National Oceanic and Atmospheric Administration (NOAA) anticipated that the gradual increase in greenhouse gases would prompt changes to the climate. The Global Monitoring Laboratory, an arm of NOAA's Earth System Research Laboratories, was created. Since then, the lab has developed increasingly sophisticated systems for long-term observation,

Power plant pollutants

data collection, and record building to understand how atmospheric gases affect climate change, extreme weather, ozone depletion, and more. Throughout the 21st century, the lab's measurement and modeling system, Carbon-Tracker, with its recent version CarbonTracker 2019, has captured intricate information on fluxes in CO_2 at the ocean surface and in the atmosphere. Updated yearly, the data give a full record of human and natural influences on rising carbon dioxide levels—the first step toward strategizing how to control CO_2 overload and return to a balanced carbon cycle, a healthy ocean, and a thriving planet.

directly below the seafloor along continental margins. There it exists as a gas hydrate—a crystalline solid lattice of water and methane that looks and acts much like ice. This ice, however, can be ignited, thus its name: "fire ice"—the masses of methane ice that crackle and fizz as fishermen dragging trawls in deep water bring them up in their nets, dissipating the methane into the air.

As Hobart King notes in a review of methane hydrate in *Geoscience News and Information*, "Methane hydrate deposits are believed to be a larger hydrocarbon resource than all of the world's oil, natural gas, and coal resources combined." There is understandable concern that warming of the ocean might trigger the release of methane now held stable by the pressure and low temperature of the deep sea. Geophysicist Carolyn Ruppel suggests that methane emitted from the deep sea may mostly oxidize before reaching the surface, but concerns remain about large expulsions that might be released as the ocean warms.

By 2019 the United Nations Conference of the Parties (COP25) climate conference in Madrid, Spain, was known as the Blue COP, and the ocean is front and center at COP26 in Glasgow, Scotland, in 2021, in recognition of the growing appreciation for the inextricable connections between the ocean and climate. Roger Revelle and Hans Suess would likely think this to be a logical outcome of looking at the evidence.

OPPOSITE: On the seafloor, methane worms colonize ice-like methane deposits, or hydrates, and dine on bacteria.

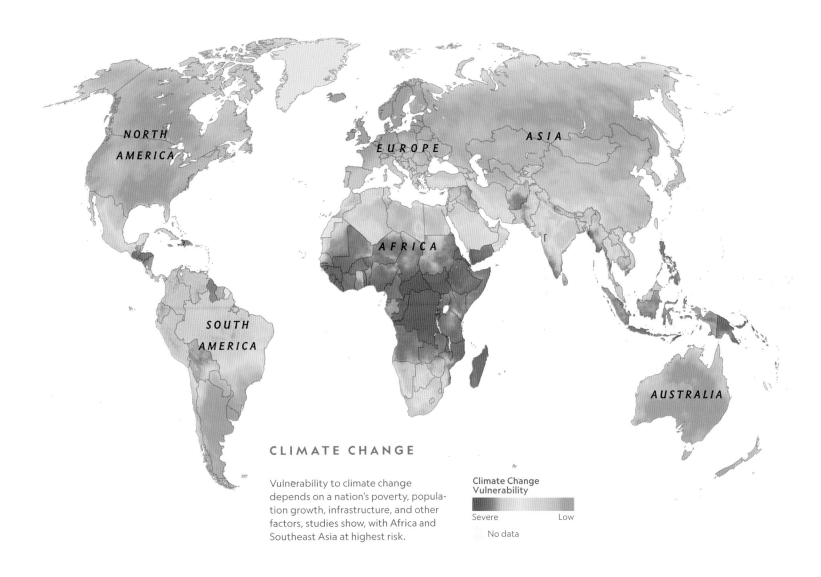

CLIMATE CHANGE

Vulnerability to climate change depends on a nation's poverty, population growth, infrastructure, and other factors, studies show, with Africa and Southeast Asia at highest risk.

Climate Change Vulnerability

Severe   Low

No data

Blue Carbon

Carbon, one of the most abundant elements in the universe, may sparkle as a polished diamond, give luster to a lump of coal, be invisible with oxygen as CO_2 and with hydrogen as methane (CH_4), or be inconspicuous but vital as an ingredient in all living things. Most of Earth's carbon is stored in rocks, soil, and sediments; the rest is in the atmosphere and ocean, and in every form of life. Like water, carbon is necessary for our existence. Captain Kirk and Spock in the science-fiction series Star Trek articulate a scientific fact when looking for "carbon-based units" as evidence of life elsewhere in the universe. But carbon combined with oxygen as CO_2 and released into the atmosphere by burning fossil fuels is altering the climate suitable for our existence.

To cope with climate change, reducing carbon emissions from human activity is a top priority, but "nature-based solutions" are gaining traction as a means of mitigating and reversing the buildup of atmospheric CO_2, by protecting intact and restoring damaged natural systems that underpin a favorable climate. The UN-REDD Programme (United Nations Programme on Reducing Emissions from Deforestation and Forest Degradation) gives incentives to countries to preserve their forests as a way of pulling massive quantities of carbon dioxide out of the atmosphere. Neglected until recently are the forests of phytoplankton, benthic algae, and coastal vegetation that are doing the heavy lift of capturing carbon dioxide from the atmosphere and passing it along as nutrition for animals large and small, eventually cycling carbon into the depths of the sea for long-duration sequestration.

"Blue carbon" is a term coined in 2009 to recognize and account for life in the ocean with respect to climate change. As marine policy expert Steven Lutz explains in a 2018 report by the environmental communications center GRID-Arendal, "Oceanic blue carbon includes carbon stored through the actions of marine life from krill to whales." David Mouillot and Gaël Mariani co-authored a 2020 analysis on how leaving more fish in the sea reduces the amount of carbon dioxide released into the atmosphere. A 2020 UNESCO report highlights mangroves, seagrasses, and marshes as systems that effectively capture and store carbon.

To define the economic benefits of protecting the ocean's living blue carbon, a group of economists looked at the value of whales as an example. International Monetary Fund economist Ralph Chami is the lead author of a 2020 study that concludes: "The carbon capture potential of whales is truly startling . . . Our conservative estimates put the value of the average great whale . . . at more than US$2 million, and easily over US$1 trillion for the current stock of great whales." If living whales have a monetary value based on their capacity to keep carbon out of the atmosphere, what about krill? And tuna, sharks, and swarms of carbon-based squid, cod, and capelin? Concern for whales and wetlands may stir awareness that the living ocean, all of it, is inextricably connected to Earth's climate and to the existence of life, humans very much included.

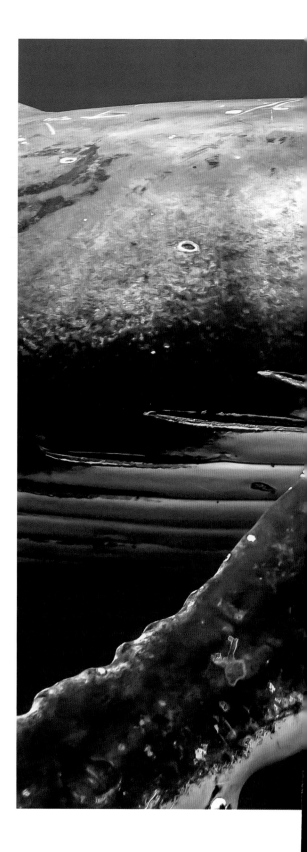

Packed with living carbon, a humpback whale and other marine life when extracted from the sea add to CO_2 overload.

Climate Change: Past

Climate change in the past is thought to have occurred at a gradual pace, sometimes much colder and sometimes much warmer than at present, marked by occasional abrupt shifts in climate and dramatic loss of life globally. In the last 500 million years, there have been five exceptionally catastrophic events resulting in the loss of 75 to 90 percent of species. The single biggest cause is thought to be volcanic eruptions that trigger major shifts in Earth's carbon cycle. Periodically, volcanoes eject massive amounts of heat-trapping gases—carbon dioxide and methane—into the atmosphere, driving global warming, ocean acidification, and a loss of dissolved oxygen in the sea.

A mass extinction occurred in a geological instant 440 million years ago, at the end of the Ordovician and dawn of the Silurian era—a time when glaciation cooled the planet and lowered sea level by hundreds of meters. The Devonian extinctions, 393 to 359 million years ago, marked the end of about 75 percent of species during a time of increased warming and low levels of oxygen in the ocean. The Permian-Triassic extinction 250 million years ago eliminated 96 percent of life on Earth. Called the "great dying," it was caused by the eruption of Siberian volcanoes that discharged an estimated 14.5 trillion tons of carbon into the atmosphere, according to a 2019 National Geographic report by Michael Greshko. The Triassic period followed, ending with a time when Earth warmed and carbon dioxide levels quadrupled.

Michael Marshall suggests in a 2019 article in *Nature* that "An extended bout of warm, wet weather 232 million years ago might have triggered the rise of the dinosaurs and completely altered the history of life on Earth," and ushered in the Jurassic period. One hundred million years ago, dinosaurs roamed what is now the realm of arctic foxes and snowy owls, amphibians slithered over Antarctica's now frozen terrain, and sharks shared space with large, predacious ichthyosaurs, long-necked plesiosaurs, and giant sea turtles.

That era ended abruptly when an asteroid burst through Earth's atmosphere, traveling more than 70,000 kilometers per hour, landing at the western edge of what is today Mexico's Yucatán Peninsula. The impact gouged a crater nearly 200 kilometers across, generating massive tsunami waves and shooting tons of debris, dust, and sulfur into the atmosphere that blocked the sun and brought about rapid planetary cooling. Volcanic eruptions at the Deccan Flats in India, possibly triggered by the asteroid, may have compounded the effects of its impact.

The end of dinosaurs marked the beginning of prosperity and diversification of mammals, including our long-ago ancestors. Since then, ice ages have come and gone, the most recent encompassing a period 115,000 to 11,700 years ago. Humans are now responsible for altering Earth's carbon cycle, with predictable consequences. At a speech in 1990, oceanographer John Knauss observed, "Nobody knows, for sure, why the dinosaurs became extinct, but it is known, for sure, that the dinosaurs are not to blame."

VISIONARIES

Ameer Abdulla
Caring for the Red Sea

....

Marine ecologist and National Geographic grantee Ameer Abdulla champions the plight of places in the Red Sea he calls "forsaken seascapes." While focusing his fieldwork there for more than a decade, he has learned to work in places of extreme political unrest, including Saudi Arabia and Yemen, among others. "I have been shot at, arrested, and banned from the remote locations I've worked in." At the same time, he has found that the places of greatest political instability "contrast beautifully with the rich marine biodiversity and habitats." The parklands around the Red Sea are unprotected, and the fishing habits of locals are undocumented. Because the area's three interdependent habitats—mangroves, seagrasses, and coral reefs—are so vital, Abdulla and his team are determining plans for their sustainable management. The corals alone, he says, may be the world's most resilient in the face of climate change. It is worth knowing more about them, to help other coral populations around the planet.

....

Over 350 million years of natural cataclysms, horseshoe crabs have survived. Here, their endangered descendants spawn in Delaware Bay.

HOPE SPOT

Spitsbergen Island, Svalbard Archipelago

Clustered between mainland Norway and the North Pole, the Svalbard archipelago is a wonderland of dramatic glaciers and mountain fjords towering above waters rich with marine species. A whaling port in the 17th and 18th centuries, the largest island, Spitsbergen, is today the base for the Svalbard Hope Spot, named in 2018. At this juncture of the Arctic Ocean, Norwegian Sea, and Greenland Sea thrive seabirds such as auks and puffins and mammals from reindeer and polar bears to walruses, harp seals, and minke whales.

A beluga whale perseveres despite changing Arctic waters.

Climate Change Consequences

To comprehend the scale of what is happening currently, it is helpful to consider how climate has changed during the history of humans. According to a summary of climate reported by NASA in 2020, there have been seven cycles of glacial advance and retreat in the past 650,000 years, attributable mostly to small variations in Earth's tilt that have changed the amount of solar radiation the northern reaches of the planet receives. The modern climate era began about 11,700 years ago, coincident with the advance of human civilization.

People have occupied Australia for some 50,000 years. Through high-fidelity oral history, according to research by Australian linguist Nicholas Reid and geographer Patrick Nunn, they have passed on memories of life before, during, and post-glacial flooding from ice melting that inundated their shorelines. Melbourne, an active port and the capital city of the state of Victoria, was established on the edge of Port Phillip Bay in the early 1800s. The bay covers nearly 2,000 square kilometers, averaging about 10 meters deep—a body of water that was dry land within the story-memories of modern Australians, a place that once was home for kangaroos and opossums.

Similarly, among California's Chumash people, Pacific coast residents for at least 13,000 years, there are cultural memories of land connecting parts of the Channel Islands. It is likely that the first people to arrive in the region of today's Santa Barbara followed the shoreline from what is now Russia across what was then a land bridge for humans and wildlife connecting to Alaska. Along Florida's west coast, during times when sea level was 100 meters lower, there was another Florida-size landmass extending into the Gulf of Mexico. Evidence of human occupation is strewn along the shores of now drowned rivers and springs that continue to spill fresh water into the sea far offshore. The island country of Singapore is largely surrounded by shallow water that floods what was once a much larger landmass. Ten thousand years ago, England was physically connected by land to Europe.

There is no evidence that humans who existed prior to the last two centuries had much to do with driving a rise in sea level. But there is a clear correlation between the actions of those who have lived since 1800 and global warming, rapid sea-level rise, and a host of other radical changes to the natural systems that maintain Earth as a habitable planet for humans.

During the past 800,000 years—including the recent, relatively stable 11,700 years—the amount of atmospheric carbon dioxide has varied, but it has never attained the levels reached in recent decades. In a NOAA report by Rebecca Lindsey in 2020, the last time atmospheric CO_2 amounts were as high as they were in 2018—407.4 parts per million—"was more than 3 million years ago, when temperature was 2°–3°C . . . higher than during the pre-industrial era, and sea level was 15–25 meters . . . higher than today." With its capacity to absorb and hold CO_2 and heat, the ocean has served as a temporary buffer against what otherwise would be more rapid and extreme atmospheric warming. The benefits are obvious, but there are

WHY IT MATTERS

Squid Are on the Move

During his two decades of work off the coast of California with jumbo squid—also called Humboldt squid—Stanford biology professor Bill Gilly noticed, in 2002, that they had started crowding into Monterey Bay. Not until an El Niño event in 2010 warmed coastal waters did they move out. Soon he and his student Timothy Frawley discovered that not only had the squid come to the bay from warmer waters to the south but they were now heading to cooler waters as far north as Alaska. Those that stayed gradually shrank in size, from one to two meters long

to about 20 centimeters long. The reason for migrations in distance and size? Likely warming waters, although gradual temperature rise, rather than a sudden El Niño event, may be the culprit. Why do we care? Because, as ecologist Bruce Robison put it in 2019, "Communities that have lived together and interacted for thousands of years are being fragmented. As they leave their traditional ecological niche, others take their place—with consequences we simply can't predict."

ABOVE: Humboldt squid migrate from Chile to feed in cooler Pacific waters. OPPOSITE: Arctic researchers measure warming effects, including turbulence.

SEA-LEVEL RISE

Coastal cities endangered by sea-level rise, including New York, Miami, and Guangzhou, by 2100 will feel the effects of an average rise of one meter—distributed unevenly around the world by winds, currents, and melting ice sheets. Warming seas will rise most in the tropics, and flooding will heighten in river-delta cities like Kolkata, Dhaka, and Ho Chi Minh City.

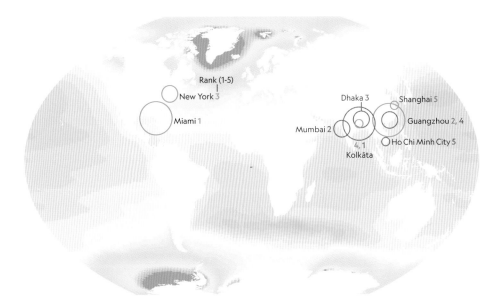

Rank (1-5)

New York 3

Miami 1

Dhaka 3

Shanghai 5

Mumbai 2

Guangzhou 2, 4

Ho Chi Minh City 5

4, 1

Kolkāta

Top five cities most at risk from rising seas (by 2070)

$3 trillion or more
$2-2.9 trillion
Up to $2 trillion
Exposed assets

14 million or more
10-13 million
Up to 10 million
Exposed population

Sea-level rise (by 2100)

+4 ft
2
0
Sea-level drop

costs, too. As the temperature rises, polar ice melts, water expands, and sea level rises.

Warmer water fuels more frequent and more powerful storms. Huracán is the Maya god of wind, storm, and fire; Huricán is the Carib Indian god of evil; and "hurricane" is a modern word for a rotating low-pressure weather system with sustained winds of 119 kilometers per hour or more. "Cyclone" and "typhoon" are other terms for the same phenomenon: a colossal, swirling, immensely powerful merger of air and water, generated over warm ocean waters. By whatever name, it is evident that storms are becoming more frequently intense and more powerful in concert with the increasing temperature of the planet.

Headlines in May 2020 announced that the Siberian town of Verkhoyansk, 10 kilometers north of the Arctic Circle, claimed a new record temperature high: 38°C. It was especially shocking because Verkhoyansk has held the world record for the lowest temperature of an inhabited place in the world, minus 67.7°C, since 1896. A single day does not clinch a trend, but temperatures measured day after day over months and years give a clear indication of climate. In 2019 Mark Fischetti reported in *Scientific American*, "In almost every year from A.D. 0 to 1950, portions of the Earth have been warmer or cooler than average. But since 1950 or so, almost all years have been overwhelmingly warmer, and the temperature rise has been far greater."

In the past, release of CO_2 from exceptional volcanic action accounted for dramatic global warming of the planet, leading to ocean acidification, ocean deoxygenation, and massive die-offs of life globally. Currently, the burning of fossil fuels by humans is simulating those processes, injecting billions of tons of greenhouse gases into the atmosphere every year.

The bleak face of rapid climate change is everywhere, from the shattered surfaces of Greenland's glaciers to the marble-white branches of bleached coral forests; from the increased number of more ferocious storms to the migrations of some marine species to new, cooler locations; accompanied, throughout, by the rapid die-off of other species already stressed by human predation, pollution, and habitat destruction. Tuna and cod may sense the shifts in ocean temperature, and polar bears may notice changes to the world they knew as cubs. But only 21st-century humans can know that what is happening to their world is reflected in changes occurring globally—and why.

A May 2019 report from the United Nations projected that a million species are at risk of extinction if current climate trends, habitat loss, pollution, and the human taking of wildlife continue unabated. Some predict that half of the world's species could disappear by the end of the century. In her book *The Sixth Extinction*, Elizabeth Kolbert warns, "There have been five great die-offs in history. This time, the cataclysm is us."

> ## "THE ECONOMY IS A WHOLLY OWNED SUBSIDIARY OF THE ENVIRONMENT."
> — HERMAN DALY, *ECONOMIST*

Ocean Circulation Is Slowing

....

Since the middle of the 20th century, experts have noticed that the ocean's global conveyor belt needs oiling. Slowing by some 15 percent over several decades, it is wreaking havoc on the ocean's health—and ours. Why? As rising carbon dioxide levels in the atmo-

sphere trigger warming temperatures, the Greenland ice sheet melts, and fresh water flows into the ocean. Normally, heavy, cold seawater south of Greenland would sink, driving the conveyor belt circulation deep into the ocean and southward. But lighter, fresh meltwater cannot sink. This causes the conveyor system to slow, affecting ocean currents and climate worldwide. For instance, this weakening fuels superstorms like the East Coast's Hurricane Irene in 2011 and Sandy in 2012, which devastated coastal North Atlantic states. Scientists predict that if circulation continues to weaken at the current pace, it will alter temperatures and weather patterns in all countries bordering the Atlantic— meaning more destructive storms and sea-level rise along the U.S. coast, melting icebergs that jam North Atlantic shipping lanes, colder winters and hotter summers in Europe, and changing rainfall patterns around the Equator.

....

ABOVE: In 2012 Hurricane Sandy ravaged the New Jersey coast. OPPOSITE: Hurricane Irma's 119 km/h gusts pummeled Florida in 2017.

DEEPER DIVE

Wildlife to the Rescue

Wearing a monitor reminiscent of a child's party hat, one elephant seal is helping us understand climate change. Tagged by NASA Jet Propulsion Laboratory (JPL) researcher Lia Siegelman in 2019, the marine mammal swam nearly 5,000 kilometers south across the Pacific to the Southern Ocean's turbulent Antarctic Circumpolar Current. There it made 80 dives per day, as deep as 1,000 meters, yielding data showing how heat moves vertically between ocean layers in the current's eddy-rich waters and warms the surface so that more heat from the sun can be absorbed. Such information brings scientists closer to understanding the ocean's capacity to absorb heat

NASA/JPL-tagged elephant seal

and thus offset global warming. Animals also help us determine their fate in a warming ocean. The Census of Marine Life's decade-long Tagging of Pacific Predators (TOPP) gathered data from 4,300 electronic tags on 23 predator species. Joined with satellite measurements of ocean temperature and surface productivity, the data gave new insights into migration patterns and habitat areas, and how both might change as the ocean warms. Some species might lose 35 percent of their range—or move more than 965 kilometers away. Such disruption would be a "heavy blow" to populations already stressed by human impacts, says NOAA ecologist Elliott Hazen.

Ocean Climate Action

NASA scientist James Hansen warned in 2020: "If humanity wishes to preserve a planet similar to that on which civilization developed and to which life on Earth is adapted, paleoclimate evidence and ongoing climate change suggest that CO_2 will need to be reduced … to at most 350 ppm [parts per million]." Severe consequences of global warming could be avoided if global warming could be held to an increase of 1.5°C relative to preindustrial levels, according to a 2018 report published by the UN Intergovernmental Panel on Climate Change (IPCC). That would require a net human-caused carbon dioxide emissions decline of 45 percent by 2030 compared with 2010 levels, and net zero by mid-century. This translates to less than 10 years to do what it takes to hold the planet's temperature below a point where recovery may not be possible.

"This is the first time in history that we have known enough to see the magnitude of the problems we face as a species, and this is the last time we will be able to do something about it," said Sebastián Piñera, president of Chile, in January 2019 as he was about to depart for a visit to Chile's Antarctic research station. In Antarctica, ice shelves are losing mass faster than they refreeze—four trillion metric tons since the mid-1990s, according to satellite data from 1994 to 2018. In the Arctic, summer ice has shrunk by more than half since 1980. Since polar regions have magnified importance in maintaining climate within a range favorable to humans, the rapid melting is commanding the rapt attention of oceanographers and climate scientists alike.

"The Ocean as a Solution to Climate Change" is part of the High Level Panel for a Sustainable Ocean Economy, an ambitious multinational program convened in 2018. Fourteen leaders, presidents or prime ministers, backed by more than 250 advisers and experts, set out to develop a road map aimed at reversing the decline of the ocean with special reference to the ocean's role in climate change. Among the proposed solutions is restoration and protection of kelp forests, mangroves, and seagrasses to absorb carbon and offset global emissions, coupled with support for global protection of 30 percent of Earth's land and sea by 2030. Following up, in December 2020, the 14 nations—Australia, Canada, Chile, Ghana, Indonesia, Japan, Kenya, Mexico, Namibia, Norway, Portugal, Fiji, Jamaica, and Palau—committed to protect 30 percent of their respective exclusive economic zones by 2025 and to manage 100 percent sustainably.

There is progress on other fronts, including widespread adoption of renewable energy sources that have helped "flatten the curve" of carbon dioxide emissions, according to a February 2020 report by the International Energy Agency (IEA). Reduced energy use owing to the COVID-19 pandemic temporarily lowered carbon emissions. But sustained reduction is needed to attain stability, and the ocean has a fundamental role in doing so. John Kerry, appointed climate envoy for the United States, has astutely observed, "You cannot protect the ocean without solving climate change, and you cannot solve climate change without protecting the ocean."

VISIONARIES

James Balog
Man of Ice

An environmental photographer and mountaineer with degrees in geography and geomorphology, James Balog has spent four decades merging art and science through innovative imaging, giving a "visual voice" to Earth's changing ecosystems. In 2007 Balog founded the Extreme Ice Survey, an unprecedented study documenting the retreat of glaciers due to rising temperatures—and the subsequent rise of sea levels. "I am fascinated by ice: the beauty of it, the mutability of it, the malleability of it," he has said. "It is also the canary in the global coal mine," allowing us to see climate change in action. A substantial majority of Earth's glaciers are shrinking because "Earth has a fever," he explains. His images show the fever's impact, including Greenland's Jakobshavn Glacier calving a chunk of ice 700 meters deep and five kilometers wide within minutes. His images in books, films, and more promote awareness. In 2012 he founded the Earth Vision Institute, combining art and science to teach about the impact of environmental change.

Its image etched in stone, a shrimp that swam in prehistoric waters resembles today's descendants.

DEEPER DIVE

As the Ice Melts

The polar ice caps are melting due to warming from greenhouse gas emissions. Between 2002 and 2019, NASA's Gravity Recovery and Climate Experiment (GRACE) satellites charted an Antarctic land-based ice loss of 147 gigatons a year, with Greenland losing 280 gigatons a year. Since 1900, global sea level has risen between 13 and 20 centimeters, and the pace is picking up. By 2100, researchers project, Earth's surface will warm by three to four degrees Celsius above preindustrial levels, enhancing that melting. It's a vicious cycle. The melt from Greenland may slow the circulation of the global conveyor belt, prompting more melting; and melt from Antarc-

Eroding Arctic iceberg

tica's land traps incoming warm water beneath its ice sheets, causing melting from below. Simulations published in *Nature* in 2019 show that if ice melt continues at this rate, global temperatures will continue warming and sea levels will rise by as much as 25 centimeters by 2100. What if we stopped emissions tomorrow? Could we go back to solid ice, steady global circulation, and less intense storms? No, say experts. Warming will continue because atmospheric levels of CO_2 are already so great. But reducing emissions and protecting natural systems can slow, level, and possibly reverse the trend, while we learn to adapt to life in a changed world.

GREAT EXPLORATIONS

Sea From Space

When Russian cosmonaut Yuri Gagarin orbited Earth in 1961, he became the first of a privileged few individuals who would view the extent and power of Earth's ocean from space. Nonhuman eyes in space can bring even more revealing insights. Since the launch of the Soviet Union's Sputnik in 1957, and the United States' Explorer 1 a year later, scientists have developed increasingly accurate satellite sensors to study Earth's land, atmosphere, weather—and the ocean surface.

Oceanography in the 21st century is an integrated science that combines data from many satellite sensors with that of ocean-based instruments. These data are analyzed by powerful supercomputers, with a global scientific community guiding the process. Today, space agencies in the United

A cloud-bathed ocean is captured by European Space Agency astronaut Alexander Gerst in 2014.

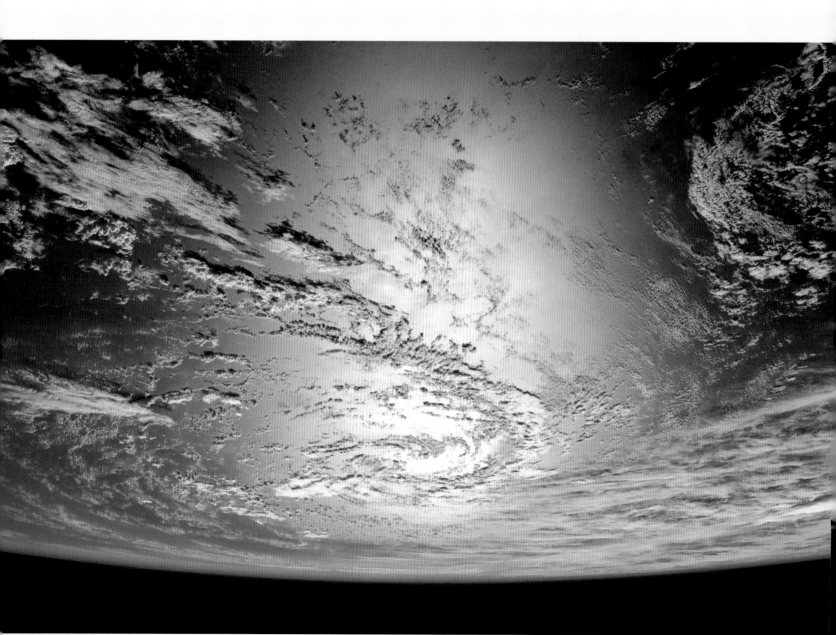

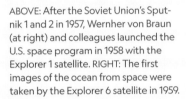

ABOVE: After the Soviet Union's Sputnik 1 and 2 in 1957, Wernher von Braun (at right) and colleagues launched the U.S. space program in 1958 with the Explorer 1 satellite. RIGHT: The first images of the ocean from space were taken by the Explorer 6 satellite in 1959.

States, Japan, China, and the 22 nations of the European Space Agency (ESA) share satellite technology, data, and data-based solutions to care for the ocean and support its role in cooling our warming planet.

Not only do satellites help capture the pulse of the planet, but they alert us when the pulse spells danger. Early warnings of severe superstorm events help communities prepare and respond more effectively. And scientists use data to make seasonal forecasts and predictions of climate cycles like El Niño, helping agricultural societies adapt and avoid crop failures that can lead to economic disaster.

As climate change speeds up, satellite technology is keeping pace. For example, in 2018 NASA scientist Ved Chirayath developed FluidCam. This camera uses a technique called fluid lensing; its supercomputer software allows accurate imaging of shallow underwater landforms and coral reefs down to 15 meters depth, usually distorted by light refracting through water. In addition, Chirayath has designed software that helps satellites accurately identify coral reefs, allowing for reef mapping with 95 percent greater accuracy than ever before.

Where do we go from here? Seeking out new ocean worlds beyond Earth will open doors to greater understanding of our own ocean's past, its effect on life, and what may lie ahead. In 2020, NASA scientist Lynnae Quick used data from the 1997–2017 Cassini-Huygens Saturn probe and a mathematical model to hypothesize that dozens of exoplanets may hold water, have similar structure and geological activity to Earth's, and be potentially livable.

Jupiter's ice-covered moon Europa may be, too. The Europa Clipper mission—the first dedicated study of a water world beyond ours—is due to launch in the 2020s. The spacecraft will conduct a detailed survey of Europa by making some 45 close flybys over the moon. With each pass, scientists on Earth will shift its flight path so that the craft eventually scans nearly the entire surface. The resulting data may confirm evidence that Europa is covered with a salty sea holding more liquid water than Earth's ocean.

TIME LINE

The View From Above

1957: Soviet Union launches Sputnik.
1958–59: U.S. launches Explorers 1–7, first satellites to carry weather instruments.
1960: U.S. launches Television Infrared Observation Satellites (TIROS), world's first polar-orbiting weather satellites.
1975: NASA-NOAA launch first Geostationary Operational Environmental Satellite (GOES) to study Earth's environment and weather.
1978: NASA launches Seasat, first oceanographic satellite; collects in 105 days as much data as ships had collected in previous 100 years. TIROS-N produces first maps of sea surface temperature, chlorophyll, photosynthetic action.
1990: Hubble Space Telescope is deployed; images other planets, stars, galaxies.
2001–present: NOAA's Jason satellites measure temperature, sea-level rise, tides, current; forecast tropical storms, track El Niño; collaborate with European agencies; create first global map of sea surface height.
2011–2038: NOAA-NASA Joint Polar Satellite System (JPSS) provides global coverage twice daily to monitor and predict atmospheric, land, ocean, and storm conditions.
2016–2036: GOES R-U series of NOAA-NASA satellites monitor storms, fog, flash floods, volcanic eruptions, and wildfires.
2020: NASA's Ocean Color Instrument (OCI) uses electromagnetic spectrum to image ocean ecology.
2022: NASA to launch PACE (Plankton, Aerosol, Cloud, ocean Ecosystem) spacecraft with OCI to observe global ocean biology, aerosols, and clouds.
2020s: James Webb Space Telescope to replace Hubble; will add to information about the building blocks of life—including water—elsewhere in the universe.
Mid- to late 2020s: Europa Clipper Mission scheduled to study Jupiter's moon Europa, an ocean world.

THE FUTURE OCEAN

Chapter Ten

"WE HAVE A LONG WAY TO GO TO MAKE PEACE WITH THIS PLANET,
AND WITH EACH OTHER . . . IT IS NOT TOO LATE FOR US TO COME AROUND,
WITHOUT LOSING THE QUALITY OF LIFE ALREADY GAINED."

— EDWARD O. WILSON, *THE CREATION*

Predicting the future of the ocean a million years from now is, in some ways, easier than imagining what it will be like in 2030 or 2050. Geological processes established over the ages will continue, with ocean basins spreading or shrinking and continents gliding across the globe, driven by tectonic forces deep within the Earth. Sea life will continue to evolve and be affected, sometimes catastrophically, by shifts in the ocean and the atmosphere. Climate will change in ages-old cycles of cooling and warming; ice ages will come and go; sea level will rise and fall. But given the swift and unprecedented impact of human actions on Earth's temperature, chemistry, and the diversity and abundance of life, the state of the ocean in the coming decades will strongly reflect what we do—or fail to do—now. The laws of nature are predictable, but human behavior is not. As never before in history, there is a species—21st-century humans—with the capacity to consciously, and maybe even conscientiously, shape the future of its own existence.

Geologists refer to the epoch of time starting 11,700 years ago as the Holocene. However, the dramatic impact wrought by humankind on the planet inspired biologist Eugene Stoermer and chemist Paul Crutzen to champion the designation of a new epoch: the Anthropocene. The term is derived from the Greek words *anthro,* for "man," and *cene,* for "new." Journalist Mark Lynas observes: "We are already into a new geological era, the Anthropocene, where human interference is the dominant factor in nearly every planetary system."

To officially qualify as a new epoch, sanctioned by the International Union of Geological Sciences (IUGS), changes must be reflected in rock strata. Some consider 1800 to be the point when human actions began to have an impact of geological magnitude. Others propose 1945, marking the first atomic weapons testing and bombing during World War II. Radioactive particles left an enduring signal in soil samples globally, together with the widespread appearance of plastics and other synthetic materials that did not previously exist. After much deliberation, when the IUGS Anthropocene Working Group met in 2016, they agreed to propose the middle of the 20th century—1950—as the point at which human activity increased sharply and distinctly enough, noted as the Great Acceleration, to warrant geological definition. It is unsettling but true: Our signature on the Earth is written in stone.

RIGHT: Galápagos sharks, at home in warm tropical waters globally, gather here in Bassas da India atoll, near Mozambique. PAGES 384-385: Vibrant shallow reefs of Port Elizabeth, South Africa, attract shimmering schools of small fish.

Changing the Ocean Changes Everything

Strolling as he spoke at a conference in 1990, Joseph Allen—mission specialist on space shuttle flights STS-5 and STS-51A—described the fundamental training required of astronauts. "First, you learn everything you can about your life-support system," he said. "Then you do everything you can to take care of it." He paused and pointed to an image of Earth from space. "There it is," he continued. "Our life-support system. Learn everything you can about it and do everything you can to take care of it." So, aboard our spacecraft Earth, how are we doing in the 21st century? What have we learned? And what are we doing to take care of the systems that keep us alive as we hurtle through the magnificently beautiful but largely lethal universe?

The view astronauts have shared of our home planet from afar catalyzed a shift in the way many people see themselves relative to one another and to the natural world. In 1968 Apollo 8 command module pilot James Lovell observed from his perch high in the sky, "The vast loneliness up here on the moon is awe-inspiring, and it makes you realize just what you have back there on Earth. The Earth from here is a grand [oasis] to the big vastness of space." Yet, remarkable complacency continues to drive conversion and destruction of the natural systems that make Earth uniquely suited to our prosperity. We could not survive on Earth throughout most of its existence any more than astronauts could fly in a disassembled spacecraft. Arriving on Earth as it was two billion years ago would have been as dangerous for us as setting foot on Mars today with no spacesuit. Yet, even now, in a climate that supports life, we know that Earth's engine, the ocean, is sputtering—and we continue to behave as if that is a minor issue, not an urgent, life-threatening matter. Throughout human history, we have displaced, rearranged, lost, and deliberately annihilated many of the pieces of systems that keep us alive without understanding the consequences. "Conquering" and "marketing" nature have been, and continue to be, persistent themes, largely overwhelming the importance of keeping life-support functions intact by protecting natural processes and the diverse life-forms that power them. In effect, we are altering Earth's *biogeochemistry* in ways that may be suitable for microbes and mushrooms but not for humans.

Until the 1950s, the ocean remained much less disturbed than the land. In the several decades since, destructive changes comparable to thousands of years of human encroachment on the land have occurred in the sea. Environmental scientist Johan Rockström and photographer Mattias A. Klum suggest in their 2012 book *The Human Quest,* "We can no longer exclude the possibility that, in pursuit of wealth and well-being, we may inadvertently trigger catastrophic outcomes at the planetary level." They urge "a mind-shift for a transformation to global sustainability" and cite planetary boundaries: critical thresholds in Earth's natural systems that

VISIONARIES

Lynn Margulis
The Great Connector

Lynn Margulis's "great gift," said fellow biologist Steve Goodwin, was "making connections that others just couldn't make." A proponent of British biologist James Lovelock's Gaia theory—that all life, air, water, and land form a vast self-regulating system—Margulis focused on the origins of life. She proposed that eukaryotic cells—those with a nucleus—originated when certain single-celled organisms engulfed other cells and formed a symbiotic relationship. New organelles, organs, and even new species are now thought to arise by symbiogenesis, in which independent entities merge to form composites. Darwin emphasized competition as the driver of evolution, and Margulis demonstrated the complementary and likely more significant power of cooperation, interaction, and mutual dependence: networking at the cellular level. Controversial in the 1960s when Margulis introduced the concept of symbiogenesis, today it is broadly accepted.

OPPOSITE ABOVE: Vulnerable on the IUCN Red List, Atlantic walruses range Arctic waters. OPPOSITE BELOW: Nicaragua's Caribbean shelters threatened spotted eagle rays.

must be respected to maintain habitability. Thresholds already crossed include climate change, biodiversity loss and extinction, ocean acidification, and ozone depletion. Environmental science professor emeritus from the University of Arizona Guy McPherson put it succinctly: "If you really think the environment is less important than the economy, try holding your breath while you count your money."

In the distant past, before humans existed, thresholds that held up to a certain point and were then exceeded triggered many phases that were inhospitable for life as we know it. The same natural forces continue to shape the world. Earth's tilt, the sun's flare, and the arrival of asteroids are beyond our control, but armed with knowledge of how our own actions are now pushing the current safe boundaries, we can steer a pathway that recognizes and stays within the limits. With the knowledge that we now have, should we not be asking how our past and current actions have impacted the stability of the systems we have heretofore taken for granted? Should we not have an accounting of how humans are altering Earth's life-support systems? Since everything we care about hinges on a planet that works in our favor, should not maintaining critical functions top our list of highest priorities? And most of all, shouldn't we be motivated to act before, not after, the boundaries have been transgressed?

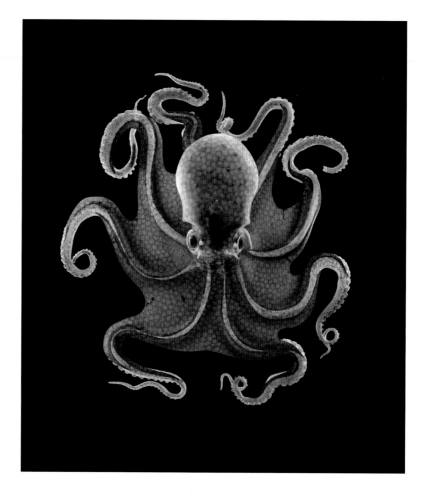

The increasing temperature and acidity of the ocean, decline of polar ice, sea-level rise, increased frequency of intense storms, mounting pollution, declining wildlife, devastating diseases, and other changes are signals that should put us on full alert about reaching tipping points that will be difficult or impossible to overcome. There is time, but not a lot, to respond to the knowledge that the ocean is in trouble, and therefore, so are we. Ocean policy and governance—including international treaties that attempt to control the uses and abuses of the ocean and its resources—are evolving, but not as quickly as the swift changes that threaten human health; global, regional, and local economies; security; and, ultimately, our survival.

The first administrator of the National Oceanic and Atmospheric Administration (NOAA), meteorologist Robert White, was mindful of the growing dangers to the oceans, the atmosphere, and the living systems that everyone uses but no one owns. He observed in 1994: "What is at stake in all of this is the fate of the global commons. We are all dependent on maintaining the habitability of the planet . . . This is the quintessential challenge for mankind in the next century."

And here we are.

ABOVE: Despite its 530-million-year-old ancestry, the common octopus is threatened by overfishing, acidic waters, and aquaculture. OPPOSITE: Small fish clustered in close formation are nonetheless vulnerable to hungry barracuda.

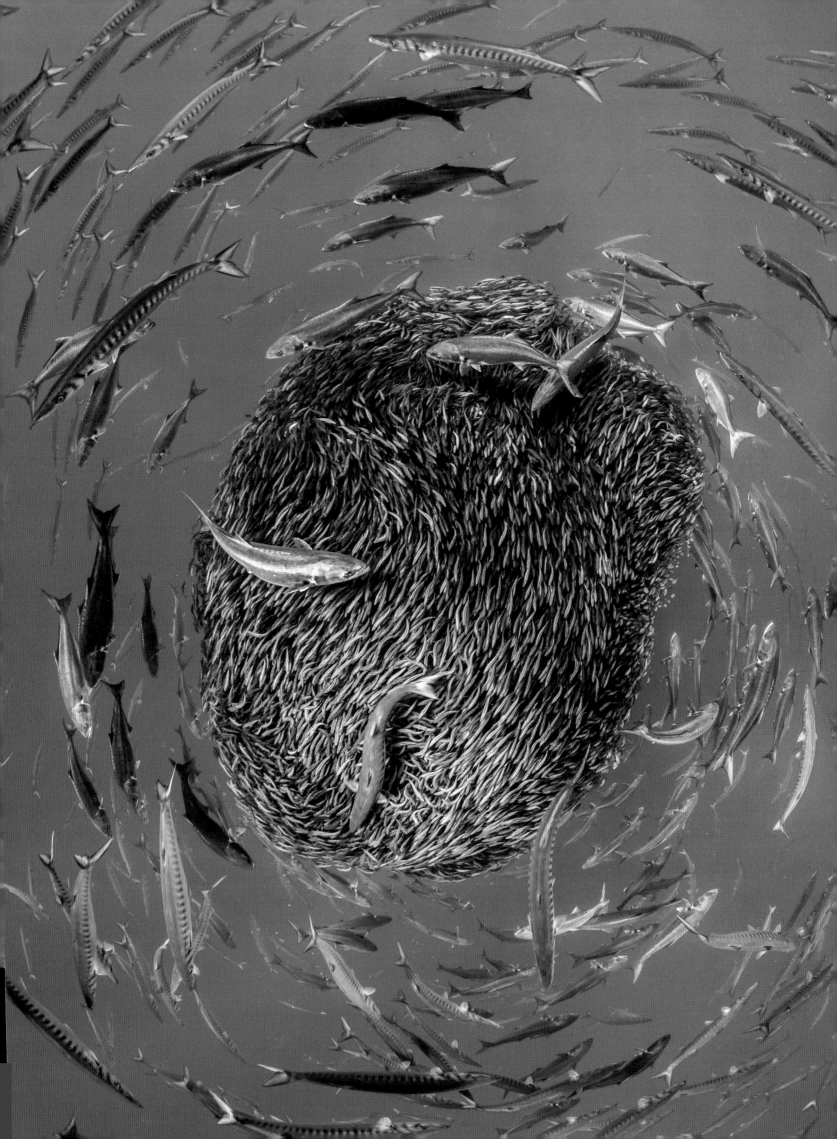

Governance: Who Determines the Ocean's Future?

Ferociously territorial, finger-length damselfish guarding their cultivated gardens of algae and broods of babies take on intruding groupers, snappers, and scuba divers hundreds of times their size and weight with repeated open-mouthed, head-on attacks. Oceanic triggerfish are notorious for vigorously defending nesting sites, sometimes extracting walnut-size pieces of flesh from fish—or humans—who venture over what the fish determine to be the no-go zone. Male sea lions make it clear that certain parts of their beachfront property are off-limits to other male sea lions—and any other large creature that ventures nearby. Yet vast areas are peacefully shared by large whales, tuna, cod, salmon, eels, sharks, and sea turtles that travel over hundreds, sometimes thousands, of kilometers in a year along traditional routes between feeding, breeding, and nursery areas. Whether in small patches on a reef or in large open spaces, sea creatures have had the ocean to themselves until a few thousand years ago, and, mostly, until about 500 years ago. The sea reflects the history of life flowing through millions of generations of incomprehensibly diverse life-forms, fully at home in their magnificent liquid space. Now *Homo sapiens,* terrestrial by nature, claim the ocean—all of it—as their own. No other species has more influence on what the future of the sea—and of the world—will be.

By the 1500s various nations, notably Portugal and Spain, were laying claim to trade routes that they defended against seafarers from other countries. Spain ambitiously sought to protect exclusive access to the Pacific Ocean by guarding the Strait of Magellan. Tensions increased as the ocean became increasingly important for transportation and trade globally. A Dutch jurist and philosopher, Hugo Grotius, reflecting on the problems imposed by increasing conflicts, articulated the concept of *Mare Liberum,* "Freedom of the Seas," in 1609. His thinking was that like air, the sea was not "susceptible to occupation," and that its use is "common to all men"—so limitless that it should be adapted for the use of all, whether for navigation or extracting wildlife. Jurisdiction over the sea developed largely as a European concept. Traditional sailing canoes in the Pacific connected, rather than divided, Polynesian cultures, and in Nordic countries Vikings did not think they needed permission from anyone to explore the far reaches of the North Atlantic. The Chinese trading fleets in the 1400s were free to sail into the Pacific and Indian Ocean as they pleased.

Conflicts mounted as more nations developed the capacity to access and use the ocean for transportation and trade and to capture fish, seabirds, seals, and other wild marine animals. Eventually, maritime states began to limit their protective stance to defensible near-shore waters—usually defined as three nautical miles—and adopt the concept of free passage on the High Seas beyond. By the 1930s, there was growing pressure to extend jurisdiction

In an ancient cycle, eels hatched in the Sargasso Sea mature in Maine's Pemaquid River, and decades later spawn at sea.

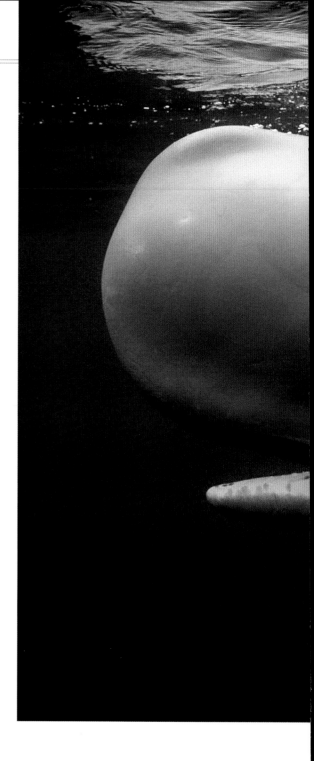

farther offshore. Actions during traumatic battles at sea during World War I brought into focus the need to formalize the concept. In a speech before Congress, U.S. President Woodrow Wilson proposed that there should be "absolute freedom of navigation upon the seas, outside territorial waters, alike in peace and in war, except as the seas may be closed in whole or in part by international action for the enforcement of international covenants." The League of Nations convened in 1930 to work out ocean governance, but no action was taken until 1945, when President Harry S. Truman declared U.S. jurisdiction over offshore oil and fishing rights to the edge of the continental shelf to protect the nation's commercial interests.

Post–World War II was a somewhat more peaceful time for people; but for life in the sea, industrial-scale fishing, already well under way, moved into high gear. As rock-and-roll music began to sweep the world of humans, most of the voices of the eight species of great whales had already been silenced, millions eliminated from their traditional ocean spaces. More than half of the population of sperm whales *(Physeter macrocephalus)*—animals with the largest brain in the universe, regional dialects, tight social bonds, traditions, habits, genetic diversity, and learned knowledge—had been converted to domestic animal feed, oil for industry, and fertilizer. Whale songs that reverberated across hundreds of kilometers of open ocean were replaced or masked by the deafening throb of engines and, occasionally, the overwhelming thunder of a nuclear explosion. During nuclear testing in the South Pacific Ocean from 1946 to 1958—justified as a cost of global security—the people of the Marshall Islands suffered the loss of their homes, culture, history, and identity. It was also the end of the world for the damsel-fish of the region and thousands of unique species occupying Pacific reefs that were vaporized.

Currently, the international policy and governance mechanism for the behavior of nations with respect to the ocean is primarily the United Nations Convention on the Law of the Sea (UNCLOS). It was drafted by many nations in the 1970s; opened for signature in 1982; signed by 119 countries that year;

················· **MISSION BLUE** ·················

United Nations: Leading Ocean Policy

From 2021 to 2030, the United Nations is planning to pull out the stops to address "defining issues of our time" impacted by the ocean: climate change, food insecurity, diseases and pandemics, diminishing biodiversity, economic inequality, conflicts, and strife. In 2015 the UN had already set a 2030 Agenda for Sustainable Development, with ocean conservation the 14th of 17 transformative goals. In 2017 UN Ambassador Peter Thomson of Fiji, named UN Special Envoy for the Ocean, began galvanizing new efforts. By 2020 these included a foundation supporting ocean science and policy, and nine communities of ocean action coordinated globally by high-

Snake sea cucumber, Bikini Atoll

level volunteers. Months before the COVID-19 pandemic postponed the 2020 UN Ocean Conference in Lisbon, leaders declared a "Blue Call to Action." Six "RISE UP" targets defined in 2020 provide a focused, streamlined list for tending to the ocean's health. They are: Restore ocean life; invest immediately in a net-zero carbon emissions future; speed the transition to a circular and sustainable economy; empower and support coastal people; unite for stronger global ocean governance; and protect at least 30 percent of the ocean by 2030. Despite the global pandemic, the focus on these goals has remained a priority for the future.

ratified by the requisite 60 nations needed to bring the treaty into force in 1993; and is now binding for the more than 160 nations that have ratified it. UNCLOS provides a comprehensive international legal framework governing activity on and in the world's ocean. It includes the rights of military mobility on the High Seas, through international straits, and in coastal waters; ensures the free movement of global commerce; clarifies High Seas freedoms for laying cables and pipelines; establishes an international framework for maritime law enforcement, marine environmental protection, and marine scientific research; and creates a mechanism for settling international disputes. A provision of the convention with profound significance allows coastal states to declare jurisdiction over living and nonliving marine resources in their exclusive economic zone (EEZ), up to 200 nautical miles off their shorelines. It also created the International Seabed Authority, to organize and control all mineral-related activities on the seabed beyond national EEZs.

In rising cacophony, shipping noise and seismic surveys drown out age-old voices of sperm whales and other marine life.

Sustainable Fisheries

It's an ancient theory: Take from the ocean only what you need to eat, and fish populations will naturally recover, even thrive. A 2007 study praised the traditional Tagbanua people of the Philippines for doing just that. The people also set aside coral reefs as protected areas and used hook-and-line methods rather than wide-sweeping nets to capture only enough fish to feed their communities. With minimal impact, they could "have their fish and eat them, too." They also protected the endangered Irrawaddy dolphin, which is often entangled in modern nets and traps. While laws in the past half-century aim to stop illegal catches, they are often not enforced. Although taken legally, most commercially caught ocean wildlife is used for currency and for products, not local domestic use, and high levels of extraction have made sustainability unrealistic. Campaigns such as the American chefs' "Take a Pass on Chilean Sea Bass" are helping consumers to recognize their role in the decline of ocean life.

What remains outside the jurisdiction of individual countries is the High Seas, about 64 percent of the ocean surface, nearly half of the planet's surface and much more than half of the biosphere. Over the years since UNCLOS's signing in 1982, international fishing interests have laid claim to life in the High Seas by declaring a network of large multinational regional fisheries management organizations (RFMOs) authorized under an UNCLOS Straddling Fish Stocks Agreement. While theoretically abiding by conservation guidelines, in reality, as award-winning journalist Ian Urbina observes in *The Outlaw Ocean*, "The vastness of the ocean makes it easy for poachers to thwart government quotas, enter forbidden areas, and pillage sanctuaries. As a result, pirate fishing boats are responsible for over 20 percent of the wild-caught seafood imported into the United States, and the percentage is likely higher abroad." Urbina has not only documented devastating accounts of pirate fishing, but also High Seas fishing as a conduit for trafficking humans, arms, drugs, and wildlife. Illegal, unregulated, and unreported catches by thousands of High Seas vessels cause serious problems for ocean governance, but legal fishing has challenges, too. Only five nations—China, Taiwan, Japan, South Korea, and Spain—take most of the tens of millions of tons extracted from the High Seas annually, and although the markets are global, a 2018 report in *Science Advances* states that as much as 54 percent of the offshore fishing would be unprofitable were it not for government subsidies.

Mindful of the critical importance of "half of the world" to planetary security, participants from 20 countries at the 2003 Defying Ocean's End conference in Los Cabos, Mexico, strongly recommended the United Nations take action to embrace the High Seas as a global public trust where no extractive activities would be allowed. That vision has not gained traction, but starting in 2014, the United Nations began to take under consideration a process that could enable the establishment of large protected areas in the High Seas with special reference to "marine biodiversity of areas beyond national jurisdiction," or BBNJ. If at least 30 percent of the High Seas is safeguarded, much of Earth's biodiversity might endure.

TOP LEFT: After years at sea, sockeye salmon return to their river birthplace to spawn and die. ABOVE: An Irrawaddy dolphin spy-hops to look around. OPPOSITE ABOVE: Sri Lankan fishers atop stilts search for fish. OPPOSITE BELOW: Fish waste feeds giant Japanese scallops in a Pacific Coast farm.

Saving the Ocean & Eating It, Too

An enduring problem concerning gauging the risk of extinction of animals in the sea is that we like to eat them. Many countries, including the United States, spend large sums to support fishery agencies that assist in finding, killing, and marketing ocean wildlife, while allocating relatively small sums for its protection. Billions of dollars annually subsidize industrial fleets extracting wildlife for commercial sale. While it is known that the abundance and diversity of life in the ocean is fundamental to the nature of planetary chemistry—affecting the cycles of water, oxygen, carbon, nutrients, minerals, and other vital biogeochemical processes—animals in the sea are generally treated as commodities. They are referred to as "stocks," measured by the ton, and taken using methods comparable to bulldozing forests to capture songbirds. Collateral damage—habitat destruction and unwanted animals killed and discarded—magnify the loss. There has been progress in fisheries reform, but the pace of ocean life decline is currently exceeding efforts to restore their numbers, let alone safeguard their role in maintaining Earth's most essential functions.

A 2019 United Nations review, the "Global Assessment Report on Biodiversity and Ecosystems Services," determined that at least a million species are in immediate threat of extinction, with consequences to the habitability of the planet. Terrestrial species are most obviously at risk, but since only a small percentage of the ocean has been explored, the number of species discovered, as well as those on the edge of survival, is certain to grow. Most marine species, including wide-ranging kinds, are faithful to specific places in the sea, and many are endemic "homebodies," making them vulnerable to extinction when their part of the ocean is damaged or destroyed. Sandra Díaz, one of Global Assessment's co-chairs, characterized the tremendous variety of living species as our "life-supporting safety net." In a 2019 National Geographic online interview by Stephen Leahy, Díaz observed, "Not only is our safety net shrinking, it's becoming more threadbare, and holes are appearing." The ocean, primary home for the largest, most substantial and comprehensive library of genetic diversity in the universe, continues to be open for business, largely unencumbered by concerns about shredding Earth's vital fabric of life.

In October 2019, a research article in *Science* by Brett Scheffers and several co-authors reviewed the heavy toll on vertebrates "across the tree of life" and determined that of the tens of thousands of species considered, one in five is affected by trade of some sort. Of the 31,500 terrestrial bird, mammal, amphibian, and scaled reptile species, about 24 percent are traded globally, a multibillion-dollar industry that is driving species toward extinction. However, fish—the most numerous of vertebrates and by far the most widely traded—were not taken into account among the wild animals at risk.

Top predators and among the fastest of fish, yellowfin tuna are vital links in the cycling of carbon and nutrients in the sea.

Ninety percent of many commonly consumed fish have been extracted, and some—notably bluefin tuna, American and European eels, all species of sturgeon, oceanic whitetip sharks, cod, halibut, and various deep-sea fish species including orange roughy and Chilean sea bass—have been pushed to levels so low that recovery may not be possible.

An example is the fate of the North Atlantic cod *(Gadus morhua),* a mainstay of food and commerce for North Atlantic countries for 500 years. In the early 1990s, Canadian scientists calculated the species to be at one percent of earlier levels, and the Canadian government declared a moratorium on catching them. Other nations followed, but by then, notable "cod highways" had been upended, knowledge lost of the migration routes led by the oldest, most experienced fish. A 2012 headline in *The Atlantic* claimed, "There Are Just 100 Fully Grown Cod Left in the North Sea," meaning that decades-old fish were rare, and the youngsters being taken were not reaching their prime. Currently, the web of ancient connections of life in the sea maintained by cod hangs by a few threads, and it is rare to see cod anywhere—except, ironically, in a restaurant.

Marine birds and mammals, vertebrates heavily exploited since the 1500s, are more often recognized as "wildlife," and national laws and international agreements are now aimed at protecting what remains of their greatly depleted populations. In a 2018 BirdLife International report, Margaret Sessa-Hawkins noted that of the 360 species of seabirds, nearly half—47 percent—are experiencing population declines, and 31 percent are threatened with extinction. Few currently are killed for food or products, but loss of coastal nesting sites; pollution; decline of krill, squid, and fish required for food; and entanglement in fishing lines, nets, and trash have made their existence increasingly perilous. Marine mammals face similar environmental issues, but in the 1970s, to stop commercial killing, the United States passed legislation protecting all marine mammals in the country's waters, followed in 1986 by an international moratorium on commercial whaling. The last whaling station in the United States closed in 1971 when the Del Monte Fishing Company in Richmond, California, discontinued its line of "Moby Dick 100 Percent Whale Meat" products—used mainly for cattle feed.

In the 21st century, legal, illegal, and unintentional killing of hundreds of thousands of seals, whales, dolphins, and other marine mammals by ship strikes and pollution persists globally. Currently, U.S. fishermen can legally kill marine mammals as "incidental catch," and permits are given to kill birds and mammals deemed to be competing with fishermen and aquaculture operations. Japanese fishermen take thousands of dolphins to market annually, and in Canada the legal kill quota for harp seals alone in 2020 was 400,000.

There are numerous state, national, regional, and international laws governing where, how many, and what kind of marine animals, mostly fish, squid, and crustaceans, can legally be taken, but given the trend of decline for most commercially captured species, the definition of "sustainable catch" has proved to be elusive. Large-scale extraction continues, however, largely based on the rationale for food security. The UN Food and Agriculture Organization is cited for claims that range from "Fish are a key source

VISIONARIES

Joel Sartore
The Photo Ark

. . . .

W hen Joel Sartore's wife, Kathy, was diagnosed with cancer in 2005, the high-energy, high-profile National Geographic photographer and Fellow put his career aside. While Kathy recovered, Sartore grew increasingly aware of life's fragile foundations. "One question continued to haunt me," he said. "How can I get people to care that we could lose half of all species by the turn of the next century?" His solution: the Photo Ark. Founded in 2006, its mission is to chronicle endangered species—"not just to create a record of what we've squandered, but to get people to change how they think and act in order to save these species." Sartore's goal is to capture studio portraits of Earth's 15,000 species in zoos and wildlife reserves, far from their natural homes. Sartore has made more than 11,000 portraits as of early 2021. The work is about humans, too. "It's folly to think that we can destroy one species and ecosystem after another and not affect humanity," he says. "When we save species, we're actually saving ourselves." Explore the Photo Ark at National Geographic's website and in Sartore's books *National Geographic Photo Ark*, *Photo Ark Vanishing*, and *Photo Ark Wonders*.

. . . .

OPPOSITE ABOVE: Pacific reef triggerfish face threats from human collectors and habitat loss. OPPOSITE BELOW: Cloaked with spines, the urchin *Arbacia punctulata* deters ocean predators but not humans with a taste for its soft interior.

of protein for millions of people" and "More than three billion people depend on fish as a major source of protein" to "4.3 billion people are reliant on fish for 15 percent of their animal protein intake." So, the question is, of the millions of tons of wild-caught animals taken annually, how much serves the needs of people who have few food options, and how much goes to global markets where dining on sea life is usually a choice and often a luxury? How much is converted to food for farmed fish, cows, pigs, chickens, and domestic pets? How much of the human diet comes from animal protein, how much protein is derived from plants, and what balance is needed for healthy humans? Beyond protein, what constitutes "food security"?

World Wildlife Fund scientist Jason Clay notes in a July 2011 *Nature* article, "Freeze the Footprint of Food," that by 2050 we may need "three Earths to meet the demands of our consumption." He identifies several measures—eliminating food waste, advancing plant breeding, and rehabilitating degraded land—that together can enable farming to feed 10 billion people and keep Earth habitable. Aquaculture and wild animals from the sea are not on the menu.

Whether your preference is for seafood on your plate or sea life in the sea, growing concerns about the declining numbers of many species have spurred increasing interest in ocean farming. For centuries, cultivation of freshwater carp, crabs, and other plant-eating species has been common in Asia, especially China, and pond farming of marine species has a long history in Hawaii and various Pacific island countries. In recent decades, some marine fish have been grown in coastal areas, including Atlantic salmon, striped bass, mahi mahi, and several kinds of shrimp, oysters, clams, and scallops. Some operations are moving offshore with netted enclosures for fish. Most marine fish that are being farmed are carnivores, however, and feeding them usually involves catching four or five tons of small wild fish for every ton of farmed animals raised. Other problems include buildup of wastes around the farms, introduction of antibiotics and parasites to native populations, and the escape of nonnative species to nearby ecosystems.

Most of these thorny problems have solutions, however. Closed systems that recycle water and use biological filtration achieve "more crop per drop" and also avoid concerns about escapes and contamination. Selecting fast-growing plant-eating species rather than slow-growing carnivores follows the model established for terrestrial farm-raised animals. In New York, the Billion Oyster Project has mobilized the public behind an effort to restore the bivalves that once were abundant enough to form mountainous reefs that interfered with ship traffic and served as a culinary staple until early in the 20th century. The Wild Oyster Project in San Francisco and similar efforts in the Chesapeake Bay and in the Gulf of Mexico are not only showing promise as a potential food source but also are engaging the public in restoring a powerful, natural filtration system, a robust habitat for marine life, and a living fortress against erosion and storms. There is enormous potential for cultivating microbes and certain marine algae for food, pharmaceuticals, and edible oils. However, the most promising future for extracting wealth from the ocean may be embodied in the ocean's enormous library of life available to those with the foresight and wisdom to read its volumes and also read between the lines.

Personal Anti-Plastic Campaign

....

Trapped in the bellies of whales and wrapped around the necks of sea turtles, plastic bags, six-pack rings, bottles, and toys are strangling ocean wildlife, poisoning water, and forming island-size gyres. Ocean plastics must go—or bring irreparable damage to millions of species, including humans. While

solutions flow forth from global governments, NGOs, and private entrepreneurs, individual grassroots efforts can cause a ripple effect. If everyone did these six simple things, writes journalist and plastics expert Laura Parker in her 2018 *National Geographic* article "Planet or Plastic," they would "feel no pain." The ocean's health—and ours—would be better for it.

1. Give up plastic bags. Take your own reusable ones to the store.
2. Skip straws. Use paper or metal ones instead.
3. Pass up plastic bottles. Invest in a refillable water bottle.
4. Avoid plastic packaging. Buy bar soap instead of liquid, in bulk.
5. Recycle what you can.
6. Don't litter.

....

ABOVE: Like an invasive species, plastic waste floats amid Indo-Pacific sergeants. OPPOSITE: Giant clams on Australia's Great Barrier Reef are the world's largest bivalve mollusks, sporting hues from royal blue to deep gold.

HOPE SPOT

Outer Seychelles

Off the east coast of Africa and north of Madagascar, the Outer Seychelles grace the Indian Ocean in five coralline island groups, home for numerous endemic land and sea plants and animals. The Seychelles were named a Hope Spot in 2014. That year just 0.02 percent of the exclusive economic zone (EEZ) was protected, but by 2020, with support from the Nature Conservancy and other organizations, and consecutive presidents, James Michel (2004–2016) and Danny Faure (2016–2020), the government of the Seychelles increased protection to 30 percent, embracing 1,035,995 square kilometers of ocean.

The Seychelles' Aldabra is Earth's second largest coral atoll.

2020:
The Turning Point

A new term was created in 2020—"Anthropause"—to describe the impact of the novel coronavirus pandemic, COVID-19, on human society since early that year. Proposed by Christian Rutz and co-authors in a June 2020 report in *Nature Ecology & Evolution,* they note, "The reduction in human mobility on land and at sea during the Anthropause is unparalleled in recent history," with effects "drastic, sudden, and widespread." To control infection, many countries went into lockdown. People stayed home, did not travel or gather in restaurants or other public places for weeks and then months, and, with cautious reopenings, followed special protocols to cope with ongoing infections.

In October 2019, biosecurity experts conducted a simulated event whereby a hypothetical virus jumped from a wild animal to a domestic animal to humans, setting off an epidemic that within weeks ballooned into a pandemic devastating millions of lives and causing global economic collapse. The simulated pandemic was called "Event 201," according to Ryan Morhard, a biosecurity specialist at the World Economic Forum in Geneva. Morhard wrote that "because we're seeing up to 200 epidemic events per year, . . . we knew that, eventually, one would cause a pandemic." In the story, reported in an August 2020 issue of *Nature,* Morhard remarked that world leaders have not been taking the threat of a global pandemic seriously enough, including the immense human and economic toll.

Two months after the simulation, a few people in Wuhan, China, became ill with an actual virus that behaved remarkably like the modeled event. A virus benign in infected animals transitioned into a novel form uniquely suited for humans as its host. Techniques long recommended by health experts—rapid detection, rapid response—were delayed long enough for the virus to be transmitted by people traveling globally by land, air, and sea, escalating the spread exponentially.

A United Nations Development Report dated June 8, 2020, noted that demand for oil had plummeted and oil prices that month had traded below zero, and that "Sizeable reductions in fossil fuel consumption are already resulting in measurable reductions in greenhouse gas emissions which benefit the ocean by slowing the impacts of climate change." Many ships, including recreational boats, stayed in ports long enough for there to be a measurable change in ocean noise and activity; in turn, dolphins, whales, seals, sea lions, and seabirds came closer to shore than usual in coastal areas from Seattle, Washington, to Trieste, Italy. In Spain, home of the largest fishing fleets in the European Union, about half of the ships stayed in port for months. On August 15, 2020, the *Japan Times* reported, "Plummeting global demand for fish and seafood as a result of the coronavirus crisis is likely to create an effect similar to the halt of commercial fishing during both world wars, when the idling of fleets led to the rebound of fish stocks."

Protection of nesting sites at Costa Rica's Ostional beach is helping the gradual comeback of highly endangered olive ridley sea turtles.

Bathed in sunlight that penetrates
Norway's Gulen fjord, a starfish
and plumose anemone share space
on a carpet of red algae.

On July 12, 2020, as COVID-19 wrought chaos across the world, a Japanese-owned, Panama-flagged, bulk-fuel-oil carrier went off course and grounded on a coral reef in Mauritius in the Western Indian Ocean. More than 1,000 tons of oil spilled, with devastating consequences for the region's renowned beaches and reefs, and the country's tourism-based economy. It was not the first time that human error resulted in a massive oil spill disaster. But this one happened at a time and place and under circumstances that riveted the attention of people who were already reeling with the effects of a global health crisis. Shipping policies, long overdue for reform, may henceforth change.

The year 2020 had been planned as a big year for the ocean. The United Nations Ocean Conference and the Climate Change Conference (UNFCCC COP 26), which was to feature major ocean themes, along with many other critical ocean meetings, were put on hold while the reality of a changed world began to sink in. There is no substitute for personal engagement, but as ocean scientist Robert Ballard demonstrated years before, much can be accomplished through "telepresence." Whether beaming images of the deep sea from a ship to a classroom or holding a conference with participants seated across the globe, there was proof during the 2020 Anthropause that meaningful meetings could be held and communication achieved vicariously.

What might be learned from these exceptional circumstances that could accomplish a paradigm shift in human behavior? As human society convulsed, nature awakened in a powerful demonstration of cause and effect. According to Rutz and his co-authors, research on Anthropause effects can reveal better understanding of human-wildlife interactions. We may find that minor changes in lifestyles can have major benefits for ecosystems and for humans, now that we can see what happens when we give nature a break. Might pulling together to combat a lethal disease demonstrate that the power exists to do what it takes to combat climate change, planetary security, and harmony with nature? Quite possibly, 2020 will prove to have been a very big year for the future of the ocean and for the future of humankind.

> "NATURE ALWAYS BALANCES HER BOOKS."
>
> —ARTHUR C. CLARKE, *2061: ODYSSEY THREE*

DEEPER DIVE

Top Conservation Tools for the 21st Century

Tools with a 21st-century twist are ramping up ocean conservation. Through increasingly savvy GIS technology, computer simulation, and other tools, organizations like California-based ESRI (Environmental Systems Research Institute) crystallize into view the ocean's structure and ecosystems, helping experts find trouble spots and set priorities for a return to health. Planning software like Marxan models solutions for challenged ecosystems. Digital tools, including SeaSketch, incorporate ideas from multiple contributors to build management plans for marine protected areas. For deep-diving research, personal subs are evolving fast, and a gentler generation of robots, includ-

Dexterous Squishy Robot Fingers

ing National Geographic grantee David Gruber's Squishy Robot Fingers, safely collect study specimens. As of 2020, digital apps and ocean hubs on the internet from groups like Ocean First Education offer lesson plans and virtual field trips. There is the power of the camera. Since 1989, National Geographic Explorer Robert Ballard's JASON project has streamed underwater habitats into classrooms through ROV cameras. Today, James Balog's time-lapse images of disappearing Greenland ice, Joel Sartore's photographs of the critically endangered hawksbill sea turtle, and Louie Psihoyos's video of dolphins being clubbed to death in his documentary *The Cove* are urgent calls to action.

2030: The Tipping Point

A Mother Goose nursery rhyme might be the right cautionary message for Mother Earth in 2030.

Humpty Dumpty sat on a wall.
Humpty Dumpty had a great fall.
All the King's horses and all the King's men
Couldn't put Humpty together again.

In 2020, progress had been made in abating pollution and other reforms outlined in Life Below Water, United Nations Sustainable Development Goal 14, but results fell well short of the targets. Time to choose a different course was slipping away, and new goals for "the next ten years" were set as the best chance humankind would ever have to reverse trends leading to a planet far less hospitable than what it has been for many preceding generations. Global warming is on the rise, driven by the increasingly high emissions of CO_2, methane, and nitrous oxide generated by burning coal, oil, and gas, coupled with release of these gases from agriculture, thawing permafrost, and the destruction of forests and wildlife, on land and in the sea. Human dominance of the planet is on track to engulf the remaining wild places, even on the High Seas, with industrial fishing, floating farms, and deep-sea mining stripping life from the seafloor.

In 2009 William McKibben's climate change initiative aimed to keep us from exceeding 350 parts per million (ppm) of CO_2 in the atmosphere. Many scientists thought 400 ppm was a red line that represented a perilous tipping point that could set in motion processes beyond our control. It was widely believed that cutting emissions by 45 percent between 2020 and 2030 would hold global temperature rise to 1.5°C, a level thought to be both achievable and tolerable. But in 2009, atmospheric CO_2 concentration had already reached 390 ppm; by 2014 it exceeded 400 ppm, and by 2020 it was at 410 ppm. The difference of 10 parts per million in the atmosphere does not seem like much, but even that increase over usual CO_2 concentrations has major consequences. The last time concentrations were above 400 ppm was more than two and a half million years ago, when the planet was warmer overall, sea level was about 20 meters higher, no ice sheet existed in the Arctic, and forests greened the Antarctic terrain. There were no large port cities or massive coastal communities to be devastated by such sea-level rise.

"NOAA and NASA confirmed that 2010 to 2019 was the hottest decade since record keeping began 140 years ago," reported Alejandra Borunda in a National Geographic science report on January 15, 2020. Ocean temperatures were the highest they've ever been, with increasing heat driving storms of increasing intensity. It will take a while—maybe until the end of the century, maybe sooner—for the Arctic Ocean to be ice free and the Antarctic continent to lose much of its glittering shield of ice, but the process is inexorable when sea and air temperatures become increasingly warmer than the ice.

VISIONARIES

Johan Rockström & Mattias Klum
Planetary Boundaries

Johan Rockström and Mattias Klum are partners for the planet. Director of the Potsdam Institute for Climate Impact Research, and founding director of the Stockholm Resilience Centre, Rockström leads the research for Planetary Boundaries Initiative, first published in 2009. The research defines nine boundaries that keep Earth "a safe operating space for humanity" and pursues ways to keep them intact. The boundaries include maintaining a healthy global water cycle; limiting the release of pollutants into the air and ocean; and stopping ozone depletion, ocean acidification, ecosystem destruction, biodiversity loss and extinctions, and climate change. Called "one of the most important natural history photographers of our time" by National Geographic, Klum joined Rockstrom to bring these boundaries alive through books, articles, and films. Sometimes the two return to places they've documented—not always to good results. "It is so utterly depressing when everything you fought for and were moved by is leveled," says Klum. "It makes you want to fight harder because there is so much to fight for."

Global warming accelerates the rush of meltwater from an ice cap on Nordaustlandet Island, in Norway's Svalbard archipelago.

Edward O. Wilson observes in his thoughtful 2016 book *Half-Earth*, "Like it or not, we remain a biological species in a biological world, wondrously well adapted to the peculiar conditions of the planet's former living environment, albeit tragically not this environment or the one we are creating. In body and soul we are children of the Holocene, the epoch that created us, yet far from well adapted to its successor, the Anthropocene." It is not possible to return to conditions that existed 200 years ago, or even 10 years ago, but in 2020, plans developed to veer from the catastrophic consequences of human impacts on planetary functions. The importance of protecting what remains of the intact systems that hold the planet steady, land and sea, motivated unprecedented action with critical goals set for 2030. An ambitious international effort, the UN Decade of Ocean Science for Sustainable Development, was organized in 2020 to mobilize action to explore and document the ocean and thereby reverse the decline in ocean health by 2030. Seabed 2030, an international program to "map the gaps" that exist in our knowledge of more than 80 percent of the ocean floor, gained momentum as new technologies increased destructive access for fishing and mining in previously undisturbed places.

The Convention on Biological Diversity (CBD), launched at the Rio de Janeiro UN Conference on Environment and Development in 1992, began a process to stem the loss of species. In 2010, at a meeting in Aichi, Japan, the process was crystallized as a plan for protecting at least 10 percent of the ocean and 17 percent of the land by 2020. As that milestone was reached, about 15 percent of the land and less than 4 percent of the ocean had deliberate, high levels of protection. British marine scientist Callum Roberts calculated that protecting at least 30 percent of the ocean could significantly restore health to depleted ocean species and benefit the ocean overall. In 2014, those attending the once-in-a-decade World Parks Congress in Sydney, Australia, endorsed the need to protect at least 30 percent of the entire natural world by 2030; and in 2016, dozens of conservation organizations globally reaffirmed this goal. In *Half-Earth,* Wilson made a compelling case for committing half of the planet's surface to nature to save the diversity of life-forms, and thereby save humankind—a concept long championed by the WILD Foundation's Nature Needs Half campaign. Half of the planet protected by 2050 has become the new goal to maintain vital biodiversity and the processes they govern.

Concerned about the impact of deforestation and forest fragmentation, ecologist Thomas Lovejoy set in motion in 1979 a long-term study in Brazilian rainforests near Manaus to determine the "minimum critical size" of forested area needed for individual species and ecosystems to persist. Would it be better to have many small areas or concentrate on large tracts to protect the species and systems involved? Not surprisingly, large areas maintain their integrity and diversity longer than smaller pieces, but the right small pieces may be critical for plants and animals that have a highly restricted range. The same principles apply to the ocean, naturally composed of a quilted network of distinctly different systems populated by both wide-ranging and locally anchored creatures. These spaces are increasingly fragmented by conversion of coastal regions for human use, by shipping traffic, ocean dumping, destructive fishing, and other human pressures in the open sea.

How much of the planet must be safeguarded in a natural state to secure an enduring future for humankind? What matters most: quality or quantity? Is there a balance? Can the needs and desires of not just eight billion people, but likely many more, be accommodated on half of the land and half of the sea? There is growing evidence that prosperity can improve for people and nature if such protective action is taken, mindful that there is only so much planet to go around. And while our numbers are on track to grow, Earth will not.

According to the Global Footprint Network, if all of Earth's eight billion people consumed as much as citizens of the United States, five Earths would be needed to support 2020 lifestyles. People elsewhere in the world would need only about three. Although scientists have identified at least 50 extrasolar planets believed to be capable of sustaining life, getting there and making them livable for humans is a far greater challenge than making peace with the small blue planet that is our home.

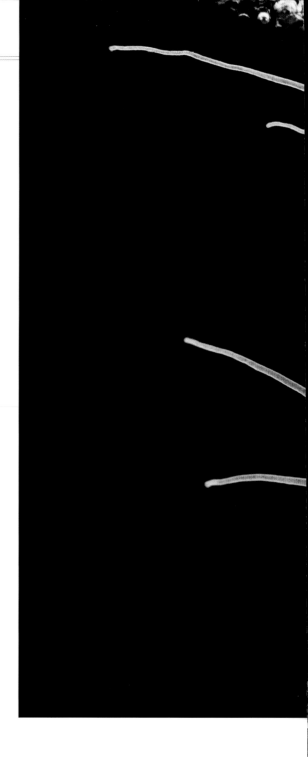

Its tentacles clinging to the underside of a melting ice floe, an anemone perseveres as the globe warms and sea level rises.

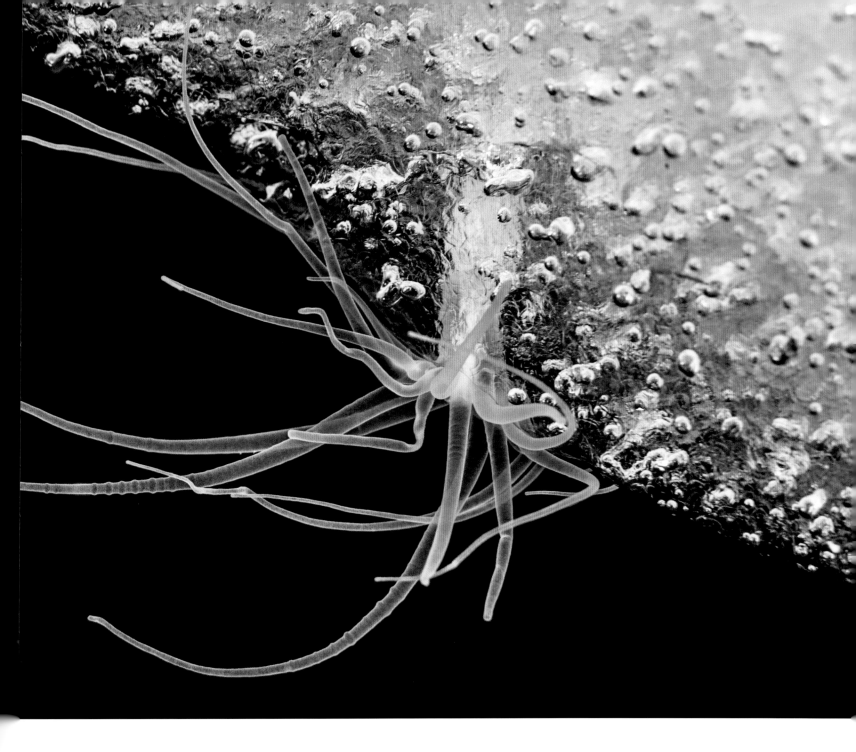

Entrepreneurs & the Plastics Challenge

The human ingenuity that brought us plastic is now working overtime to take it away. "We believe that private entrepreneurship has a huge role to play in ocean plastic cleanup and prevention," wrote Andrew Rummens of the policy research group Strata in 2019. That same year National Geographic, with the U.K.'s Sky Ocean Ventures, announced the Ocean Plastic Innovation Challenge, inviting creative suggestions to rid the ocean of plastic. Qwarzo, a French team, won for its 100 percent recyclable, biodegradable, and compostable paper technology; Chile's Algramo team for its low-cost, reusable product packaging; and #PerpetualPlastic of Germany for

Ocean volunteers clean up.

its sculpture representing the paths and fate of all plastics ever produced—made from flip-flops washed up on the beaches of Bali. There are more. Aiming to become the "Tesla of biodegradable products," the U.K.'s Polymateria develops chemical additives to break down plastic products quickly and safely. The Netherlands' Ocean Cleanup transports ocean plastic to build an artificial coastline, while protecting local marine life. And U.S. companies like 4ocean, Norton Point, and Parley for the Oceans first clean beaches of plastic trash, then upcycle it into bracelets, sunglasses, swimwear, and running shoes— cutting-edge style for a clean ocean.

2050: What Is the Future We Want?

What will the ocean be like in the middle of the 21st century?

The question was put to experts advising on the proposed content for the Living Seas Pavilion at Disney World's Epcot Center in Orlando, Florida, in 1982. According to Kym Murphy, leader of the project, the distinguished oceanographer William Nierenberg was skeptical. "As a young scientist I was on a panel convened to anticipate important new developments for the decades ahead. I have now lived long enough to see some of the things we missed." He continued, "Going to the moon and the space program generally, the significance of nuclear energy, lasers, the electronics revolution, breakthroughs in genetics, biotechnology, the impact of computers, plastics, ceramics, carbon fiber . . . So much has happened so fast, and the pace is accelerating. I don't want to pretend that I know what might happen next!"

After deliberating, the panel agreed that, by conveying to the public what was already known about the ocean and why it matters to everyone everywhere, and by sharing the thrill of exploring the largest unknown part of the planet, the ocean exhibit might inspire and excite people. Visitors coming to experience Disney's Magic Kingdom and Epcot Center could fulfill a desire to dive vicariously into the magic kingdom of life in the sea. Jacques Cousteau said, "People protect what they love." Knowing is the key to caring, and all things considered, the greatest threats to the ocean are largely caused by those who do not know why they should care.

In the 1850s, most people who lived in San Francisco did not worry about the fate of seabirds living on the Farallon Islands, 48 kilometers offshore from the Golden Gate Bridge. Half a million blue-speckled, pointy-tipped eggs of the murre *(Uria aalge)* were taken every year from the islands to help feed the region's growing gold-mining population. The birds were nearly exterminated before protective measures saved the few remaining pairs, and in nearby Petaluma, chicken farming yielded a reliable egg supply. Early in the 20th century, so many migratory ducks, geese, and other waterfowl were killed by market hunters in North America that laws were enacted to keep them from having the same fate as the passenger pigeon, hunted to extinction by 1911. A lucrative industry based on adorning hats with bird feathers led to the near extermination of snowy egrets and hundreds of other species valued for their decorative plumage.

Only when public opinion reached a tipping point favoring live birds over tastes in food and fashion trends were laws passed that came just in time to save most of the species involved. Supreme Court Justice Oliver Wendell Holmes declared in 1918 that the protection of birds was in the national interest and that without such measures, one could foresee a day when no birds would survive for any power—state or federal—to regulate. Similarly, only when the weight of public opinion shifted from enjoying soup and steaks derived from sea turtles to valuing the turtles as living treasures at

VISIONARIES

Katy Croff Bell
The Ocean's Future Is Technology

When National Geographic Fellow Katy Croff Bell hosted the "Here Be Dragons" event in 2018 at the Massachusetts Institute of Technology Media Lab, fire-breathing reptiles were not on the agenda. As an oceanographer and maritime archaeologist, Bell was launching an Open Ocean initiative. Its mission: to use technology to pinpoint "dragons"—gaps in ocean exploration, innovation, access, and diversity—and find solutions to fill them. The ideal solutions are low-cost and open-source, so that ocean awareness can reach historically excluded communities and more people can be empowered to explore their own ocean environments. Now, as president of the Ocean Discovery League, Bell is building an ocean exploration accelerator—an organization that offers support services for ocean entrepreneurs, helping them find faster, cheaper, and more accessible ways to explore the deep ocean. At the same time, she's training a diverse team to lead those global efforts. And she's just getting started.

OPPOSITE ABOVE: Active by day, Indo-Pacific sunset wrasses retire early at night. OPPOSITE BELOW: A rare gathering of scalloped hammerhead sharks, endangered victims of the shark-fin trade

risk of becoming extinct was it possible to bring about serious protection. For whales, after centuries of being treated as products, there finally emerged sufficient awareness—and caring—about their importance beyond tons of meat and oil for public attitudes, behaviors, and laws to change. As with birds, sea turtles, and whales, might we come to see beyond commodity values in octopus, squid, krill, crabs, lobsters, Chilean sea bass, and the thousands of diverse forms of life we lump together ignominiously as "fish"? Bluefin tuna, sailfish, marlin, swordfish, and mako sharks are the fastest fish in the sea, but along with some of the slowest—seahorses, manta rays, and grouper—they currently are in a race for survival.

In a world where the human population has increased from about one billion in 1800, to two billion in the 1930s, four billion in the 1980s, and eight billion early in the 21st century—with 10 billion anticipated by 2050—there has been a growing recognition that human demands on the land, fresh water, air, and sea are creating stresses beyond what scientists recognize as the planet's "carrying capacity." There is optimism that technological advances will keep pace, solving problems as they arise. A February 2020 World Economic Forum analysis reports that technologies exist to switch to wind, water, and solar energy generation worldwide that could first slow, and then reverse, the effects of global warming by 2050. According to planning for the 2020 UN Ocean Conference, offshore wind that now supplies 0.2 percent of global electricity is forecast to match or exceed the energy supplied by offshore oil by 2050. Solar energy is increasingly being tapped directly for domestic and industrial uses, and some project safely mastering hydrogen, geothermal, and nuclear fusion for energy by mid-century. Food security will likely be attained from more efficient, nutritious, less wasteful, and increasingly plant-based habits with less reliance on farmed animals and ocean wildlife. As respect for the immeasurable and irreplaceable value of naturally biodiverse systems grows, expanded protection for them could increase significantly.

But there are also projections that by 2050 global temperature will still be rising, ocean acidification increasing, wildlife diminishing, and wild places collapsing. Poverty, health, hunger, and security concerns haunt even the most prosperous countries. According to a 2017 assessment by the Ellen MacArthur Foundation in partnership with the World Economic Forum, by 2050 the ocean could have more plastic than fish if present rates of removal of fish and input of trash continue. Pieces of plastic break down into smaller sizes—destructive to ocean and human life—and they never go away. Meanwhile, life in the ocean, indifferent to human affairs, must live or succumb to radically changed circumstances. Nothing in the strategies for survival of tuna or whales or cod or krill has prepared them for shipping traffic that slices through their traditional pathways, or for the noise, pollution, and unparalleled levels of human predation that increasingly threaten them. Nothing in the inherent strategies for survival of humans has prepared us for the new menacing realities we have created for Earth and ourselves.

The future we want may no longer be ours to choose. Still, there is reason for hope that a long and prosperous future is possible if we treat the living ocean as if our existence depends on it, because—now we know—it does.

Basking in sunbeams, an orca bridges two realms, air and water, both vital for its existence.

"BIG SCARY PROBLEMS WITHOUT SOLUTIONS LEAD
TO APATHY, NOT ACTION. . . . SMALL STEPS TAKEN BY
MANY PEOPLE IN THEIR BACKYARDS ADD UP."

—NANCY KNOWLTON, SANT CHAIR FOR
MARINE SCIENCE, SMITHSONIAN INSTITUTION

GREAT EXPLORATIONS

Cleaning Up the Ocean

An endangered Australian sea lion, the continent's only native pinniped, surfaces for air in the Indian Ocean.

The Keep America Beautiful program—launched in 1953 after cows in Vermont were found eating glass bottles tossed in haystacks—catalyzed anti-littering action across the United States. Other campaigns, from Johannesburg to Paris to Sydney later followed, including Sweden's Keep Nature Tidy in 1963. But trash in the ocean was another thing: If not broken down by the water, at least it would be out of sight. The ocean would still support travel, provide food and recreation, and underpin all of Earth's vital systems. Right?

Wrong. Over time, scientists shared their growing knowledge and concerns. While vast, the ocean is a sensitive, finely honed machine responsible for regulating air, climate, and the health of every land-based life-form—including humans. That regulation depends on the care of its underwater

ABOVE: A moray eel adopts a discarded rubber tire off the Galápagos Islands. RIGHT: The Philippines' pristine Tubbataha Reefs Natural Park, a UNESCO World Heritage site, thrives with diverse marine life.

ecosystems, species, and forces. Pollution is its insidious disrupter. Almost two decades after the first land campaigns, awareness of ocean trash turned into action against it—on international, national, and grassroots levels. An early catalyst, in 1971, was biologist Ed Carpenter's warning after sighting flecks of plastic in Sargasso seaweed during a research trip—a harbinger of what toxins lay deeper and unseen.

In 1972 the so-called Dumping Act—the U.S. government's Marine Protection, Research, and Sanctuaries Act—authorized the EPA and NOAA to find sources of ocean-bound sewage and industrial waste and curtail their activity. Soon U.S. laws also monitored coastal seafloor dredging, with its destruction of life and habitats.

On the global front, in 1973 the United Nations formed the International Convention for the Prevention of Pollution from Ships. Then its arm for education, science, and culture—UNESCO—joined the IUCN and other global organizations to create guidelines for worldwide marine protected areas (MPAs). In 1982, the UN's Convention on the Law of the Sea outlined rules for all nations to use the High Seas sustainably.

In 2017 the UN Clean Seas campaign engaged 60 countries to clean up plastics. By 2018, the U.S. government's Subcommittee on Ocean Science and Technology (SOST) had set tangible goals to "safeguard human health" and "improve forecasts of marine contaminants and pathogens." By 2020, Japan's Ministry of the Environment was allocating billions of yen annually for ocean waste disposal and plastics recycling.

Today, many private-sector scientists, conservationists, and locals support "rewilding"—restoring nature to its wild state through community engagement. Mission Blue alone has enlisted thousands of citizens in support of their local Hope Spots. And since 1986, the Washington, D.C.–based Ocean Conservancy's volunteers have collected millions of metric tons of trash during International Coastal Cleanup Day—10,433 metric tons by 122 nations in 2018 alone. Such statistics are both sobering and hopeful.

TIME LINE

Ocean Clean-up Efforts

1971: Marine biologist Ed Carpenter reports flecks of plastic on seaweed in the Sargasso Sea.

1972: Eighty-seven nations via London Dumping Convention begin guidelines to protect marine environment (effective 1975); U.S. passes "Dumping Act"; Ocean Conservancy founded in Washington, D.C.

1973: UN establishes MARPOL: International Convention for the Prevention of Marine Pollution from Ships.

1975: UNESCO joins IUCN and others to set guidelines for marine protected areas (MPAs).

1977: EPA sets ocean dumping rules.

1982: UN Convention on the Law of the Sea signed by some 160 nations.

1984: California surfers start Surfrider Foundation.

1986: Ocean Conservancy launches International Coastal Cleanup Day.

1988: U.S. Congress bans dumping of medical, industrial, and construction wastes and sewage sludge.

1990: U.K. surfers in Cornwall found Surfers Against Sewage.

1997: U.S. captain Charles Moore identifies Great Pacific Garbage Patch.

2006: London Protocol updates 1972 London Dumping Convention; 53 nations prohibit toxic dumping.

2009: Roberta Dixon-Valk, Amanda Marechal, Tim Silverwood start Australia's Take 3 for the Sea. Marcus Eriksen, Anna Cummins found California's 5 Gyres to reduce plastic trash; aim to ban plastic beads in cosmetics by 2022.

2014: Kristal Ambrose founds Bahamas Plastic Movement and "Adopt a Beach."

2016: UN officially sets goal of protecting 30 percent of the world ocean by 2030.

2017: UN founds Clean Seas Campaign with 64 nations.

2018: U.S. congressional subcommittee sets goals to curtail ocean pollution.

2019: UN World Wildlife Day focuses on rewilding.

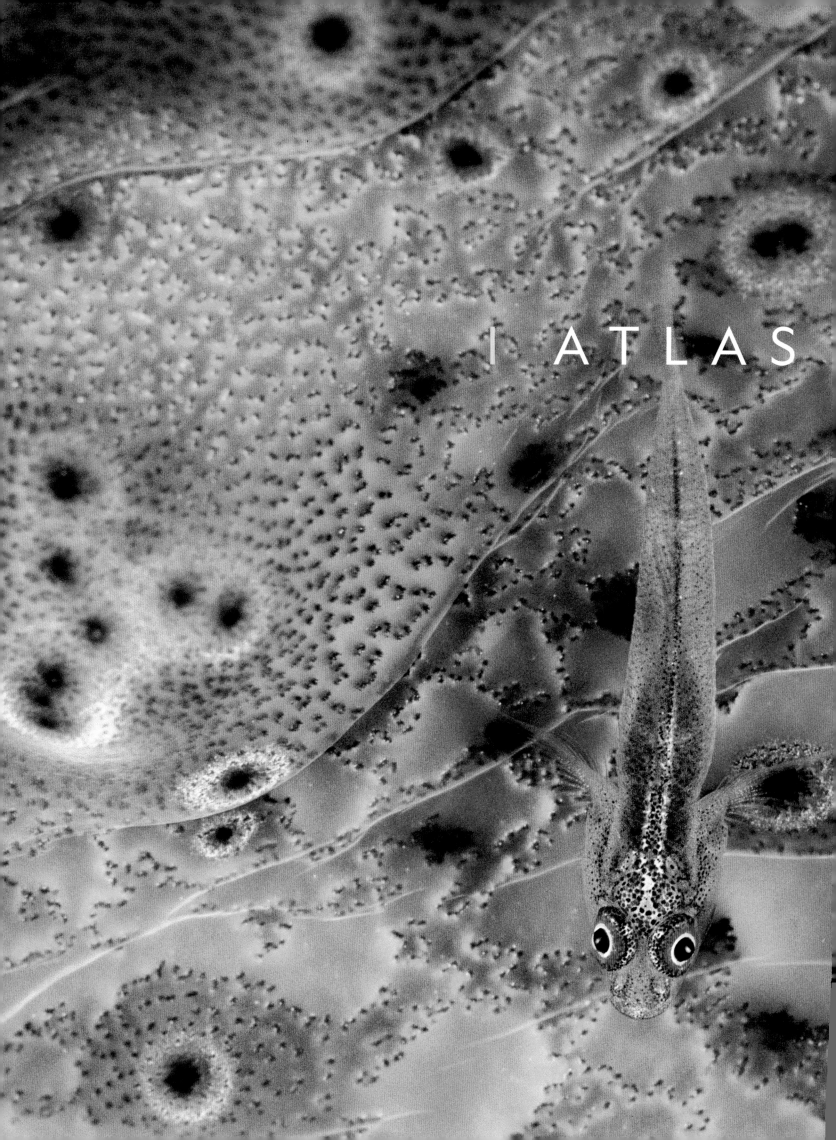

ATLAS

PART FOUR

OF THE OCEAN

4

"STANDING ON THE SHORE AND LOOKING OUT TO SEA, THE BOY SAID, 'THERE'S A LOT OF WATER OUT THERE.' AND THE WISE OLD OCEANOGRAPHER RESPONDED, 'AND THAT'S ONLY THE TOP OF IT.'"

—RICHARD ELLIS, *SINGING WHALES AND FLYING SQUID*

For most of the history of humankind, knowledge of what is beneath the surface of the sea has remained largely unknown. Even now, large areas of the seafloor have not been mapped with resolution comparable to what is known of the moon, Mars, and Jupiter. And knowledge of the dynamic currents, waves, and boundaries between the surface and the seafloor—the living, liquid ocean—is even less well documented. We are at the point now when we are at least, and at last, beginning to recognize the magnitude of what remains to be discovered.

In the 1990s, efforts to create detailed maps of the ocean floor were projected to take at least 150 years given the technologies available at the time—and the pace at which they were applied. A new project, Seabed 2030, a collaboration between the Nippon Foundation (NF) and the General Bathymetric Chart of the Oceans (GEBCO), aims to produce an accurate, detailed map of the worldwide seafloor by 2030. By combining current information into a unified database and encouraging efforts to collect and submit new data, it will generate maps that will be freely available to the public at a resolution never before possible. The completed maps will be released on the NF-GEBCO Seabed 2030 official website and will also be made available through online sources such as Google Earth and the Environmental Systems Research Institute's (ESRI's) Ocean Basemap.

In 2019, winners were announced for a Shell Oil– and NOAA-sponsored XPRIZE competition to build an autonomous, high-resolution seafloor mapping system that could operate from an uncrewed boat. The leading project was submitted by an international team from the University of New Hampshire, led by Rochelle Wigley and Yulia Zarayskaya. Other proposals were rewarded for creative approaches that can accelerate seafloor mapping.

But what of the water column? Scientists at ESRI, the company that pioneered global information systems in the 1960s, are now focusing on how to document and effectively portray layers of data in the sea, including the occurrence of living systems. Fishermen who began using sonar to locate fish underwater have said that now fish have no place to hide. Soon, whatever is out there, down there, may at last be made visible.

RIGHT: Hugging the coast of Australia's Queensland for some 2,000 kilometers, the Great Barrier Reef is Earth's longest and largest reef system. PAGES 420-421: A goby shelters on the mantle of a giant clam in Papua New Guinea's Milne Bay, charted by the Spanish in 1606 and later named for British admiral Alexander Milne.

Why We Map the Ocean

n the first volume of *National Geographic* magazine in 1888, Gardiner G. Hubbard wrote, "When we embark on the great ocean of discovery, the horizon of the unknown advances with us and surrounds us wherever we go. The more we know, the greater we find is our ignorance. " He might have added, "and that is why we keep exploring!" The first people to see the ocean many thousands of years ago likely wondered how big it was, how deep, and what unknowns were beyond the horizon.

Some 3,500 years ago seafarers from islands in Polynesia stepped into outrigger canoes to find out, equipped with only their understanding of the stars, winds, currents, and the flights of birds. Gradually they learned to read the currents and to integrate their knowledge to document directions and the distances to other islands, and then continents. They recorded this information on stick charts, the world's first maps of the ocean. The placement of shells and fibers indicated islands, waves, and currents.

Early Pacific navigators passed their knowledge, discoveries, and instruments from generation to generation, spawning lines of master Pacific navigators who seemed born with an inherent understanding of the dimensions of the sea. Meanwhile, other seafarers across the globe were finding their own ways to document their known ocean.

Careful logs kept by early Mediterranean ship captains gave detailed sketches of coastlines, distances from port to port, currents, winds, and trouble spots. But it wasn't until the 13th century that all the elements were combined, with highly progressive mathematical calculations, into the first real chart, intricately engraved into a length of calf hide. Called a portolan chart—from the Italian word for "a collection of sailing directions"—it allowed seafarers to actually visualize the area they traveled.

By the mid-1500s, the Belgian cartographer Gerardus Mercator had made maps with his Mercator projection, which is still widely used today. Although areas are stretched or compressed, basically to fit a spherical Earth onto a flat map, this projection presented navigators with straight-line compass routes from points A to B. Together, portolan charts and Mercator maps "revolutionized how people perceived space, much like Google Earth has done in our lifetimes," says Library of Congress geography specialist John Hessler.

While mapping a route from A to B laid the framework for crossing the ocean, once on the journey the crew needed to know where they were at all times. Longitude and latitude lines for showing precise locations on land and sea had been drawn into primitive maps since the time of Eratosthenes in the third century B.C. At sea, mariners could tell

> "A MAP IS THE GREATEST OF ALL EPIC POEMS. ITS LINES AND COLORS SHOW THE REALIZATION OF GREAT DREAMS."
>
> —GILBERT H. GROSVENOR

ABOVE: On a stick chart from Micronesia's Marshall Islands, sticks represent Pacific wave patterns and shells, atolls. OPPOSITE: An 18th-century astrolabe, or "star taker," helped navigators calculate latitude.

their latitude by the stars, but because the Earth rotates, they could not use the sun and stars to determine their east/west longitude without knowing the time. Knowing time at sea was difficult because the pendulum that kept time in old clocks was disrupted by a ship's motion. In 1761, after more than 30 years of work, British clockmaker John Harrison finally developed the seagoing chronometer that revolutionized ocean navigation.

Mapping currents to aid ships in speedy travel and fishermen in identifying grounds for plentiful catches was still new territory. In 1769 American statesman and polymath Benjamin Franklin was the first to plot a major current, the Gulf Stream, which runs from the Gulf of Mexico north along America's eastern seaboard. And Alexander von Humboldt in 1802 identified the powerful Peru Current, today named for him.

There was also the challenge of mapping the ocean below its surface—nearly 11 kilometers at the deepest depth. Prompted by a closer look at the Gulf Stream in 1843, the U.S. Coastal Survey began exploring and recording ocean depth and varying temperatures, the makeup of the ocean floor, and its life-forms. And in 1855, a former American naval captain, Matthew Fontaine Maury, mapped wind and ocean currents around the world in his *Physical Geography of the Sea.* From 1872 to 1876, the first global ocean scientific expedition, H.M.S. *Challenger,* with six scientists, recorded temperature, currents, and marine life across 68,890 nautical miles of world ocean. Using wire, sinkers, and an 18-horsepower steam engine, they also measured depth and retrieved samples of seafloor sediments. Half a century later, during World War I, seafloor soundings took on new meaning as sonar was developed for battleships to send out sound signals to locate enemy submarines. Sonar soon was used to map seafloor mountains, canyons, and trenches.

It was not until 1968 that oceanographer Bruce Heezen and cartographer Marie Tharp used contemporary sonar readings, along with previous scientific records, to structure the first map of the Atlantic Ocean floor, followed by the Pacific map, and then a complete world view. Their efforts represented a culmination of attempts to record what was there. It was also just the beginning.

Residents of Tonga, named the "Friendly Islands" by Captain James Cook, visit the H.M.S. *Challenger* during its 1872–76 global expedition to chart the world ocean.

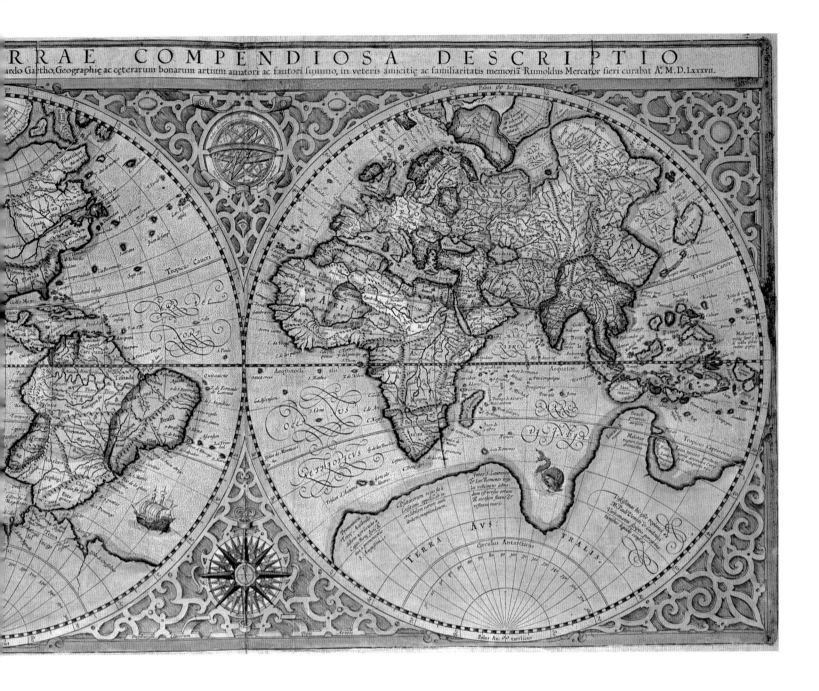

RRAE COMPENDIOSA DESCRIPTIO

Today, the international initiative Seabed 2030 is promoting compilation of bathymetric data into a 100-meter-resolution digital model of the world seafloor over the coming decade. High-resolution geographic information systems (GIS) technology uses various data to create 3D visualizations that show magnificent geologic formations, ocean currents, sea life distribution, environmental impacts, mass animal migrations, volcanic action at plate boundaries, and more.

That vital mapping, as evidenced throughout this book, does not always bring good news: It records declining coral reefs, gyres filled with plastic, melting ice shelves, pockets of dying sea life, burgeoning toxic algae, and precious seafloor minerals as well. Still, we need it to continue, to be as accurate as possible. Because it is important for us to know this mysterious 70 percent of our planet. Because the ocean must be celebrated and protected. Because the ocean is the origin of our life—and because our lives depend on it.

Although stretched and compressed, Gerardus Mercator's 16th-century map provided navigators straight-line compass routes from point to point.

ARCTIC OCEAN

World Ocean

SEAFLOOR

ARCTIC CIRCLE

ASIA

NORTH
AMERICA

NORTH

PACIFIC

OCEAN

TROPIC OF CANCER

EQUATOR

INDIAN

OCEAN

SOUTH

TROPIC OF CAPRICORN

AUSTRALIA

PACIFIC

OCEAN

SOUTHERN OCEAN

ANTARCTIC CIRCLE

ANTARCTICA

ARCTIC OCEAN

ARCTIC CIRCLE

EUROPE

ASIA

NORTH

ATLANTIC

TROPIC OF CANCER

OCEAN

AFRICA

EQUATOR

SOUTH
AMERICA

SOUTH

INDIAN

TROPIC OF CAPRICORN

ATLANTIC

OCEAN

OCEAN

SOUTHERN OCEAN

ANTARCTIC CIRCLE

SCALE 1:86,475,000
1 CENTIMETER = 865 KILOMETERS; 1 INCH = 1365 STATUTE MILES

| 0 | 2000 | 4000 |
KILOMETERS

| 0 | 2000 | 4000 |
STATUTE MILES
Scale at the Equator

ANTARCTICA

World Ocean

··

POLITICAL MAP AND DEPTH CONTOURS

ARCTIC OCEAN

QUEEN ELIZABETH

BEAUFORT
SEA

Wrangel I.   CHUKCHI SEA

BROOKS RANGE

VICTORIA
ISLAND

ARCTIC CIRCLE

S I B E R I A   Lena   ALASKA
                        U.S.

Great Bear
Lake

R U S S I A

Lake
Baikal

Amur

KAMCHATKA
PENINSULA   Central Range   BERING SEA   Alaska
                                            Peninsula

60°

Date Line

Monday
Sunday

Aleutian Islands

ALEXANDER
ARCHIPELAGO

ROCKY MOUNTAINS

Great Slave
Lake

C A N A D A

SEA OF
OKHOTSK

Haida Gwai

Lake
Winnipeg

Sakhalin

Kuril Islands
Russia

Vancouver
Vancouver I.

Ulaanbaatar   Qiqihar   Harbin   Sapporo

MONGOLIA   Changchun   Jilin

GOBI   Shenyang

Baotou   Beijing   Pyongyang   NORTH
                                  KOREA

Sea of
Japan
East Sea

N O R T H   P A C I F I C

Salt Lake City   UNITED
                  STATES

Lanzhou   Tianjin   Seoul   SOUTH   Tokyo
                               KOREA

Denver

Xi'an   Qingdao   Ōsaka

C H I N A   Nanjing   Yellow
                       Sea   East
                              China
                              Sea

Los Angeles

Phoenix

30°N   Wuhan   Shanghai

Chongqing   Izu Islands
             Japan

San Diego   Dallas

Midway Is.
U.S.

Houston

Guiyang   Fuzhou   Bonin Is.
                    Japan

Guadalupe I.
Mexico

TROPIC OF CANCER   Xiamen   Taipei   Daito Is.   Volcano Is.   Marcus   HAWAIIAN ISLANDS   Honolulu

Guangzhou   TAIWAN   Japan   Japan   Japan   U.S.   Maui

Guadalajara

Hanoi   Hong Kong   Ryukyu Islands   Hawai'i   MEXICO

LAOS   Hainan   PHILIPPINE   Northern   Wake I.   Johnston Atoll   Revillagigedo Islands   Mexico City   GULF

THAILAND   South   SEA   Mariana Islands   U.S.   U.S.   Mexico

Bangkok   China   Saipan   GUATEMALA
                                                                                                         EL SALVADOR

MYANMAR   VIETNAM   Sea   Quezon City   Guam U.S.

(BURMA)   CAMBODIA   Manila   MICRONESIA   Clipperton
                                                            France

Kuala Lumpur   PHILIPPINES   World's greatest ocean depth   Bikini Atoll   MARSHALL
                              (36037 ft) 10984 m   ISLANDS

BRUNEI   PALAU   FEDERATED STATES OF MICRONESIA

SINGAPORE   MALAYSIA   Davao City   CAROLINE ISLANDS   O C E A N

Sumatra   Borneo   Sulu
                    Sea   Celebes
                          Sea   M   E   L   A   N

EQUATOR   Palmyra Atoll
           U.S.

I N D O N E S I A   Howland I.
                     U.S.

GREATER SUNDA ISLANDS   NAURU   Gilbert
                                  Islands   Baker I.   Jarvis I.   GALÁPAGOS ISLANDS
                                                                     U.S.   Ecuador

Jakarta   Java Sea   NEW GUINEA   PAPUA   KIRIBATI

Java   Surabaya   LESSER SUNDA IS.   NEW GUINEA   Phoenix   Marquesas Is.
                                                      Islands   France

Christmas I.   TIMOR-LESTE   TUVALU

Australia   Timor   MOLUCCAS   Timor
                                 Sea   SOLOMON   Tokelau
                                        ISLANDS   N.Z.   TUAMOTU ARCHIPELAGO

Arafura Sea   VANUATU   Wallis &   American

I N D I A N   CORAL   Futuna   Samoa   Tahiti
                       SEA   Fr.   U.S.

O C E A N   FIJI   SAMOA   French Polynesia
                                              France   Gambier Islands

GREAT SANDY   GREAT DIVIDING RANGE   New Caledonia   TONGA   Niue   Cook Islands
DESERT   France   N.Z.   New Zealand

Sea Level

100
250
500   A U S T R A L I A   Brisbane   Pitcairn Is.
1000   United Kingdom
1500   Great Victoria   Norfolk I.   TROPIC OF CAPRICORN   AUSTRAL ISLANDS   Pitcairn I.
        Desert   Australia

Perth   Great
         Australian   Adelaide   Sydney   Kermadec Is.   Easter Island   Salas y Gómez Island
         Bight   Canberra, A.C.T.   N.Z.   Chile   Chile

2000
2500   Darling   Melbourne   NORTH ISLAND

3000   Bass Strait   TASMAN   Wellington   NEW   S O U T H

3500   Tasmania   SEA   ZEALAND

Depth   4000   SOUTH ISLAND   Chatham Is.
contours   N.Z.
in meters   4500   P A C I F I C
5000
5500   Bounty Is.
6000   N.Z.
6500   Auckland Is.   Antipodes Is.
7000   N.Z.   N.Z.   O C E A N
7500   Campbell I.
        N.Z.
8000
8500   Macquarie I.
9000   Australia
9500
10000
10500

10984   ANTARCTIC CIRCLE   Balleny Islands

WILKES LAND   S O U T H E R N   O C E A N

SCALE 1:86,475,000
0   1000   2000
KILOMETERS
0   1000   2000
STATUTE MILES
Scale at the Equator

TRANSANTARCTIC MTS.   AMUNDSEN
                       SEA

ROSS SEA   M A R I E   B Y R D   L A N D   ELL
                                               AN

120°   150°E   180°   150°W   120°

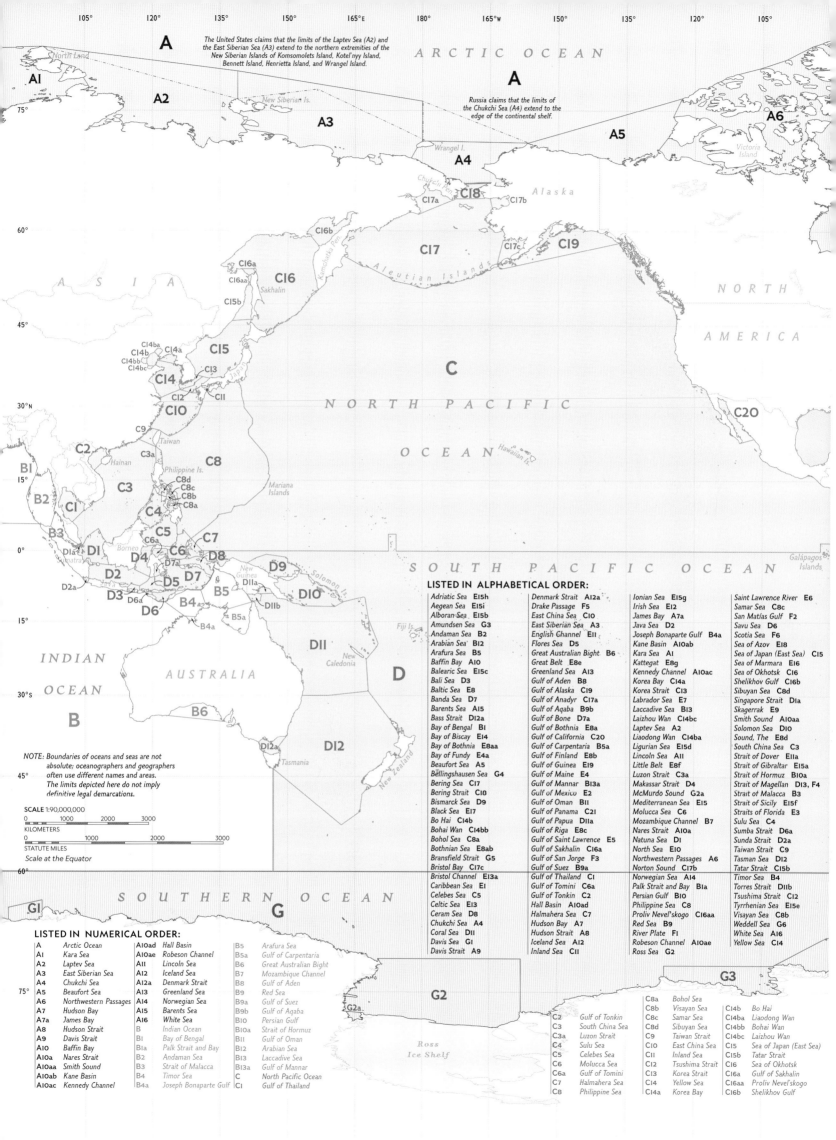

105° 120° 135° 150° 165°E 180° 165°W 150° 135° 120° 105°

A — The United States claims that the limits of the Laptev Sea (A2) and the East Siberian Sea (A3) extend to the northern extremities of the New Siberian Islands of Komsomolets Island, Kotel'nyy Island, Bennett Island, Henrietta Island, and Wrangel Island.

ARCTIC OCEAN

A

Russia claims that the limits of the Chukchi Sea (A4) extend to the edge of the continental shelf.

75°

A S I A

60°

45°

30°N

15°

0°

15°

30°S

45°

60°

NORTH AMERICA

NORTH PACIFIC OCEAN

C

SOUTH PACIFIC OCEAN

Galápagos Islands

INDIAN OCEAN

AUSTRALIA

B

NOTE: Boundaries of oceans and seas are not absolute; oceanographers and geographers often use different names and areas. The limits depicted here do not imply definitive legal demarcations.

SCALE 1:90,000,000
KILOMETERS
0   1000   2000   3000
STATUTE MILES
0   1000   2000   3000
Scale at the Equator

SOUTHERN OCEAN

G

Ross Ice Shelf

LISTED IN ALPHABETICAL ORDER:

Adriatic Sea E15h	Denmark Strait A12a	Ionian Sea E15g	Saint Lawrence River E6
Aegean Sea E15i	Drake Passage F5	Irish Sea E12	Samar Sea C8c
Alboran Sea E15b	East China Sea C10	James Bay A7a	San Matías Gulf F2
Amundsen Sea G3	East Siberian Sea A3	Java Sea D2	Savu Sea D6
Andaman Sea B2	English Channel E11	Joseph Bonaparte Gulf B4a	Scotia Sea F6
Arabian Sea B12	Flores Sea D5	Kane Basin A10ab	Sea of Azov E18
Arafura Sea B5	Great Australian Bight B6	Kara Sea A1	Sea of Japan (East Sea) C15
Baffin Bay A10	Great Belt E8e	Kattegat E8g	Sea of Marmara E16
Balearic Sea E15c	Greenland Sea A13	Kennedy Channel A10ac	Sea of Okhotsk C16
Bali Sea D3	Gulf of Aden B8	Korea Bay C14a	Shelikhov Gulf C16b
Baltic Sea E8	Gulf of Alaska C19	Korea Strait C13	Sibuyan Sea C8d
Banda Sea D7	Gulf of Anadyr C17a	Labrador Sea E7	Singapore Strait D1a
Barents Sea A15	Gulf of Aqaba B9b	Laccadive Sea B13	Skagerrak E9
Bass Strait D12a	Gulf of Bone D7a	Laizhou Wan C14bc	Smith Sound A10aa
Bay of Bengal B1	Gulf of Bothnia E8a	Laptev Sea A2	Solomon Sea D10
Bay of Biscay E14	Gulf of California C20	Liaodong Wan C14ba	Sound, The E8d
Bay of Bothnia E8aa	Gulf of Carpentaria B5a	Ligurian Sea E15d	South China Sea C3
Bay of Fundy E4a	Gulf of Finland E8b	Lincoln Sea A11	Strait of Dover E11a
Beaufort Sea A5	Gulf of Guinea E19	Little Belt E8f	Strait of Gibraltar E15a
Bellingshausen Sea G4	Gulf of Maine E4	Luzon Strait C3a	Strait of Hormuz B10a
Bering Sea C17	Gulf of Mannar B13a	Makassar Strait D4	Strait of Magellan D13, F4
Bering Strait C10	Gulf of Mexico E2	McMurdo Sound G2a	Strait of Malacca B3
Bismarck Sea D9	Gulf of Oman B11	Mediterranean Sea E15	Strait of Sicily E15f
Black Sea E17	Gulf of Panama C1	Molucca Sea C6	Straits of Florida E3
Bo Hai C14b	Gulf of Papua D11a	Mozambique Channel B7	Sulu Sea C4
Bohai Wan C14bb	Gulf of Riga E8c	Nares Strait A10a	Sumba Strait D6a
Bohol Sea C8a	Gulf of Saint Lawrence E5	Natuna Sea D1	Sunda Strait D2a
Bothnian Sea E8ab	Gulf of Sakhalin C16a	North Sea E10	Taiwan Strait C9
Bransfield Strait G5	Gulf of San Jorge F3	Northwestern Passages A6	Tasman Sea D12
Bristol Bay C17c	Gulf of Suez B9a	Norton Sound C17b	Tatar Strait C15b
Bristol Channel E13a	Gulf of Thailand C1	Norwegian Sea A14	Timor Sea B4
Caribbean Sea E1	Gulf of Tomini C6a	Palk Strait and Bay B1a	Torres Strait D11b
Celebes Sea C5	Gulf of Tonkin C2	Persian Gulf B10	Tsushima Strait C12
Celtic Sea E13	Hall Basin A10ad	Philippine Sea C8	Tyrrhenian Sea E15e
Ceram Sea D8	Halmahera Sea C7	Proliv Nevel'skogo C16aa	Visayan Sea C8b
Chukchi Sea A4	Hudson Bay A7	Red Sea B9	Weddell Sea G6
Coral Sea D11	Hudson Strait A8	River Plate F1	White Sea A16
Davis Sea G1	Iceland Sea A12	Robeson Channel A10ae	Yellow Sea C14
Davis Strait A9	Inland Sea C11	Ross Sea G2	

LISTED IN NUMERICAL ORDER:

A	Arctic Ocean	A10ad	Hall Basin	B5	Arafura Sea	
A1	Kara Sea	A10ae	Robeson Channel	B5a	Gulf of Carpentaria	
A2	Laptev Sea	A11	Lincoln Sea	B6	Great Australian Bight	
A3	East Siberian Sea	A12	Iceland Sea	B7	Mozambique Channel	
A4	Chukchi Sea	A12a	Denmark Strait	B8	Gulf of Aden	
A5	Beaufort Sea	A13	Greenland Sea	B9	Red Sea	
A6	Northwestern Passages	A14	Norwegian Sea	B9a	Gulf of Suez	
A7	Hudson Bay	A15	Barents Sea	B9b	Gulf of Aqaba	
A7a	James Bay	A16	White Sea	B10	Persian Gulf	
A8	Hudson Strait	B	Indian Ocean	B10a	Strait of Hormuz	
A9	Davis Strait	B1	Bay of Bengal	B11	Gulf of Oman	
A10	Baffin Bay	B1a	Palk Strait and Bay	B12	Arabian Sea	
A10a	Nares Strait	B2	Andaman Sea	B13	Laccadive Sea	
A10aa	Smith Sound	B3	Strait of Malacca	B13a	Gulf of Mannar	
A10ab	Kane Basin	B4	Timor Sea	C	North Pacific Ocean	
A10ac	Kennedy Channel	B4a	Joseph Bonaparte Gulf	C1	Gulf of Thailand	

C2	Gulf of Tonkin	C8a	Bohol Sea	C14b	Bo Hai
C3	South China Sea	C8b	Visayan Sea	C14ba	Liaodong Wan
C3a	Luzon Strait	C8c	Samar Sea	C14bb	Bohai Wan
C4	Sulu Sea	C8d	Sibuyan Sea	C14bc	Laizhou Wan
C5	Celebes Sea	C9	Taiwan Strait	C15	Sea of Japan (East Sea)
C6	Molucca Sea	C10	East China Sea	C15b	Tatar Strait
C6a	Gulf of Tomini	C11	Inland Sea	C16	Sea of Okhotsk
C7	Halmahera Sea	C12	Tsushima Strait	C16a	Gulf of Sakhalin
C8	Philippine Sea	C13	Korea Strait	C16aa	Proliv Nevel'skogo
		C14	Yellow Sea	C16b	Shelikhov Gulf
		C14a	Korea Bay		

World Ocean

SEA LIMITS

SOUTHERN OCEAN
There is international agreement that the icy waters around Antarctica form a distinct ocean region. While there is no consensus on its name or its extent, most countries call it the Southern Ocean and use 60° south latitude as an approximation of its northern limit.

		D11 Coral Sea	E8 Baltic Sea	E13 Celtic Sea	F South Atlantic Ocean
		D11a Gulf of Papua	E8a Gulf of Bothnia	E13a Bristol Channel	F1 River Plate
		D11b Torres Strait	E8aa Bay of Bothnia	E14 Bay of Biscay	F2 San Matías Gulf
	D1 Natuna Sea	D12 Tasman Sea	E8ab Bothnian Sea	E15 Mediterranean Sea	F3 Gulf of San Jorge
	D1a Singapore Strait	D12a Bass Strait	E8b Gulf of Finland	E15a Strait of Gibraltar	F4 Strait of Magellan
	D2 Java Sea	D13 Strait of Magellan	E8c Gulf of Riga	E15b Alboran Sea	F5 Drake Passage
	D2a Sunda Strait	E North Atlantic Ocean	E8d The Sound	E15c Balearic Sea	F6 Scotia Sea
	D3 Bali Sea	E1 Caribbean Sea	E8e Great Belt	E15d Ligurian Sea	G Southern Ocean
	D4 Makassar Strait	E2 Gulf of Mexico	E8f Little Belt	E15e Tyrrhenian Sea	G1 Davis Sea
C17 Bering Sea	D5 Flores Sea	E3 Straits of Florida	E8g Kattegat	E15f Strait of Sicily	G2 Ross Sea
C17a Gulf of Anadyr	D6 Savu Sea	E4 Gulf of Maine	E9 Skagerrak	E15g Ionian Sea	G2a McMurdo Sound
C17b Norton Sound	D6a Sumba Strait	E4a Bay of Fundy	E10 North Sea	E15h Adriatic Sea	G3 Amundsen Sea
C17c Bristol Bay	D7 Banda Sea	E5 Gulf of St. Lawrence	E11 English Channel	E15i Aegean Sea	G4 Bellingshausen Sea
C18 Bering Strait	D7a Gulf of Bone	E6 St. Lawrence River	E11a Strait of Dover	E16 Sea of Marmara	G5 Bransfield Strait
C19 Gulf of Alaska	D8 Ceram Sea	E7 Labrador Sea	E12 Irish Sea	E17 Black Sea	G6 Weddell Sea
C20 Gulf of California	D9 Bismarck Sea			E18 Sea of Azov	
C21 Gulf of Panama	D10 Solomon Sea			E19 Gulf of Guinea	
D South Pacific Ocean					

I ATLANTIC OCEAN

Atlantic Ocean

Formation of the Atlantic Ocean began 180 million years ago as Earth's shifting plates tore apart the vast continent of Pangaea, separating Europe and Africa to the east from the Americas to the west. Look closely at the puzzle piece–like coasts that face each other across the water. Their outlines would fit together if cut out with scissors on a map and pushed together. As the continents and coasts on either side filled with inhabitants and lands were mapped in their own primitive fashion, the vast blue expanse between them remained uncharted.

Some 2,500 years after the first navigators spanned the Pacific in agile outriggers, a clan of Vikings in what is now Norway looked west across their Atlantic waters, then ventured forth in sturdy wooden plank boats held together with iron rivets. It was the 10th century, 500 years before Columbus crossed the same ocean to reach the New World.

While coastal landmarks along the shore likely guided the Norsemen toward open water, once there they needed other navigational keys, certainly the position of the sun and stars, and perhaps the routes of migrating birds. Sagas, rather than written logs, were their records, leaving historians to cobble together cartographic information from their voyages. One 20th-century translation of Viking lore gave a heart-stopping account of a boat "beset by fogs and north winds until they lost all track of their course" and were supremely *hayfilla* (bewildered). Some experts think they may have had primitive instruments, perhaps crystals of calcite, called sunstones, which could read the position of the sun even behind clouds. With little other guidance, they reached Greenland, 1,609 kilometers west of their home. There is increasing evidence that in 1001 Leif Erickson, or fellow Vikings, set foot on the shores of Newfoundland in today's Canada.

In 1492 Columbus made his first voyage across the Atlantic using portolan charts. Since then, cartographers have mapped the coastal outlines on either side of the ocean basin, its currents such as the Gulf Stream, and, gradually—beginning with the H.M.S. *Challenger* expedition in 1872 and continuing with the 1968 map by Bruce Heezen and Marie Tharp—developed an overview of the depth and makeup of the seafloor with its rift valleys, fracture zones, abyssal plains, and many seamounts.

The work continues, ever more accurate—and diverse. The mapping shown throughout this book is as general as an outline bordered by continents and as specific as that of temperature pockets, hydrothermal vents, endangered deep-water coral reefs, plastic-polluted gyres, and the migration route of the endangered North Atlantic right whale. By clarifying places, positions, populations, even pollution, cartography helps us envision—and care for—what is otherwise not visible.

BY THE NUMBERS

Atlantic Ocean

....

Total Area
81,705,396 square kilometers

Total Volume
307,923,430 cubic kilometers

Greatest Depth
8,605 meters (Puerto Rico Trench)

Includes 25 percent of Earth's water
(excludes North Sea, Baltic Sea,
Mediterranean Sea, and Irish Sea)

NORTH ATLANTIC

Geographic Boundaries
Equator to 60° N / 98° W to 2° W

Average Depth
3,408 meters

SOUTH ATLANTIC

Geographic Boundaries
Equator to 60° S / 70° W to 20° E

Average Depth
3,967 meters

RIGHT: Sculpted by the Atlantic, cliffs tower 40 meters above Scotland's Isle of Lewis. PAGES 434-435: Toxin-rich tentacles of a Portuguese man-of-war may stream 50 meters from its body to capture prey.

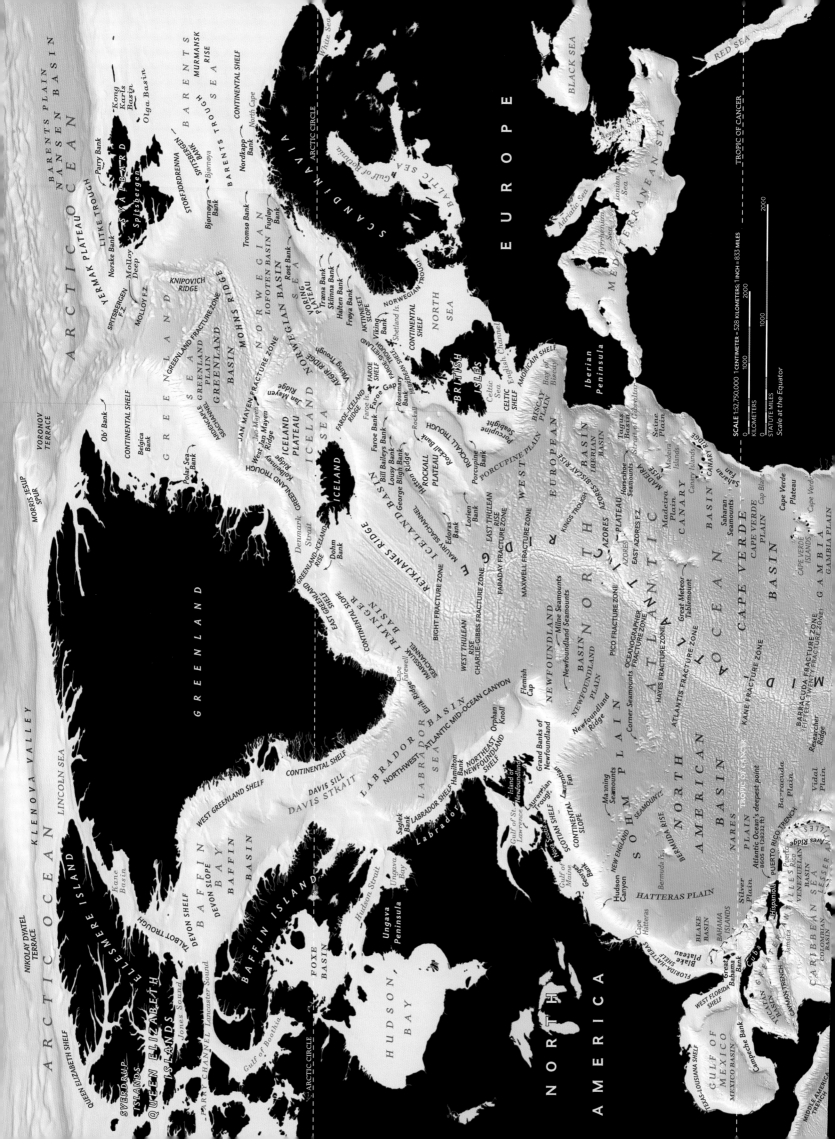

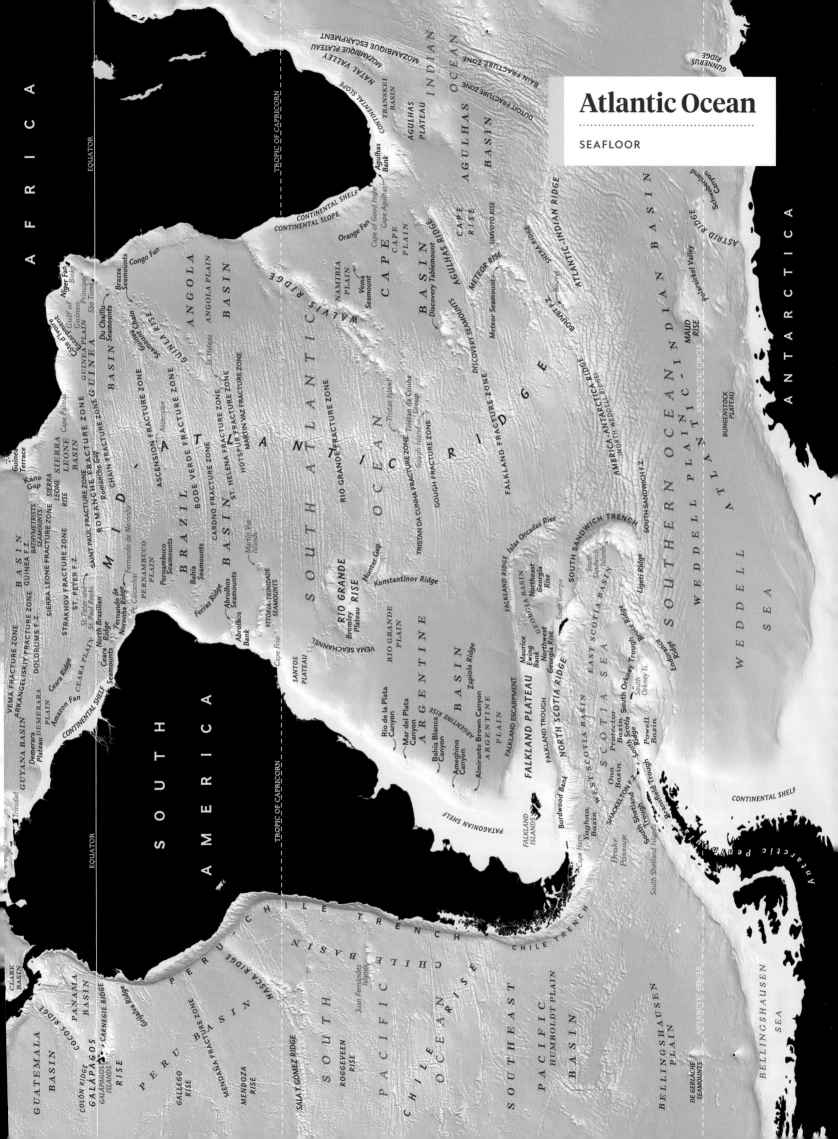

Atlantic Ocean

SEAFLOOR

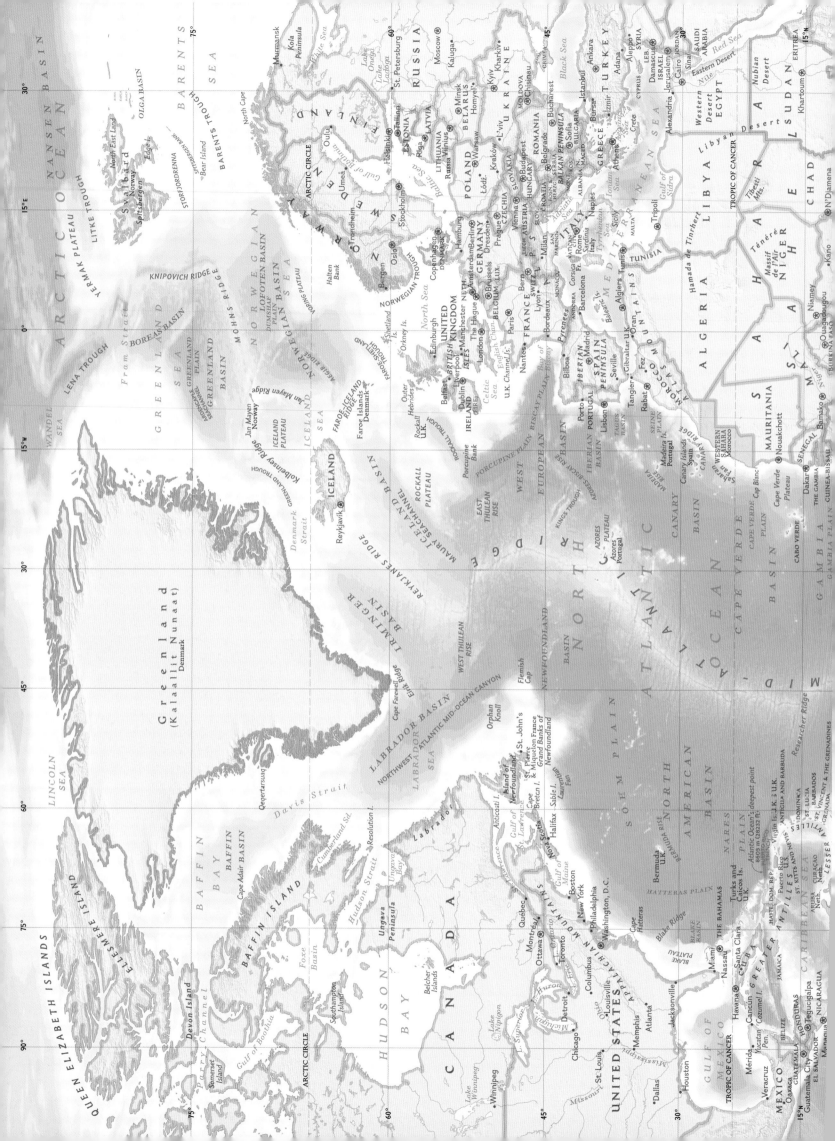

Atlantic Ocean

POLITICAL MAP AND DEPTH CONTOURS

Mediterranean Sea

Sparkling against sun-drenched, salt-embedded cliffs, the Mediterranean unfolds its languid boundary of blue between Europe to the north, Africa to the south, and Asia to the east. A niche of the Atlantic celebrated for its temperate waters, the Mediterranean's past is a stormy one.

A remnant of the Tethys Sea born 200 million years ago as the continents of Europe, Asia, and America were torn apart by plate tectonics, the Mediterranean has remained. Later plate action squeezed the Tethys between Africa and Europe, forced its seafloor beneath Europe, cut off its connection to the Atlantic Ocean, and dried it into a massive bed of salt. When yet more plate action sent the Atlantic pouring back into the salty bed through today's Strait of Gibraltar, the modern Mediterranean was born.

This descendant of Tethys has its own stormy saga, documented by the many sunken ships below its surface. The ancient Phoenicians, Greeks, Romans, and Egyptians plied the Mediterranean, reading the rocky, salt-rich coastlines, navigating swirling currents and shoals that ripped apart ships. Greek poet Homer's *Odyssey* tells of Odysseus's ships dashed on the rocks of the Illyrian Coast, today's Balkan Peninsula. In the Bible's New Testament, the Apostle Paul survived a massive storm that splintered his ship off Malta. Despite the serious challenges behind these stories, seafarers continued to fish and trade, recording the nature of the water and coastlines they traveled. From their records came the second-century world map of the Greek geographer Ptolemy, with the Mediterranean prominently displayed.

Through time, as portolan charts and then Mercator maps became more accurate, the Mediterranean has been intricately mapped with intriguing coastlines that feature many smaller seas, including the Aegean and the Adriatic. Thanks to archaeologists led by George Bass, Robert Ballard, and others, the shipwreck-strewn seafloors and the challenging rocky coasts, currents, and winds their crews navigated are being defined, providing a more comprehensive view of early life and commerce.

Over recent decades, maps show geologic features such as deep-sea hydrothermal vents, a submarine ridge off Sicily, surface currents, temperature, climate change, and biodiversity, including invasive species. Still other maps plot the ongoing march of the ancient Tethys seabed, which has not gone quietly. Plate motion continues to force it below the European continent and has for millennia been triggering major earthquakes and volcanic eruptions from Lisbon to Istanbul, not the least of them Vesuvius in A.D. 79, which buried Pompeii in hot ash.

> "THE MEDITERRANEAN DRIED UP COMPLETELY 6 MILLION YEARS AGO. DURING A . . . GLOBAL ICE AGE . . . SEA LEVEL WAS LOWERED BELOW THE NATURAL DAM AT GIBRALTAR AND THE MEDITERRANEAN WAS CUT OFF FROM ATLANTIC WATERS."
>
> —L. K. GLOVER, MARINE GEOLOGIST

Blithely pulsing through the blue, the common octopus has a name that belies its uncommon intellect and talent as a predator, escape artist, and parent.

Gulf of Mexico

Across the Gulf of Mexico birds skim and dip to feed; to the west, a glorious sunset transforms the crystalline blue surface to a shade of rich, deep rose. One can watch it all, standing several kilometers offshore, feet firmly planted on the seafloor, head comfortably above the temperate water. This is just one wonder of the Gulf. These shallow, broad, sunlit margins—a submerged ancient coast—support benthic algae, seagrass meadows, and invertebrates, as well as more than 15,000 species of fish and other life in the deep sea beyond, including chemosynthetic microbes that power diverse life around cold seeps.

This ninth largest body of water on the planet covers about 3.9 million square kilometers between North America, South America, and Cuba, with the deepest part, Sigsbee Deep, 3,658 meters down. While the basin itself was born of plate action that separated the North and South American continents some 200 million years ago, it continued shaping as seawater rushing in and out deposited extensive layers of salt, and, 65 million years ago, as tsunamis and earthquakes generated by an asteroid impact on the Yucatán Peninsula (the Chicxulub Crater) massively disrupted its seafloors and continental slopes, and brought an end to the dinosaurs.

The Gulf seafloor is also a productive petroleum field, from relatively shallow waters to depths of 2,150 meters. Depressions of salt-rich brine lakes dot the seafloor, a product of majestic salt domes thousands of meters tall, called diapirs, which sometimes break the sea surface, forming islands up to eight kilometers across. Both are residuals of excess salt trapped in the Gulf as it separated from the Atlantic.

While the Gulf's basin is mostly surrounded by land, it is connected by water to both land and sea. Not only does it receive water from the rivers of three nations, including the massive Mississippi; it also supports one of the ocean's most powerful currents—the Gulf Stream. Warm Caribbean water, called the Loop Current in the Gulf, enters through a narrow opening at the Yucatán Peninsula, moves clockwise around the Gulf, and exits through the Straits of Florida, sweeping northward along North America's eastern coast and then bending toward the United Kingdom. Its warm waters fuel tropical hurricanes, which often wreak havoc far inland.

Scientists through time have documented the Gulf's surface waters, but only over the past several decades have they mapped the details of its seafloor with its unique geological formations, world-impacting current, and chemosynthetic colonies. NOAA's *Gulf of Mexico Data Atlas* alone has 288 maps on 90 topics. They have also mapped its shipwrecks, hurricane tracks, and the petroleum deposits that provide energy for burgeoning world economies but also support construction of oil rigs, a massive network of pipelines, and some disastrous oil spills that have had serious impacts.

"JUST A SHORT DISTANCE OFFSHORE AND ALL THE WAY TO MEXICO AND CUBA ARE SOME OF THE BLUEST WATERS ON EARTH, WITH A RICH DIVERSITY OF MARINE LIFE AND HABITATS."

—JOHN WESLEY TUNNELL, JR., HARTE RESEARCH INSTITUTE

Exquisite as a jewel-encrusted brooch, the bay scallop, denizen of seagrass meadows in the Gulf of Mexico, scans seas for predators with its 20 pairs of baby blue eyes.

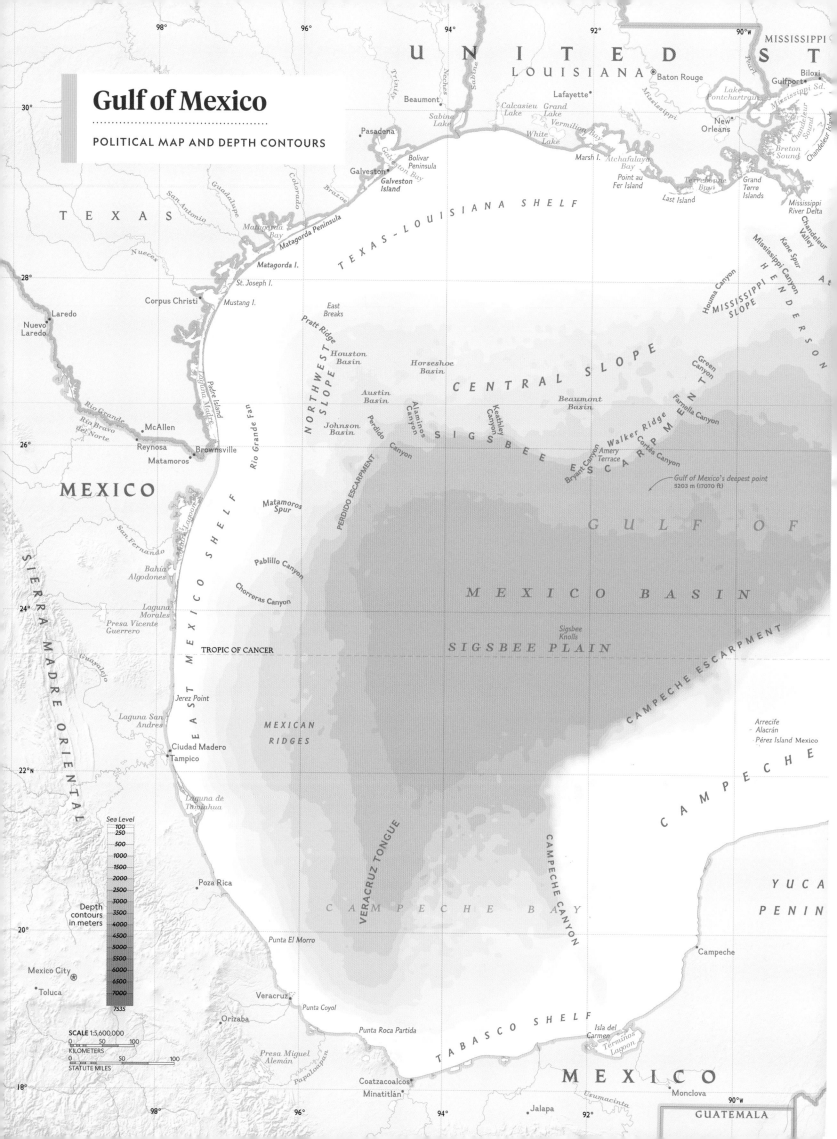

Gulf of Mexico

POLITICAL MAP AND DEPTH CONTOURS

UNITED ST

MISSISSIPPI

LOUISIANA

Baton Rouge

Biloxi
Gulfport
Lake
Pontchartrain
New
Orleans
Chandeleur
Sound
Breton
Sound
Chandeleur Is.
Mississippi
River Delta
Chandeleur
Valley

Lafayette

Beaumont
Neches
Sabine
Trinity
Calcasieu
Lake
Grand
Lake
Vermilion
Lake
White
Lake
Vermilion Bay
Marsh I.
Atchafalaya
Bay
Point au
Fer Island
Last Island
Terrebonne
Bay
Grand
Terre
Islands

Pasadena
Sabine
Lake

Bolivar
Peninsula
Galveston
Galveston Bay
Galveston
Island

Kane Spur
Houma Canyon
Mississippi Canyon
At
Mississippi
SLOPE
HENDERSON

TEXAS-LOUISIANA SHELF

TEXAS

Guadalupe
San Antonio
Colorado
Brazos

Matagorda
Bay
Matagorda Peninsula
Matagorda I.

Nueces
St. Joseph I.
Corpus Christi
Mustang I.

East
Breaks

Pratt Ridge

Houston
Basin

Horseshoe
Basin

CENTRAL SLOPE

Green
Canyon

Beaumont
Basin

Farnella Canyon

NORTHWEST SLOPE

Austin
Basin

Alaminos
Canyon

Keathley
Canyon

Walker Ridge

Cortés Canyon
Amery
Terrace
Bryant Canyon

Gulf of Mexico's deepest point
5203 m (17070 ft)

Laredo
Nuevo
Laredo

Rio Grande
Rio Bravo
del Norte
McAllen
Reynosa
Brownsville
Matamoros

SIGSBEE

E

SCARPMENT

Johnson
Basin

Perdido
Canyon

Rio Grande Fan

Padre Island

Laguna Madre

PERDIDO ESCARPMENT

GULF OF

MEXICO

MEXICO BASIN

Matamoros
Spur

San Fernando

Madre Lagoon

Bahía
Algodones

Laguna
Morales
Presa Vicente
Guerrero

TROPIC OF CANCER

Pablillo Canyon

Chorreras Canyon

EAST MEXICO SHELF

Sigsbee
Knolls

SIGSBEE PLAIN

CAMPECHE ESCARPMENT

Guayalejo

Laguna San
Andres

Jerez Point

MEXICAN
RIDGES

Arrecife
Alacrán
Pérez Island Mexico

SIERRA MADRE ORIENTAL

Ciudad Madero
Tampico

Laguna de
Tamiahua

VERACRUZ TONGUE

CAMPECHE CANYON

CAMPECHE

Sea Level
100
250
500
1000
1500
2000
2500
3000
3500
4000
4500
5000
5500
6000
6500
7000
7535

Depth
contours
in meters

Poza Rica

YUCA
PENIN

Mexico City
Toluca

Punta El Morro

CAMPECHE BAY

Campeche

SCALE 1:5,600,000
0   50   100
KILOMETERS
0   50   100
STATUTE MILES

Veracruz
Orizaba
Punta Coyol

Punta Roca Partida

Presa Miguel
Alemán

Papaloapan

Coatzacoalcos
Minatitlán

TABASCO SHELF

Isla del
Carmen
Términos
Lagoon

Usumacinta

MEXICO

Jalapa

Monclova

GUATEMALA

30°
28°
26°
24°
22°N
20°
18°

98°   96°   94°   92°   90°W

Caribbean Sea

Native to its more than 7,000 islands, the Carib Indians gave the Caribbean Sea its name, and their god Huricán became the name for the deadly storms that spawn there. Outside of hurricane season, any traveler can understand why the Caribbean's tropical waters and islands form a vacationer's paradise. But its picturesque palm trees, sparkling white sand beaches, and serene crystalline waters belie a violent past.

Once fully integrated into the Atlantic, the Caribbean began claiming its own identity about 80 million years ago. Part of the small Caribbean tectonic plate was subducted under the Atlantic oceanic plate off the island of Puerto Rico, forming the North Atlantic's deepest area, the Puerto Rico Trench. Another portion of the Caribbean and South American plates overrode the North American plate, building an island arc of active volcanoes today called the Lesser Antilles. From the northeast coast of Venezuela, these islands circle east around the Caribbean like a jeweled necklace, curving west to join the Greater Antilles of Hispaniola and Cuba, which arc southwest to the Yucatán.

The blue basin between these nations has harbored some of the most stunning coral reefs in the world—including the Great Belize Barrier Reef Reserve System, "the most remarkable reef in the West Indies," wrote naturalist Charles Darwin. A World Heritage site since 1996, it is rich with vibrant coral formations and mangrove forests that host threatened loggerhead sea turtles, manatees, and red-footed boobies. These animals and more find sanctuary in its 963-square-kilometer system stretching from the Yucatán Peninsula south to the reefs of Honduras, second only in size and length to Australia's Great Barrier Reef.

Early maps scrawled in Columbus's 1492 journal depict the island of Hispaniola, today's Haiti and Dominican Republic. A map made in 1500 by Juan de la Cosa—Columbus's pilot and navigator—gives an intricately rendered first look at the loop of islands from the Yucatán down to Venezuela. Around 1600, Englishman Sir Walter Raleigh, Queen Elizabeth I's cherished "sea dog"—both pirate and explorer—had inscribed a unique map of an area he called El Dorado ("the golden one"), today's Guyana. But all did not come together until 1680, when a complete and accurate map of the islands and mainland encircling the small sea was rendered by Dutch cartographer Johannes van Keulen, complete with latitude and longitude and realistic images of the peoples who lived there.

Since then, the Caribbean's trenches, hurricane paths, reef systems, and regions of extraordinary marine life have been carefully documented and mapped. But van Keulen's map may be the most accurate in relaying its magical quality.

"THE CARIBBEAN SEA, ONE OF THE WORLD'S MOST ALLURING BODIES OF WATER, A RARE GEM AMONG THE OCEANS, DEFINED BY THE ISLANDS THAT FORM A CHAIN OF LOVELY JEWELS TO THE NORTH AND EAST."

—JAMES A. MICHENER, *CARIBBEAN*

Caped in Caribbean sapphire and rich with marine life, the Bahamian island district of Exuma includes 365 small islands, or cays, in its 100-square-kilometer area.

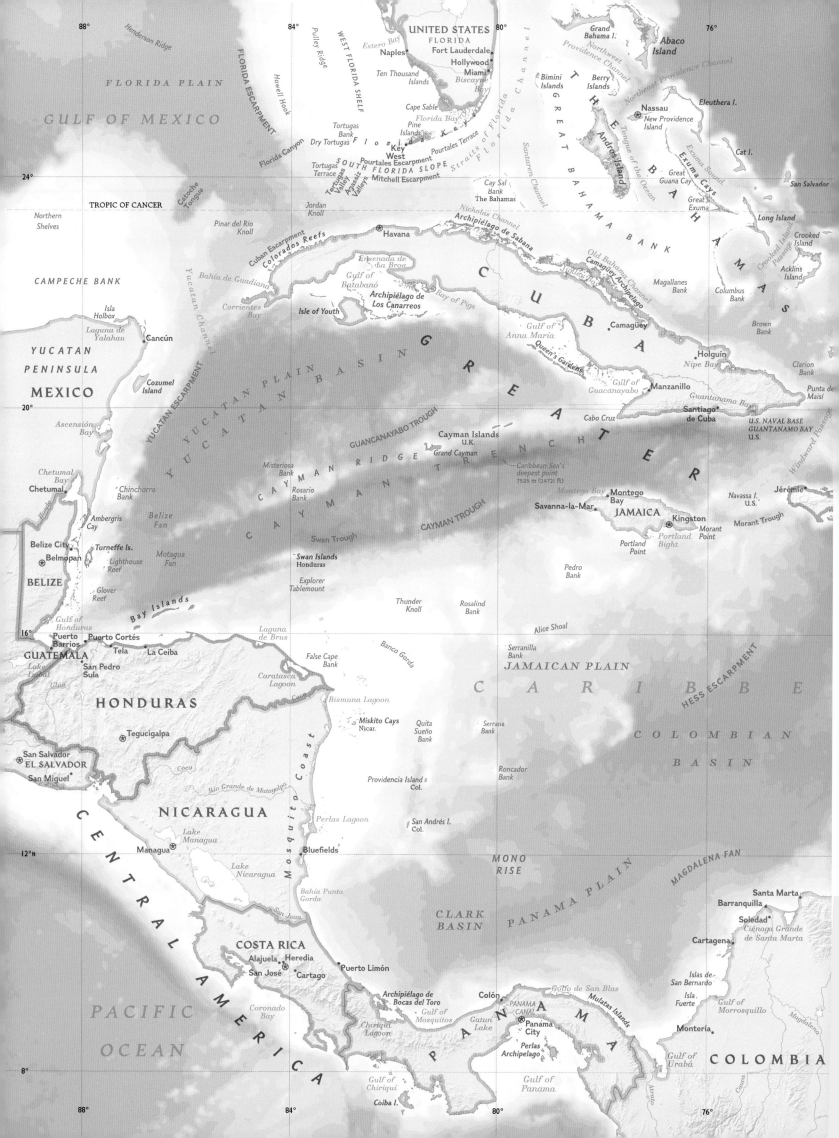

88°

84°

80°

76°

UNITED STATES
FLORIDA

Henderson Ridge

Estero Bay
Naples

Fort Lauderdale
Hollywood
Miami

Grand
Bahama I.

Abaco
Island

Pulley Ridge

WEST FLORIDA SHELF

Ten Thousand
Islands

Biscayne
Bay

*Northwest
Providence Channel*

FLORIDA PLAIN

Howell Hook

FLORIDA ESCARPMENT

Cape Sable

Florida Bay
Pine
Islands

Bimini
Islands

Berry
Islands

Northeast Providence Channel

Eleuthera I.

GULF OF MEXICO

Florida Keys

Nassau
New Providence
Island

*Tortugas
Bank*
Dry Tortugas

Key
West

Pourtales Terrace

Straits of Florida

*G
R
E
A
T*

Andros Island

Tongue of the Ocean

Exuma Sound

Exuma Cays

Cat I.

24°

TROPIC OF CANCER

*Tortugas
Terrace*

*Tortugas
Valleys*

SOUTH FLORIDA SLOPE
*Agassiz
Valleys*

Mitchell Escarpment

Cay Sal
Bank
The Bahamas

Santaren Channel

*B
A
H
A
M
A
B
A
N
K*

*T
H
E*

*B
A
H
A
M
A
S*

San Salvador

*Northern
Shelves*

Jordan
Knoll

Nicholas Channel

Archipiélago de Sabana

Old Bahama Channel

Magallanes
Bank

Long Island

Pinar del Río
Knoll

Havana

Camagüey Archipelago

Columbus
Bank

Crooked
Island

*Crooked Island
Passage*

CAMPECHE BANK

Cuban Escarpment
Colorados Reefs

Bahía de Guadiana

*Ensenada de
la Broa*

*Gulf of
Batabanó*

*C
U
B
A*

Jigüey Bay

Camagüey

Brown
Bank

Acklins
Island

Isla
Holbox

Archipiélago de
Los Canarreos

Bay of Pigs

*Gulf of
Anna María*

Holguín

Nipe Bay

Clarion
Bank

*Laguna de
Yalahau*

Cancún

*Corrientes
Bay*

Isle of Youth

*G
R
E
A
T
E
R*

20°

**YUCATAN
PENINSULA**

MEXICO

Cozumel
Island

Yucatan Channel

YUCATAN ESCARPMENT

YUCATAN PLAIN

YUCATAN BASIN

GUANCANAYABO TROUGH

Queen's Gardens

*Gulf of
Guacanayabo*

Manzanillo

Guantanamo Bay

Santiago
de Cuba

Punta
de Maisí

Cabo Cruz

U.S. NAVAL BASE
GUANTANAMO BAY
U.S.

*Ascensión
Bay*

Cayman Islands
U.K.

Grand Cayman

*T
R
E
N
C
H*

Windward Passage

*Chetumal
Bay*

Chetumal

*Chinchorro
Bank*

*Misteriosa
Bank*

CAYMAN RIDGE

Caribbean Sea's
deepest point
7535 m (24721 ft)

Montego Bay

Montego
Bay

Navassa I.
U.S.

Jérémie

*Rosario
Bank*

*C
A
Y
M
A
N*

Savanna-la-Mar

JAMAICA

Kingston

Ambergris
Cay

Belize City

*Belize
Fan*

CAYMAN TROUGH

Swan Trough

*Morant
Point*

Morant Trough

Turneffe Is.

BELIZE
Belmopán

*Lighthouse
Reef*

*Motagua
Fan*

Swan Islands
Honduras

*Explorer
Tablemount*

*Pedro
Bank*

Portland
Point

*Portland
Bight*

*Glover
Reef*

16°

Bay Islands

*Thunder
Knoll*

*Rosalind
Bank*

Alice Shoal

*Gulf of
Honduras*

Puerto
Barrios

Puerto Cortés

*Laguna
de Brus*

Serranilla
Bank

JAMAICAN PLAIN

GUATEMALA

Tela
La Ceiba

False Cape
Bank

Banco Gorda

*C
A
R
I
B
B
E
E*

HESS ESCARPMENT

San Pedro
Sula

*Caratasca
Lagoon*

Bismuna Lagoon

Miskito Cays
Nicar.

Quita
Sueño
Bank

Serrana
Bank

*C
O
L
O
M
B
I
A
N*

HONDURAS

Coco

Tegucigalpa

Roncador
Bank

*B
A
S
I
N*

San Salvador
EL SALVADOR

Coco

Providencia Island
Col.

San Miguel

Río Grande de Matagalpa

*M
o
s
q
u
i
t
o
C
o
a
s
t*

NICARAGUA

Perlas Lagoon

San Andrés I.
Col.

12°N

*Lake
Managua*

Managua

Bluefields

*MONO
RISE*

CENTRAL

*Lake
Nicaragua*

*Bahía Punta
Gorda*

**CLARK
BASIN**

PANAMA PLAIN

MAGDALENA FAN

Santa Marta

San Juan

Barranquilla

Soledad

*Ciénaga Grande
de Santa Marta*

Cartagena

COSTA RICA

Alajuela Heredia

San José
Cartago

Puerto Limón

Colón

Golfo de San Blas

Mulatas Islands

Islas de
San Bernardo

Isla
Fuerte

Montería

PACIFIC

*Coronado
Bay*

Archipiélago de
Bocas del Toro

*A
M
E
R
I
C
A*

PANAMA
CANAL

*Gatun
Lake*

*P
A
N
A
M
A*

Panama
City

*Gulf of
Morrosquillo*

Magdalena

*Chiriquí
Lagoon*

*Gulf of
Mosquitos*

Perlas
Archipelago

COLOMBIA

*Gulf of
Urabá*

OCEAN

*Gulf of
Chiriquí*

Coiba I.

*Gulf of
Panama*

Atrato

Cauca

8°

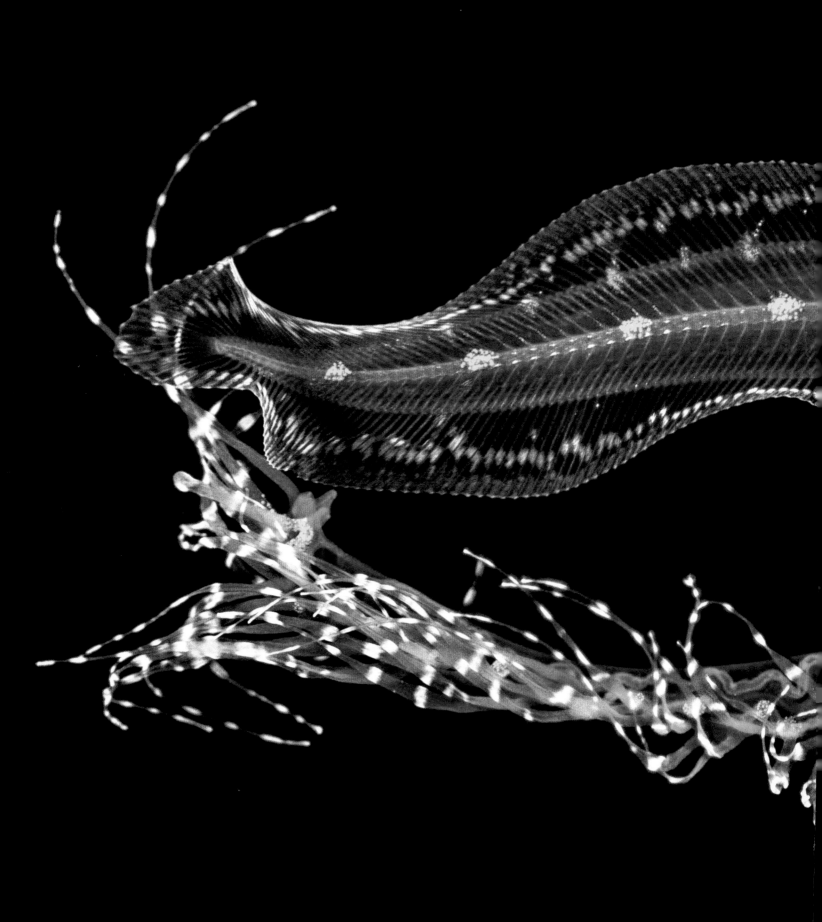

Pacific Ocean

When Portuguese explorer Ferdinand Magellan launched his round-the-world voyage beginning in 1519, it took his ships more than a month to sail from raging Atlantic waters around the tip of South America—later named the Strait of Magellan—into the Pacific. As he emerged from the stormy Atlantic into a "beautiful, quiet ocean," he called it Mar Pacifico, or "peaceful sea." It was not only peaceful but also so big that it took his crew another year to cross it, stopping in the Philippines and at other islands along the way, with Magellan himself perishing during the voyage.

Today we know that the Pacific is not only the oldest and largest ocean basin—all the others would easily fit inside it—but a body of many other superlatives as well: Its surface area is greater than all of Earth's landmasses combined; its waters can be mirror calm but can also thrash with Earth's most violent storms; its surrounding Ring of Fire supports many of Earth's most active volcanoes, with devastating earthquakes and tsunamis; its waters are the most biologically diverse, thriving with millions of species; and its seafloor holds both Earth's deepest place, the Mariana Trench, and its highest mountain from seafloor to summit, Hawaii's Mauna Kea, as well as the most islands—some 25,000—and an estimated 30,000 submerged mountains, or seamounts.

The prominent German cartographer Martin Waldseemüller's 1507 map showed that the Americas separated two distinct oceans. But only after the Magellan expedition did a full Pacific come into view in a 1529 map by Portuguese cartographer Diogo Ribeiro, who compiled the notes of the crew. It was the first to show the Pacific in accurate dimensions, and has been called the first scientific world map.

From the 1600s onward, Dutch, Portuguese, Spanish, and British sailed the Pacific to trade with spice islanders, to pirate, and to claim land for their kingdoms. From 1768 to 1780 British captain James Cook crisscrossed the Pacific three times, mapping coasts and intricately recording flora and fauna of land and sea. In 1836, aboard the H.M.S. *Beagle,* Charles Darwin explored the Pacific's remote volcanic Galápagos archipelago, 1,369 kilometers west of Ecuador, where unique life-forms had evolved.

A fuller picture of the inner Pacific began to emerge a few decades later, when the H.M.S. *Challenger* expedition circled the globe recording depth, temperature, currents, and 4,700 new species. Covering some 127,000 kilometers, its depth soundings first identified the Mariana Trench, Earth's deepest place, in the western Pacific. Over the last few decades, scientists have used increasingly sophisticated equipment, from aquatic drones and mini robots to satellite imaging and geospatial modeling, to chart even more of the Pacific's silent life, tectonic tumult, deep-coral ecosystems, and life-giving currents.

BY THE NUMBERS

Pacific Ocean

....

Total Area
152,617,159 square kilometers

Total Volume
645,369,567 cubic kilometers

Greatest Depth
11 kilometers (Challenger Deep, in the Mariana Trench, North Pacific)

Includes 46 percent of the Earth's water area, the largest water expanse
(excludes the East Asian Sea and Bering Sea)

NORTH PACIFIC

Geographic Boundaries
Equator to 64° N / 130° E to 30° W

Average Depth
4,573 meters

SOUTH PACIFIC

Geographic Boundaries
Equator to 60° S / 145° E to 70° W

Average Depth
3,935 meters

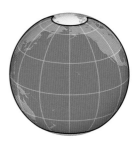

RIGHT: A cormorant dives into a galaxy of small fish off Southern California. PAGES 454-455: Off Hawaii's Kona coast at night, diaphanous veils cloak a larval cusk eel.

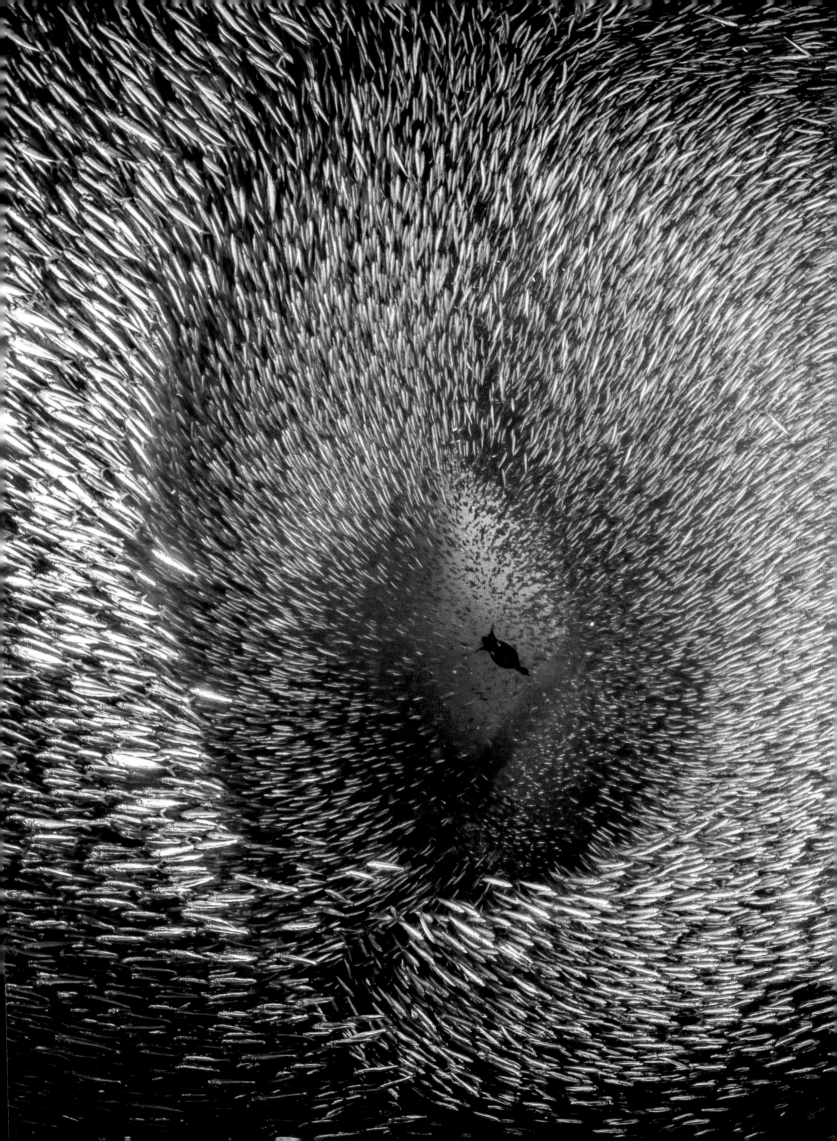

Pacific Ocean

SEAFLOOR

SCALE 1:57,300,000 1 CENTIMETER = 573 KILOMETERS; 1 INCH = 904 MILES
KILOMETERS
STATUTE MILES
Scale at the Equator

ARCTIC OC

ARCTIC CIRCLE

ASIA

Gulf of Anadyr

St. Lawrence Island

CONTINENTAL SHELF

BERING SE

Shelikhov Gulf

Kamchatka Peninsula

COMMANDER BASIN
Commander Islands

SHIRSHOV RIDGE

ALEUTIAN

Zhemchug Canyon
Pribilof Is.

BOWERS BASIN

Attu I.

Bowers Basin

Aleutian I.

BOWERS RIDGE

NORTH ALEUTIAN SLOPE
SOUTH ALEUTIAN SLOPE

ALEUTIAN RIDGE

ALEUTIAN

SEA OF OKHOTSK

OKHOTSK BASIN

Akademil Nauk Rise

KURIL BASIN

Sakhalin

Tatar Trough

KURIL-KAMCHATKA TRENCH

ZENKEVICH RISE (HOKKAIDO RISE)

EMPEROR SEAMOUNTS

EMPEROR TROUGH

CHINOOK TROUGH

Ōjin Rise

N O R T H   P A C I

JAPAN BASIN

Hokkaido

Japan Trench

Japan Rise

NORTHWEST

SHATSKIY RISE

MERCATOR BASIN

HESS RISE

Lillivokalani Ridge

SEA OF JAPAN (EAST SEA)

Honshu

Cipangu Basin

Japanese Guyots

Tamu Massif

MID-PACIFIC

H A W A I I A N

YELLOW SEA

Kyushu

IZU-OGASAWARA RISE

Nadezhda Basin

Midway Is.

NORTHWEST BASIN

N O R T H   P A C I

EAST CHINA SEA

Taiwan Strait

OKINAWA TROUGH

Ryukyu Ridge

RYUKYU TRENCH

Taiwan

Kyushu

IZU-OGASAWARA TRENCH

BONIN TRENCH

Uda Spur

Ptolemy Basin

Marcus-Wake Seamounts

KALANIOPUU BASIN

MID-PACIFIC MOUNTAINS

Johnston Atoll

Johnston Seamounts

TROPIC OF CANCER

INDIA

Hainan

SOUTH CHINA SEA

Luzon

PHILIPPINE ISLANDS

PHILIPPINE SEA

WEST MARIANA BASIN

Challenger Deep (36037 ft) 10984 m World's greatest ocean depth

MAGELLAN SEAMOUNTS

Wake Island

PIGAFETTA BASIN

EAST MARIANA BASIN

Bikini Atoll

Ratak Ridge

Marshall Seamounts

Ralik Ridge

CENTRAL PACIFIC

MAGELLAN RISE

INDOCHINA PENINSULA

BAY OF BENGAL

Andaman Islands

ANDAMAN SEA

Gulf of Thailand

SOUTH CHINA BASIN

PALAWAN TROUGH

SULU BASIN

Mindanao

Palau

YAP TRENCH

PALAU TRENCH

MARIANA TROUGH

MARIANA TRENCH

Caroline Seamounts

Ponape

CENTRAL PACIFIC

BASIN

Sri Lanka

Nicobar Islands

Malay Peninsula

Strait of Malacca

SUNDA SHELF

Natuna Sea

Borneo

CELEBES BASIN

PHILIPPINE TRENCH

WEST CAROLINE BASIN

EAUMPIK RISE

CAROLINE ISLANDS

EAST CAROLINE BASIN

LYRA BASIN

Howland I.

Baker I.

Gilbert Islands

CEYLON PLAIN

MID-INDIAN BASIN

NINETYEAST RIDGE

COCOS BASIN

JAVA RIDGE

INVESTIGATOR RIDGE

EQUATOR

INDONESIA

GREATER SUNDA ISLANDS

Celebes

North Banda Basin

Ceram Trough

NEW GUINEA TRENCH

WEST MELANESIAN TRENCH

New Guinea

Bismarck Archipelago

SOLOMON BASIN

MELANESIAN BASIN

Nauru Basin

ONTONG-JAVA RISE

Gilbert Islands

Phoenix Seamounts

Phoenix Islands

TOKELAU BASIN

Tokelau

Cocos Is.

Java Sea

Flores Basin

South Banda Basin

Weber Basin

Osborn Plateau

CHRISTMAS RISE

Christmas I.

JAVA TRENCH

Savu Basin

Lesser Sunda Islands

Timor Basin

TIMOR SEA

ARAFURA SEA

ARAFURA SHELF

Solomon Islands

SOUTH SOLOMON TROUGH

VITIAZ TRENCH

ELLICE BASIN

Ellice Ridge

Wallis Is.

Samoa Is.

FIJI

COCOS-KEELING RISE

ROO RISE

GASCOYNE PLAIN

NORTH AUSTRALIAN BASIN

SAHUL SHELF

Gulf of Carpentaria

CORAL SEA BASIN

Louisiade Plateau

CORAL SEA

D'Entrecasteaux

Vanatu

NORTH FIJI PLATEAU

NORTH FIJI BASIN

WHARTON BASIN

EXMOUTH PLATEAU

ROWLEY SHELF

Zenith Plateau

CUVIER BASIN

Cuvier Plateau

Carnarvon Terrace

Marion Plateau

Kenn Plateau

CATO TROUGH

Lord Howe Seamounts

Dampier Ridge

New Caledonia

NEW CALEDONIA

Loyalty Basin

NEW HEBRIDES TRENCH

SOUTH FIJI BASIN

SOLVILLE RIDGE

LAU RIDGE

Lau Basin

Tonga Ridge

TONGA TRENCH

TROPIC OF CAPRICORN

AUSTRALIA

East Indiaman Ridge

Lost Dutchmen Ridge

PERTH BASIN

Harold Terrace

BROKEN RIDGE

Diamantina Escarpment

DIAMANTINA FRACTURE ZONE

NATURALISTE PLATEAU

DIAMANTINA FRACTURE ZONE

DIAMANTINA TRENCH

CONTINENTAL SHELF

Great Australian Bight

CONTINENTAL SLOPE

North Norfolk Basin

NORFOLK RIDGE

LORD HOWE RISE

Lord Howe Island

NORFOLK TROUGH

South Norfolk Basin

Reinga Basin

KERMADEC RIDGE

KERMADEC TRENCH

North Island

Hikurangi Plateau

GEELVINCK FRACTURE ZONE

SOUTHEAST INDIAN RIDGE

INDIAN OCEAN

SOUTH AUSTRALIAN PLAIN

SOUTH AUSTRALIAN BASIN

TASMAN PLAIN

Bass Str.

Tasmania

TASMAN SEA

TASMAN BASIN

SOUTH TASMAN RISE

East Tasman Plateau

L'Atalante Valley

NEW ZEALAND

South Island

CHALLENGER PLATEAU

Hikurangi Terrace

CHATHAM RISE

Chatham Islands

ZEEHAEN F.Z.

HEEMSKERCK F.Z.

Australian-Antarctic Discordance

INDIAN-ANTARCTIC RIDGE

SOUTH INDIAN BASIN

SOUTH AUSTRALIAN BASIN

MACQUARIE RIDGE

EMERALD BASIN

AUCKLAND ESCARPMENT

Auckland Islands

SUBANTARCTIC SLOPE

CAMPBELL PLATEAU

CAMPBELL ESCARPMENT

BOUNTY TROUGH

Bounty Is.

Antipodes Is.

Bounty Plateau

Bollons Seamount

Chun Spur

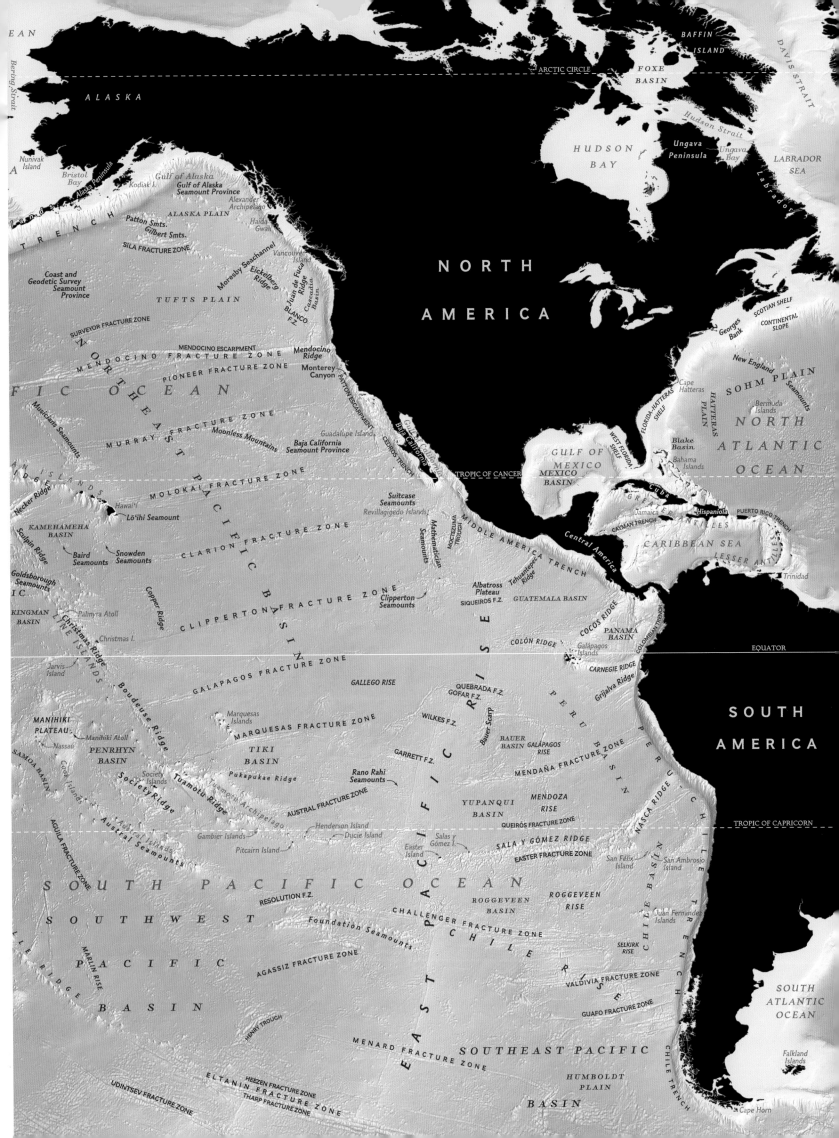

EAN

Bering Strait

ALASKA

Nunivak Island

Bristol Bay

Alaska Peninsula

Kodiak I.

Gulf of Alaska

Gulf of Alaska Seamount Province

ALASKA PLAIN

Haida Gwaii

Alexander Archipelago

TRENCH

Patton Smts.
Gilbert Smts.

SILA FRACTURE ZONE

Coast and Geodetic Survey Seamount Province

TUFTS PLAIN

Moresby Seachannel

Vancouver Island

Eickelberg Ridge

Juan de Fuca Ridge

Cascadia Basin

BLANCO F.Z.

SURVEYOR FRACTURE ZONE

MENDOCINO ESCARPMENT

MENDOCINO FRACTURE ZONE

Mendocino Ridge

PIONEER FRACTURE ZONE

Monterey Canyon

FIC OCEAN

OR T H E A S T

Musicians Seamounts

MURRAY FRACTURE ZONE

Moonless Mountains

PATTON ESCARPMENT

Guadalupe Island

Baja California Seamount Province

Gulf of California

Baja California

CEDROS TRENCH

NORTH AMERICA

HUDSON BAY

ARCTIC CIRCLE

FOXE BASIN

BAFFIN ISLAND

DAVIS STRAIT

Hudson Strait

Ungava Peninsula

Ungava Bay

LABRADOR SEA

Labrador

SCOTIAN SHELF

Georges Bank

CONTINENTAL SLOPE

New England

SOHM PLAIN

Cape Hatteras

FLORIDA-HATTERAS SHELF

HATTERAS PLAIN

NORTH ATLANTIC OCEAN

Seamounts

Bermuda Islands

TROPIC OF CANCER

WEST FLORIDA SHELF

GULF OF MEXICO

MEXICO BASIN

Blake Basin

Bahama Islands

Cuba

GREATER ANTILLES

Jamaica

Hispaniola

PUERTO RICO TRENCH

CAYMAN TRENCH

CARIBBEAN SEA

LESSER ANTILLES

Trinidad

NE C K E R RIDGE

Necker Ridge

MOLOKAI FRACTURE ZONE

Hawai'i

Lō'ihi Seamount

KAMEHAMEHA BASIN

Sculpin Ridge

Baird Seamounts

Snowden Seamounts

Goldsborough Seamounts

IC

KINGMAN BASIN

Christmas Ridge

LINE ISLANDS

Palmyra Atoll

Christmas I.

Copper Ridge

CLARION FRACTURE ZONE

NORTHEAST PACIFIC BASIN

Suitcase Seamounts

Revillagigedo Islands

Mathematician Seamounts

MOCTEZUMA TROUGH

MIDDLE AMERICA TRENCH

Central America

Tehuantepec Ridge

Albatross Plateau

SIQUEIROS F.Z.

GUATEMALA BASIN

Clipperton Seamounts

CLIPPERTON FRACTURE ZONE

COCOS RIDGE

COLÓN RIDGE

PANAMA BASIN

CARNEGIE RIDGE

Galápagos Islands

COLOMBIAN TRENCH

EQUATOR

SOUTH AMERICA

Grijalva Ridge

PERU BASIN

Jarvis Island

GALAPAGOS FRACTURE ZONE

GALLEGO RISE

QUEBRADA F.Z.

GOFAR F.Z.

WILKES F.Z.

Bauer Scarp

BAUER BASIN

GALÁPAGOS RISE

MENDAÑA FRACTURE ZONE

PERU-CHILE

Marquesas Islands

MARQUESAS FRACTURE ZONE

MANIHIKI PLATEAU

Manihiki Atoll

Nassau

SAMOA BASIN

PENRHYN BASIN

Cook Islands

Society Islands

Society Ridge

Tuamotu Ridge

Tuamotu Archipelago

TIKI BASIN

Pukapukae Ridge

Rano Rahi Seamounts

GARRETT F.Z.

AUSTRAL FRACTURE ZONE

YUPANQUI BASIN

MENDOZA RISE

QUEIRÓS FRACTURE ZONE

EAST PACIFIC RISE

AGUILA FRACTURE ZONE

Austral Islands

Austral Seamounts

Gambier Islands

Henderson Island

Ducie Island

Salas y Gómez I.

Easter Island

SALA Y GÓMEZ RIDGE

EASTER FRACTURE ZONE

San Félix Island

San Ambrosio Island

NASCA RIDGE

CHILE BASIN

TROPIC OF CAPRICORN

Pitcairn Island

SOUTH PACIFIC OCEAN

ROGGEVEEN BASIN

ROGGEVEEN RISE

Juan Fernández Islands

SOUTHWEST

RESOLUTION F.Z.

CHALLENGER FRACTURE ZONE

C H I L E

SELKIRK RISE

CHILE RISE

CHILE TRENCH

PACIFIC

MARLIN RISE

Foundation Seamounts

AGASSIZ FRACTURE ZONE

VALDIVIA FRACTURE ZONE

SOUTH ATLANTIC OCEAN

BASIN

HENRY TROUGH

GUAFO FRACTURE ZONE

Falkland Islands

UDINTSEV FRACTURE ZONE

MENARD FRACTURE ZONE

SOUTHEAST PACIFIC

HUMBOLDT PLAIN

BASIN

Cape Horn

ELTANIN FRACTURE ZONE

HEEZEN FRACTURE ZONE

THARP FRACTURE ZONE

RIDGE

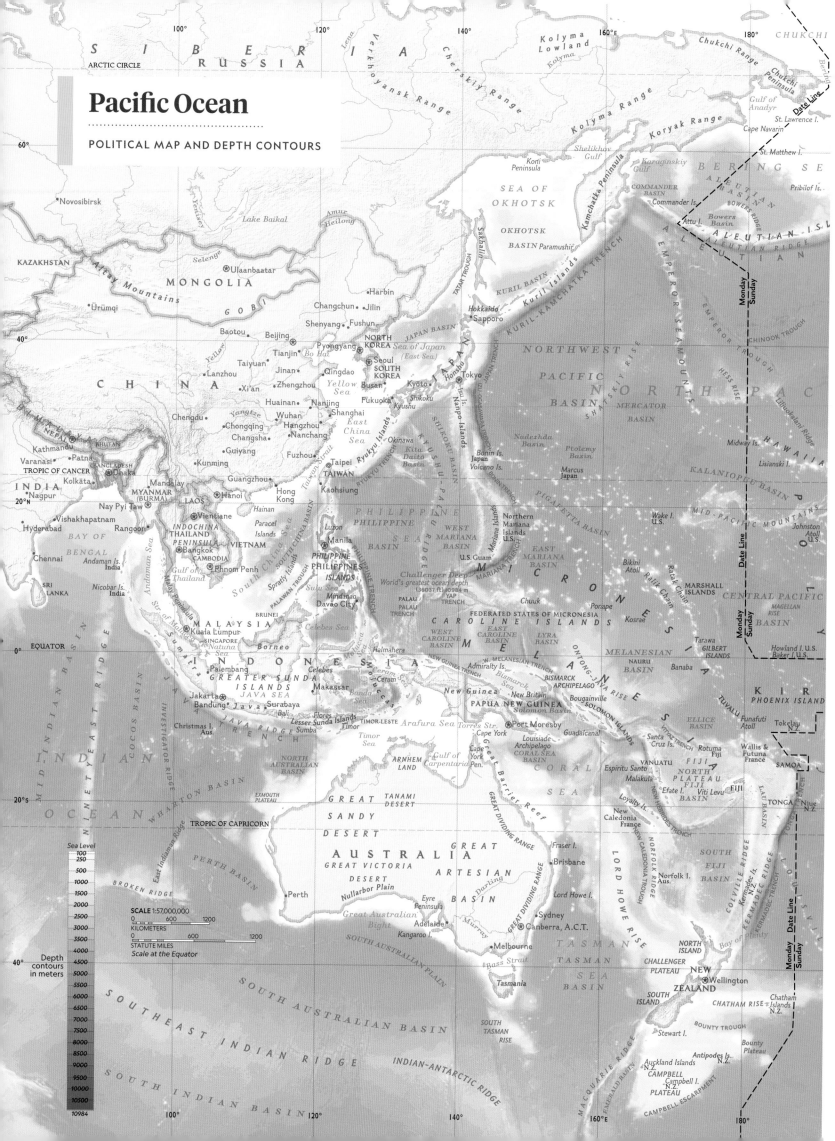

Pacific Ocean

POLITICAL MAP AND DEPTH CONTOURS

Depth contours in meters

Sea Level
100
250
500
1000
1500
2000
2500
3000
3500
4000
4500
5000
5500
6000
6500
7000
7500
8000
8500
9000
9500
10000
10500
10984

SCALE 1:57,000,000

KILOMETERS
0   600   1200

STATUTE MILES
0   600   1200

Scale at the Equator

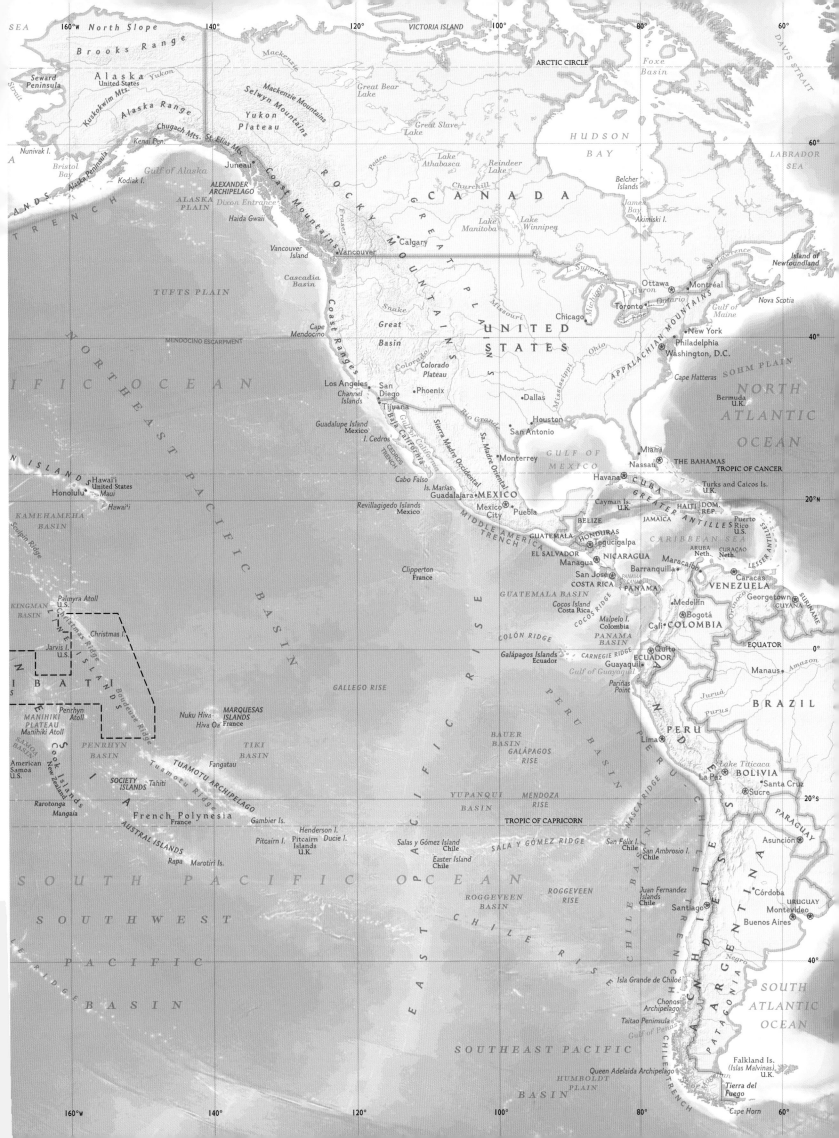

Oceania

A comprehensive Pacific Ocean map yields an extraordinary sight: some 25,000 islands that can be counted by only the most dedicated person—or by digital assistance. Even more extraordinary are their origins.

Some sit on continental shelf extensions of nearby landmasses. Many are volcanoes, built up from multiple eruptions over millennia until their peaks broke the ocean surface. Today most of them are silent and their volcanic rock "clothed in perpetual green," as author James Michener described Hawaii, supporting lush flora and fauna both on land and along their coasts, including coral atolls. But some, such as the Marianas' Pagan (eruption 2012), Papua New Guinea's Kadovar (2019), and Hawaii's Kīlauea (2018) and Mauna Loa (1984)—Earth's largest active volcano—are still fiery conduits to the interior of the planet.

In addition, 30,000 seamounts—undersea volcanoes, mostly extinct—do not break the surface. They, too, support lush systems of life, including deep-sea cold-water corals and the animals they shelter.

Named Oceania in 1812 by geographer Conrad Malte-Brun, this island-rich swath sweeps some 8.5 million square kilometers across the Pacific between about 30° N and 30° S latitudes, from Papua New Guinea in the west to Easter Island off the coast of Chile in the east, and is divided into four regions: Melanesia, Micronesia, Polynesia, and Australia.

Melanesia is the most populous of the island regions, including New Guinea, New Caledonia, Vanuatu, Fiji, and the Solomon Islands. Micronesia comprises the Federated States of Micronesia, along with the Marianas, Guam, Wake Island, Palau, the Marshall Islands, Kiribati, and Nauru. Polynesia groups New Zealand and the Hawaiian Islands together with the Midway Islands, Samoa, American Samoa, Tonga, Tuvalu, the Cook Islands, French Polynesia, and Easter Island.

As varied as their origin and placement across the ocean is the heritage of their peoples. People migrated from different parts of Asia by vessel to their far-reaching homes. By the time European navigators began charting the islands, the residents not only had been settled for many centuries but were expert navigators. Using stick charts fashioned of shells and lengths of coconut or other fiber, they plotted wind directions, ocean currents, stars, and landmarks, and then memorized them before each journey. Often a navigator's chart was personal; only he knew how to interpret it.

Westerners gradually caught up in their charting. In 1536 a map of Terra Australia was rendered by French mathematician Orontius Finus, which bears in Latin the phrase "Southern continent recently discovered but not yet completely examined." Throughout the 19th century, archipelagoes were gradually filled in.

"OCEANIA IS VAST, OCEANIA IS EXPAND-ING, OCEANIA IS HOSPITABLE AND GENEROUS, OCEANIA IS HUMANITY RISING FROM THE DEPTHS OF BRINE AND REGIONS OF FIRE DEEPER STILL, OCEANIA IS US. WE ARE THE SEA, WE ARE THE OCEAN . . . "

—EPELI HAU'OFA, *WE ARE THE OCEAN: SELECTED WORKS*

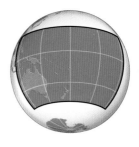

In New Guinea, a tiny emperor shrimp (*Periclimenes imperator*) tiptoes across a living carpet, the stretchy skin of its host, a sea cucumber (*Bohadschia argus*).

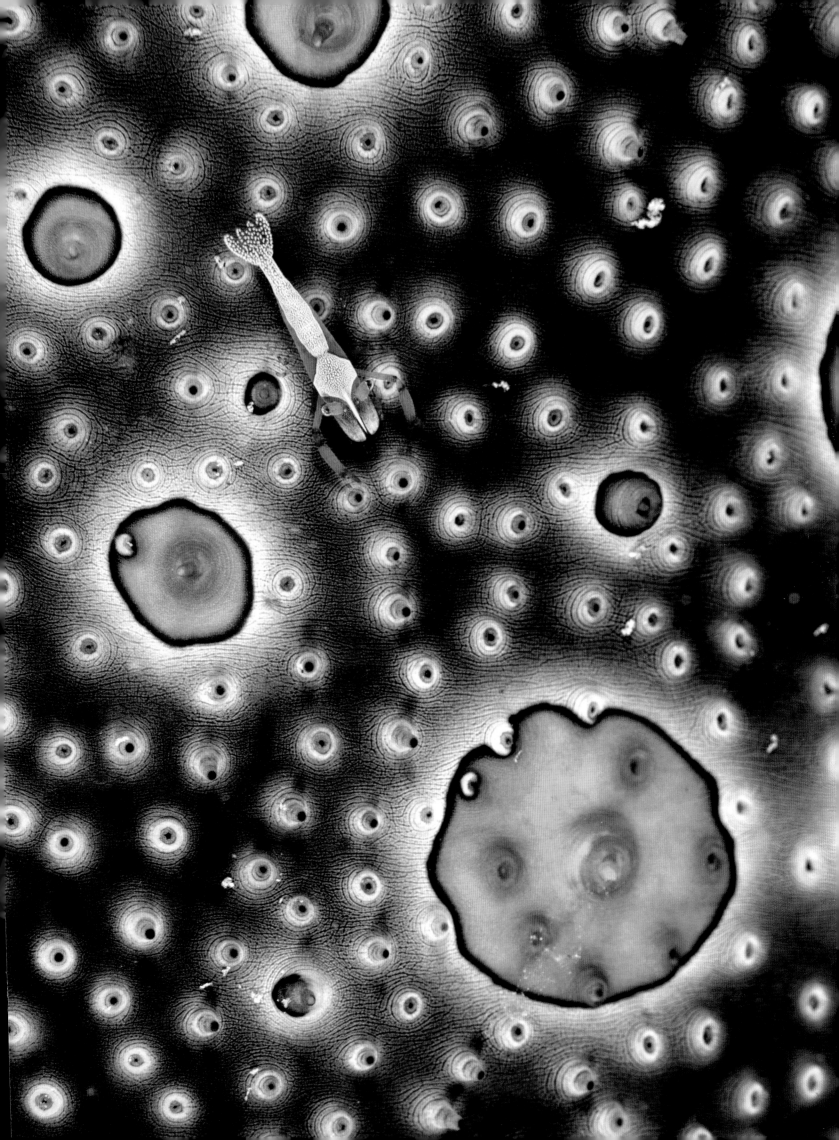

Oceania

·····································

POLITICAL MAP AND DEPTH CONTOURS

TROPIC OF CANCER

HAWAIIAN ISLANDS
NECKER RIDGE
Nihoa
Ni'ihau Kaua'i
O'ahu
Honolulu
Lāna'i Moloka'i
Kaho'olawe Maui
United States
Hawai'i

Hess
Tablemount

Johnston Atoll
U.S.

NORTH PACIFIC OCEAN

CLARION FRACTURE ZONE

NORTHEAST PACIFIC

CLIPPERTON BASIN

Magellan
Rise

Kingman Reef
U.S. Palmyra Atoll

Fanning I.
Kiribati

CLIPPERTON FRACTURE ZONE

EQUATOR

Howland Island
U.S. Baker Island

Christmas I.
Kiribati

U.S. Jarvis Island

GALAPAGOS FRACTURE ZONE

K I R I B A T I

Kanton
Birnie I. Enderbury I.
Nikumaroro Orona Manra
Malden Island

PHOENIX ISLANDS

Starbuck Island

MARQUESAS ISLANDS
France
Eiao Hatutu
Nuku Hiva Ua Huka
Ua Pou Hiva Oa
Fatu Hiva

MARQUESAS FRACTURE ZONE

unafuti Atoll
Nukulaelae

Atafu Tokelau
Nukunonu N.Z.
Fakaofu

**MANIHIKI
PLATEAU**

Penrhyn

Caroline Island
Vostok Island Kiribati

Niulakita

SAMOA
American
Samoa U.S.
Swains I.

Danger Is.
Nassau

Manihiki Atoll

Flint Island

Wallis & Futuna
France
Futuna
Arch.

Apia
Savaii I. Upolu
Tutuila Manua Is.
Pago Pago

Suwarrow
Atoll

Rose Atoll

SAMOA ISLANDS

TUAMOTU ARCHIPELAGO
Disappointment
Islands
Pukapuka

Taveuni

Lau
Group

Vava'u
Group

New
Zealand

Palmerston Atoll

Manihi
Mataiva Rangiroa Takaroa
Motu One Tikei
Bora- Takapoto
Bora Makatea Raraka
Maupihaa Fakarava Raroia
Moorea Tatakoto
Papeete Anaa Hikueru
Tahiti Ravahere Hao Pukarua
Hereheretue Reao
French Polynesia
France

Vatoa

TONGA
Ono-
i-Lau Nuku'alofa
Tongatapu
Group

Ha'apai
Group

Niue N.Z.

Aitutaki Atoll
Hervey Is.
Mitiaro
Atiu Mauke

**SOCIETY
ISLANDS**

Tureia Actaeon
Group
Duke of Marutea
Gloucester Islands Tematagi
Moruroa
Fangataufa Morane Mangareva
Pitcairn Islands
United Kingdom

LAU RIDGE
Lau Basin

Rarotonga

Mangaia

Maria
Islands Rimatara
Rurutu
Tubuai
Raivavae I.

A U S T R A L

TROPIC OF CAPRICORN

Gambier Islands Temoe
Oeno Island

Henderson Island

Pitcairn Island
Ducie Island

Kermadec
Islands
N.Z.

President Thiers
Bank

Neilson Reef
Rapa Marotiri
Islands

I S L A N D S

**SOUTH
PACIFIC**

OCEAN

KERMADEC TRENCH

TONGA ISLANDS

TONGA TRENCH

Date Line
Monday | Sunday

LOUISVILLE RIDGE

SOUTHWEST PACIFIC BASIN

RISE
Chatham Is.
N.Z.

Depth contours in meters	
Sea Level	
100	
250	
500	
1000	
1500	
2000	
2500	
3000	
3500	
4000	
4500	
5000	
5500	
6000	
6500	
7000	
7500	
8000	
8500	
9000	
9500	
10000	
10500	
10984	

SCALE 1:31,400,000

KILOMETERS
0   400   800

STATUTE MILES
0   400   800
Scale at the Equator

Date Line
Monday | Sunday

COOK ISLANDS

LINE ISLANDS

South China Sea

The South China Sea, called "Southern Sea" in ninth-century China, earned its current name in part from 16th-century Portuguese navigators who used the Mar de China as a trade route to the nation. Today, its bordering nations use a range of names, including China's South Sea, Vietnam's East Sea, and the Philippines' West Philippine Sea—and they all maintain competing claims to it, largely because of oil reserves beneath the seafloor.

Ringed by China and Taiwan to the north, the Philippines to the east, Borneo and Sumatra to the south, and Malaysia, Thailand, Cambodia, and Vietnam to the west, the South China Sea today carries one-third of the world's shipping. Much of it traverses the Strait of Malacca between Sumatra and Malaysia, where pirates have been on the watch for centuries and are still a regular threat.

This sea is home to hundreds of small islands and seamounts that support flourishing ecosystems. Some seamounts near the Spratly Islands rise so precipitously from the seafloor to just below the sea surface that the area is named "Dangerous Ground" for sailors.

Covering some six million square kilometers, the Coral Triangle accounts for 30 percent of the world's coral reefs, with 500 species of reef-building corals. The Triangle shelters the highest marine biodiversity on Earth. The treasures, while vast, have been under siege from agricultural and industrial runoff, overfishing, and climate change.

While Charles Darwin was forming his theory of evolution in the Galápagos Islands, British naturalist Alfred Russel Wallace was doing similar work in the South China Sea islands. The life he found on Bali was completely different from that of its neighbor just 32 kilometers away, Lombok. Balinese natural life resembled that of Java and Sumatra to the west and north, and life in Lombok looked like that of New Guinea and Australia to the east and south. The line that separates the two is called Wallace's Line, and one theory for such distinct species variations is that due to plate movement and sea-level rise, animals isolated on each side of the line followed their own evolutionary path.

As eager as he was to discover this exciting anomaly, Wallace was equally eager to protect it from human opportunism. Even in the mid-1800s, he worried: "Future ages will certainly look back on us as a people so immersed in the pursuit of wealth as to be blind to higher considerations." Steps are being taken, through the United Nations, the Association of Southeast Asian Nations, and other organizations working toward protection.

"BIODIVERSITY STARTS IN THE DISTANT PAST AND IT POINTS TOWARD THE FUTURE."

—FRANS LANTING, ARTIST AND PHOTOGRAPHER

Shrouded in mist, wooded mountains and limestone pillars embrace South Vietnam's Ha Long Bay in the South China Sea.

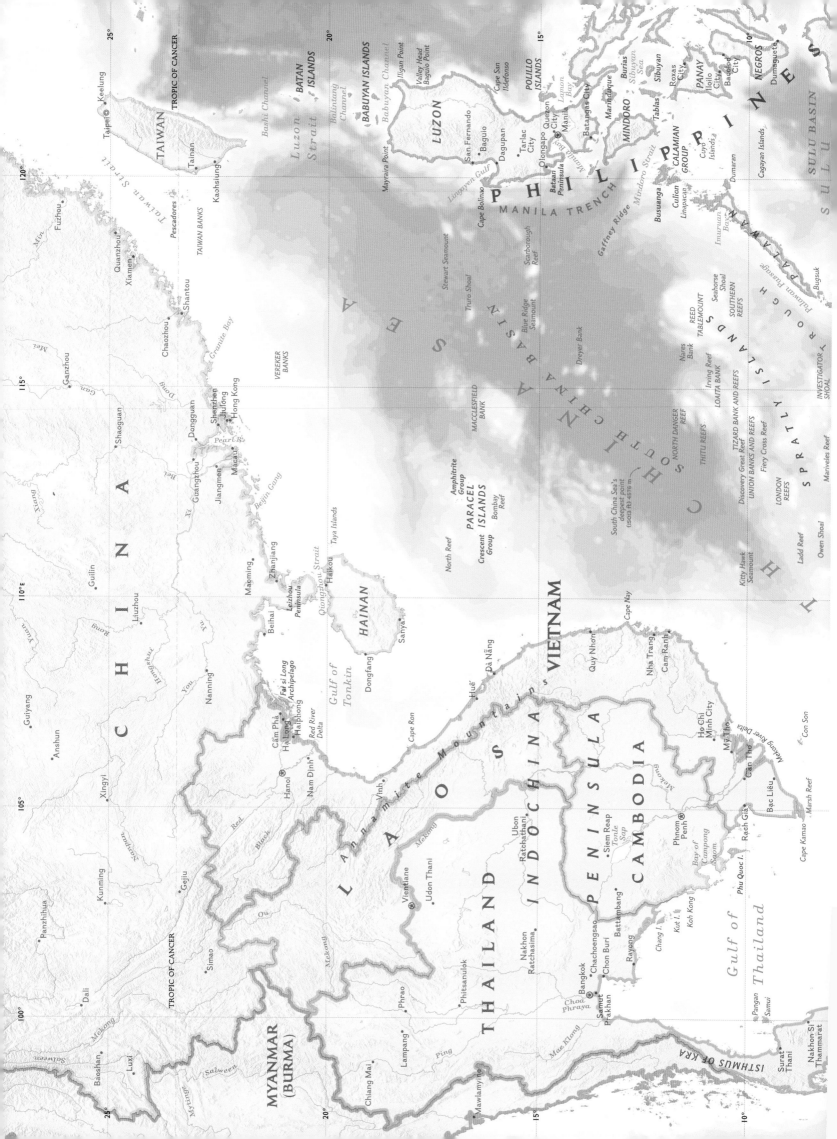

South China Sea

POLITICAL MAP AND DEPTH CONTOURS

SCALE 1:10,000,000

Depth contours in meters

Sea Level
100
250
500
1000
1500
2000
2500
3000
3500
4000
4500
5000
5500
6000
6500
7000
7125

KILOMETERS
STATUTE MILES
Scale at the Equator

I | INDIAN OCEAN

Indian Ocean

The youngest and third largest of the ocean basins, the Indian Ocean was formed as the Indian subcontinent broke from Gondwanaland 200 million years ago and, moved by plate tectonics, eventually slammed into Eurasia to the north.

While these waters tend to be calm from November to April, warm, moist rising air during monsoon season from May through October triggers powerful, destructive tropical storms.

Although largely unexplored, the Indian Ocean is known to host a great diversity of marine life. The West Indian Ocean alone shelters between 11,000 and 20,000 species according to estimates in 2018 by the World Wildlife Fund.

Abyssal plains, seamounts, and hydrothermal vents harbor lush deep-sea communities, and coral reefs and shallow seagrass meadows line its rocky shores. The Mid-Indian Ridge, in a massive inverted Y, connects to southern ridges. As it spreads, adding new seafloor, it also spawns earthquakes. So do its trenches, where one plate is forced beneath the other. In 2004 a Sunda Trench quake of 9.1 magnitude triggered tsunami waves as high as 51 meters, devastating Sumatra.

Monsoons (from the Arabic *mawsim*, for "season") as early as 2000 B.C. propelled merchant ships from the southwest in spring and from the northeast in winter. In the 1400s, the Chinese treasure fleet admiral Zheng He plied these waters in the name of his emperor in 60 extravagantly outfitted ships, opening the coast to Chinese settlement. By the 1500s Portuguese navigator Vasco da Gama had rounded Africa's Cape of Good Hope on the "Route of Spices," opening the door for European trade.

Each new voyage added to ever more detailed maps of geological formations and knowledge of wildlife, trade routes, winds, and currents—mainly the North Equatorial Current, which reverses direction with monsoon season in the Indian Ocean, and the South Equatorial Current, which sustains a steady counterclockwise motion.

In 2014 new discoveries resulted from the disappearance of Malaysian Airlines flight 370 into the sea. While covering 120,000 square kilometers in search-and-rescue efforts, experts uncovered many new seamounts and ecosystems, adding significantly to the ocean's cartographic profile.

Of the major ocean basins, the Indian has the fewest radiating arms, but they are significant for humans and wildlife—including the Gulfs of Aden and Oman and the Bay of Bengal. Highlighted on the following pages, the Red Sea and Persian Gulf are vital places, ecologically and economically.

BY THE NUMBERS

Indian Ocean

....

Total Area
67,469,539 square kilometers

Total Volume
261,519,545 cubic kilometers

Greatest Depth
7,125 meters (Sunda Trench, south of the Indonesian arc of islands)

Includes 20 percent of Earth's water area
(excludes major seas around Indonesia and northern Australia)

Contains the world's longest linear feature, the Ninety East Ridge, a chain of submerged peaks.

Underwater temperatures 1.6 kilometers deep are Earth's hottest, 22°C, in the Red Sea, an active spreading center.

Geographic Boundaries
25° N to 60° S / 20° E to 145° E

Average Depth
3,960 meters

RIGHT: Swimming in synchrony, longfin bannerfish cruise past a giant Gorgonian coral. PAGES 470-471: Coral plates rim a turquoise lagoon off the Maldives archipelago.

Indian Ocean

SEAFLOOR

I A

EAST
CHINA
SEA

JAPAN

Kyushu

Japanese
Guyots

NADEZHDA
BASIN

IZU-OGASAWARA RISE

IZU-OGASAWARA TRENCH

Two Jima Ridge

Uda Spur

BONIN TRENCH

PIGAFETTA BASIN

TROPIC OF CANCER

OKINAWA TROUGH

Ryukyu Islands

Taiwan Strait

Taiwan

RYUKYU TRENCH

Ryukyu Ridge

Kita Daitō
Basin

Daitō Ridge

SHIKOKU BASIN

KINAN Seamounts

KYUSHU-PALAU RIDGE

Kinan Ridge

Gulf of
Tonkin

Hainan

PHILIPPINE

SEA

PHILIPPINE
BASIN

WEST
MARIANA
BASIN

West Mariana Ridge

MARIANA TROUGH

Mariana Islands

MARIANA TRENCH

INDOCHINA
PENINSULA

Macclesfield
Bank

SOUTH CHINA SEA BASIN

Luzon Ridge

Luzon
Plateau

PACIFIC   OCEAN

Y OF
NGAL

Ganges Fan

Andaman
Islands

ANDAMAN BASIN

SOUTH   CHINA   SEA

MANILA TRENCH

PHILIPPINE
ISLANDS

PHILIPPINE TRENCH

Palau

PALAU TRENCH

YAP TRENCH

WEST CAROLINE
RISE

EAST
MARIANA
BASIN

Caroline Seamounts

LYRA
BASIN

Gulf of
Thailand

PALAWAN TROUGH

SULU
BASIN

AYU TROUGH

PALAU TRENCH

EAURIPIK RISE

CAROLINE   ISLANDS

WEST CAROLINE TROUGH

Nicobar
Islands

ANDAMAN SEA

Malay Peninsula

Strait of Malacca

SUNDA SHELF

NATUNA
SEA

Sibutu-Basilan Ridge

CELEBES
BASIN

Sangihe Ridge

WEST
CAROLINE
BASIN

EAST
CAROLINE
BASIN

NINETY EAST RIDGE

COCOS BASIN

COCOS PLAIN

Kuenen Rise

Sumatra

Mentawai Basin

MENTAWEI RIDGE

Borneo

EQUATOR

I N D O N E S I A

Celebes

CERAM TROUGH

NORTH
BANDA
BASIN

NEW GUINEA TRENCH

WEST MELANESIAN TRENCH

Manus
Island

New Guinea
Basin

New Guinea

SOLOMON
BASIN

Cocos Is.

Enggano

JAVA

Java Sea

GREATER

SUNDA   ISLANDS

BANDA
SEA

WEBER BASIN

ARU BASIN

Moresby Valley

PAPUA
PLATEAU

Christmas I.

Vening Melnesz
Seamounts

CHRISTMAS RISE

Monsoon
- Rise

INVESTIGATOR RIDGE

JAVA TRENCH

JAVA RIDGE

Sunda Trough

Lesser Sunda Islands

Bali

FLORES
BASIN

SOUTH BANDA
BASIN

Savu
Basin

Timor

TIMOR TROUGH

ARAFURA SEA

ARAFURA SHELF

Torres
Strait

Great Barrier Reef

CORAL SEA
BASIN

CORAL SEA

Lombok Basin

Indian Ocean's deepest point
7125 m (23376 ft)

ROO RISE

NORTH
AUSTRALIAN
BASIN

TIMOR
SEA

SAHUL SHELF

Gulf of
Carpentaria

QUEENSLAND
PLATEAU

TOWNSVILLE TROUGH

Cocos Is.

COCOS-KEELING
RISE

GASCOYNE
PLAIN

Platypus
Spur

EXMOUTH
PLATEAU

Rowley Shelf

OSBORN
PLATEAU

WHARTON   BASIN

ZENITH
PLATEAU

Lost Dutchmen Ridge

CUVIER
BASIN

CUVIER
PLATEAU

Carnarvon Terrace

MARION
PLATEAU

TROPIC OF CAPRICORN

EAST INDIAMAN RIDGE

Batavia
Seamount

Guilden
Draak Seamount

PERTH   BASIN

HARTOG RIDGE

NATURALISTE FZ

AUSTRALIA

BROKEN RIDGE

DIAMANTINA ESCARPMENT

OB' TRENCH

NATURALISTE
PLATEAU

Naturaliste Trough

EUCLA SHELF

Great Australian
Bight

Eyre Terrace

Ceduna
Terrace

CEDUNA SLOPE

CONTINENTAL SLOPE

TASMAN PLAIN

DIAMANTINA TRENCH

DIAMANTINA   FRACTURE   ZONE

CONTINENTAL SLOPE

Beachport
Terrace

SOUTH   AUSTRALIAN   PLAIN

Bass Strait

TASMAN
SEA

FRACTURE ZONE

Tasmania

TASMAN
SEA

East Tasman
Saddle

EAST
TASMAN
PLATEAU

SOUTH TASMAN RISE

SOUTH TASMAN ESCARPMENT

L'Atalante Valley

S T

I N D I A N   RIDGE

SOUTH   AUSTRALIAN   BASIN

BASIN

Australian-Antarctic

Discordance

Indian Ocean

POLITICAL MAP AND DEPTH CONTOURS

HOPE SPOT

Maldives Atolls

The turquoise waters of the Maldives Atolls—a chain of 1,200 coral islands sweeping across the Indian Ocean— shelter some 250 species of scleractinian corals and 55 genera of hermatypic corals, as well as threatened species of turtles, whales, whale sharks, pearl oysters, eels, and pufferfish. But paradise is in trouble. Due to climate change, these waters are rising, overwhelming shorelines and displacing human and marine life. The government of the Maldives is working overtime to reinforce coasts and relocate residents.

Humpback snappers school near the Maldives' North Ari Atoll.

Red Sea

ts renown as a diver's paradise of coral reefs, glittering fish, and violet blue waters may cloud the full significance of the Red Sea on the world stage.

As Arabia began to diverge from Africa some 35 million years ago, the seafloor beneath it also began pulling apart, bubbling up with molten lava that cools and builds more seafloor and driving the new tectonic plates apart at a rapid clip of about 1.5 centimeters a year—half the average growth of a human fingernail. Already 250 kilometers wide, the Red Sea is on its way to rivaling the Atlantic in size some 150 million years from now.

From the north end of the sea down through Africa, the split is linked to the East African Rift Valley and a similar spreading of Lake Victoria some 2,400 kilometers to the south, slowly on its way to becoming a new ocean.

A 1947–48 scientific expedition aboard the Swedish ship *Albatross* first discovered unusually high temperatures and salinity in deep waters at the middle of the sea. The central rift area is dotted with at least 20 brine lakes dense with salt and dissolved minerals from hydrothermal springs spewing up from inner earth. While other such vents likewise emit chemical cocktails, these were different, with an added ingredient: salt-rich sediments from an ancient sea that likely evaporated during the Miocene epoch some five million years ago.

The chemicals appeared to be supporting life similar to that in other deep-sea hydrothermal vent areas. These life-forms offer only a hint of this natural aquarium, which shelters 1,500 kilometers of more than 300 species of coral—some of it more than 5,000 years old. Also, islands, mangroves, and marshes host 10 species of sharks and about a thousand other species of fish, 10 percent of which live only in its waters. Dugongs, sea turtles, manta rays, and dolphins also flourish here.

The Red Sea was a trade route of the Egyptians as early as 2500 B.C., and a Suez conduit from the Nile to the Red Sea was opened around the sixth century B.C.

It is not good news for Red Sea denizens that their magical realm is also one of Earth's most traveled seaways, with some 20,000 ships passing through the Suez Canal each year. Waste disposal, anchors, oil spills, and climate change all make life hazardous for marine populations. But there is hope. As the bordering countries and conservation organizations become more mindful of the problems, some are taking measures to protect the natural values while accommodating commercial interests.

"TO MY LEFT WAS ASIA, TO MY RIGHT, AFRICA. BETWEEN THEM, I HAD AT MY BACK THE RICHEST REEF I'D EVER SEEN."

—DAVID DOUBILET, *LIGHT IN THE SEA*

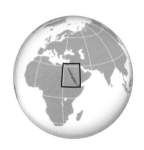

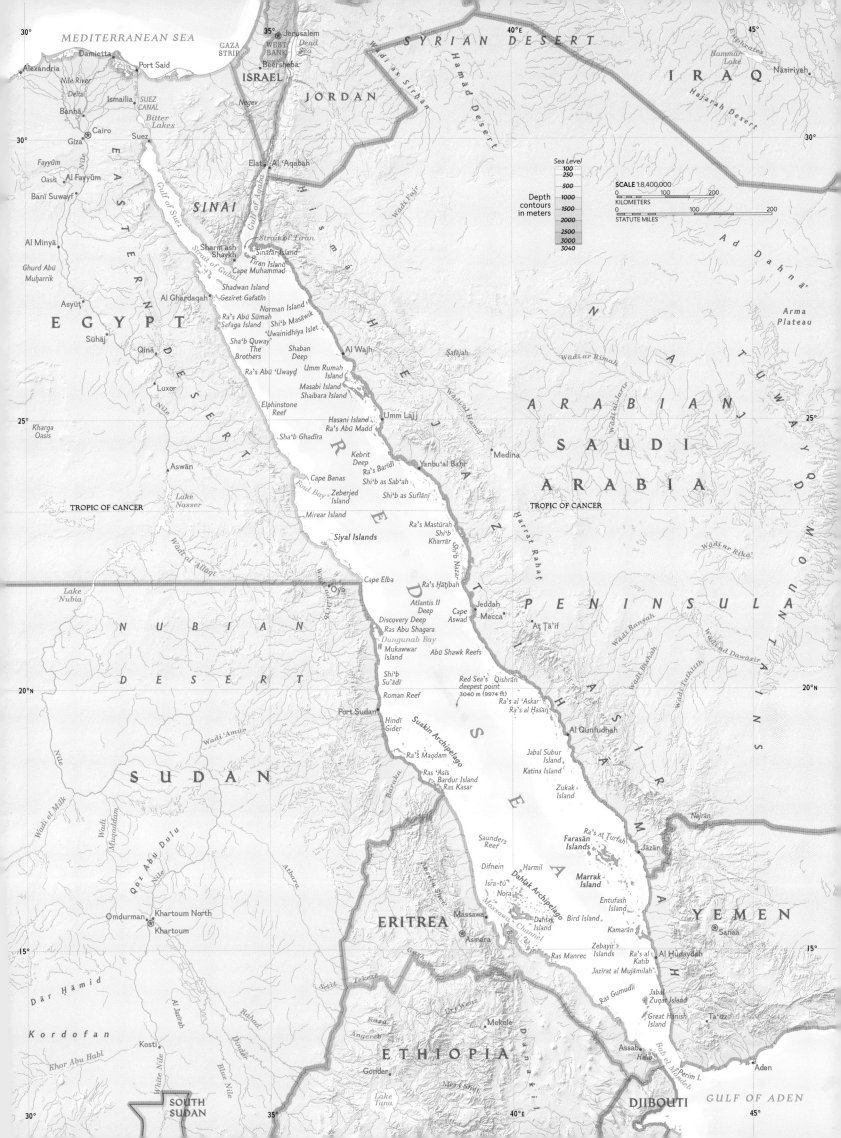

MEDITERRANEAN SEA

30°

Alexandria
Damietta
Port Said
Nile River
Delta
Ismailia
SUEZ
CANAL
Banhā
Bitter
Lakes
Cairo
Giza
Suez

GAZA
STRIP
WEST
BANK
Jerusalem
Dead
Sea
Beersheba

ISRAEL
JORDAN

35°

SYRIAN DESERT

Wadi as Sirhān

Hamād Desert

40°E

IRAQ

Euphrates

Hammār
Lake

Nasiriyeh

Hajārah Desert

45°

30°

Fayyūm
Oasis
Al Fayyūm
Banī Suwayf

Al Minyā

Ghurd Abū
Muḥarrik

Asyūṭ

Sūhāj

Qinā

Luxor

Aswān

Kharga
Oasis

Lake
Nasser

TROPIC OF CANCER

Wadi al ʿAllāqī

Lake
Nubia

NUBIAN

DESERT

Wadi el Milk

Wadi
Muqaddam

SUDAN

Omdurman
Khartoum North
Khartoum

Qoz Abū Dulu

Nile

Dār Ḥamid

Al Jazīrah

Kordofan

Khor Abu Habl

Kosti

White Nile

Blue Nile

SOUTH
SUDAN

30°

EGYPT

EASTERN

DESERT

SINAI

Gulf of Suez

Gulf of Aqaba

Elat
Al ʿAqabah

Strait of Tiran
Sinafar Island
Tiran Island
Cape Muhammad
Sharm ash
Shaykh

Shadwan Island
Gezīret Gafatīn
Al Ghardaqah
Norman Island
Raʾs Abū Sūmah
Safaga Island
Shiʿb Masāwik
ʿUwainidhīya Islet
Shaʿb Quway
The
Brothers
Shaban
Deep
Raʾs Abū ʿUwayḍ
Umm Rumah
Island
Masabi Island
Shaibara Island
Elphinstone
Reef
Hasani Island
Raʾs Abū Madd
Shaʿb Ghadīra
Kebrit
Deep
Raʾs Barīdī
Cape Banas
Foul Bay
Zeberjed
Island
Shiʿb as Sabʿah
Shiʿb as Suflānī
Mirear Island
Siyal Islands
Cape Elba
Oyo
Wadi Dīb

H
E
J
A
Z

Al Wajh

Umm Lajj

Yanbuʿal Baḥr

Ṣafājah

Wadi al Ḥamḍ

R
E
D

S
E
A

Raʾs Mastūrah
Shiʿb
Kharrār
Shiʿb Nazar
Raʾs Ḥāṭibah
Atlantis II
Deep
Discovery Deep
Ras Abu Shagara
Dungunab Bay
Mukawwar
Island
Shiʿb
Suʿādī
Roman Reef
Port Sudan
Hindī
Gider
Raʾs Maqdam
Ras ʿAsīs
Bardur Island
Ras Kasar
Cape
Aswad
Jeddah
Mecca
Abū Shawk Reefs
Red Sea's
deepest point
3040 m (9974 ft)
Qishrān
Raʾs al ʿAskar
Raʾs al Ḥasan
Al Qunfudhah
Suakin Archipelago
Jabal Subur
Island
Katina Island
Zukak
Island

Wadi Fajr

Wadi al Ḥamḍ

35°

A R A B I A N

SAUDI

ARABIA

Medina

T
I
H
A
M
A

Ḥijāz Mts

Wadi ar Rimah

Wadi ar Rika

Wadi Ranyah

Wadi Bishah

Wadi ad Dawāsir

Wadi Tathlīth

Harrat Rahaṭ

At Ṭāʾif

Arma
Plateau

N
A
F
U
D

T
U
W
A
Y
Q

A
S
I
R

M
O
U
N
T
A
I
N
S

Ad Dahnāʾ

TROPIC OF CANCER

PENINSULA

25°N

20°N

Wadi ʿAmur

Baraka

Wadi Amūr

Port Sudan

Saunders
Reef

Difnein

Isra-tú
Nora

Massawa

Asmara

Harmil

Dahlak Archipelago

Dahlak
Channel

Dahlak

Bird Island

Ras Manrec

Entufash
Island

Kamarān

Marrak
Island

Farasān
Islands

Raʾs at Turfah

Jāzān

Najrān

Zebayir
Islands

Jazīrat al Mujāmilah

Raʾs al
Katib

Al Ḥudaydah

Jabal
Zuqar Island

Great Hanish
Island

YEMEN

Sanaa

Taʿizz

15°

Gash

Tekezē

Setit

Atbara

ERITREA

Barka

Anseba

Rahad

Dinder

Angereb

Kaza

Gonder

Lake
Tana

Mereb

ETHIOPIA

Mekele

Danakil

D
A
N
A
K
I
L

Assab
Halib

DJIBOUTI

Bab el Mandeb
Perim I.

Aden

GULF OF ADEN

45°

15°

40°E

35°

30°

Persian Gulf

A rich landscape of freshwater marshes and rivers during the last global ice age 18,000 years ago, the Persian Gulf is a different place today. As that ice age ended and glaciers melted, sea level rose and salt water from the Indian Ocean flooded the freshwater plains thriving with lions, elephants, and zebras. Gradually, new citizens—dugong, sea turtles, shrimp, fish, and birds—inhabited this submerged realm thick with seagrasses, algae, mangroves, and coral reefs.

Some 241,000 square kilometers in area, the Persian Gulf—also known as the Arabian Gulf and the Gulf of Iran—stretches between Iran to the east and Iraq, Kuwait, Qatar, and the United Arab Emirates (UAE) to the west. Access to the Gulf comes from the north through the Shatt al-Arab, the confluence of the fertile Tigris and Euphrates Rivers, and from the south through the Strait of Hormuz—gateway to the Indian Ocean.

Early Mesopotamia, the cradle of civilization, thrived here, and the tools of Paleolithic nomadic tribes have been found along its shores. Under Persian king Darius the Great around 500 B.C., the coastal culture began to develop with a strong navy, thriving ports, and profitable trade reaching as far as China. The Gulf's shallow waters, no deeper than 110 meters, long harbored the world's most thriving pearl-diving industry.

While the islands of Bahrain, an Arab state, and Iran's Qeshm in the Strait of Hormuz are populated, smaller islands throughout the Gulf are uninhabited salt domes and coral reefs jutting above the surface.

The land on either side of the Gulf—a product of tectonic plate rifting, subduction, and collision over eons—holds ancient aquifers that carry fossil water, now being accessed for a growing human population. Also, over time, a 12,000-meter-thick buildup of limestone, quartz, salt deposits, and sediments from marine life under the land and seafloor fostered the Gulf's stores of oil. These were first discovered in 1908 in Iran, then exploited after World War II, with the Gulf floor a prime drilling site today.

A major supplier of the world's petroleum, the Gulf's oil and gas deposits form the backbone of the area's economy—and strife, notably the Iraqi invasion of Kuwait in 1990–91 to assume the smaller nation's oil reserves.

Drilling, shipping, runoff, oil spills, and climate change all threaten healthy Gulf ecosystems. At the seventh World Government Summit in Dubai in 2019, scientists projected a 33 percent loss in marine biodiversity by 2090.

"IN THE 'CRADLE OF CIVILIZATION' . . . THE CHEMISTRY, PHYSICAL ENVIRONMENT, AND GEOLOGICAL HISTORY OF THE AREA HAVE SHAPED . . . ONE OF THE WORLD'S MOST PRODUCTIVE BODIES OF WATER."

—FRANCESCA CAVA, *OCEANUS*

Aglow in a NASA satellite image, the Persian Gulf radiates activity as a premier petroleum producer, and the site of rapid environmental change.

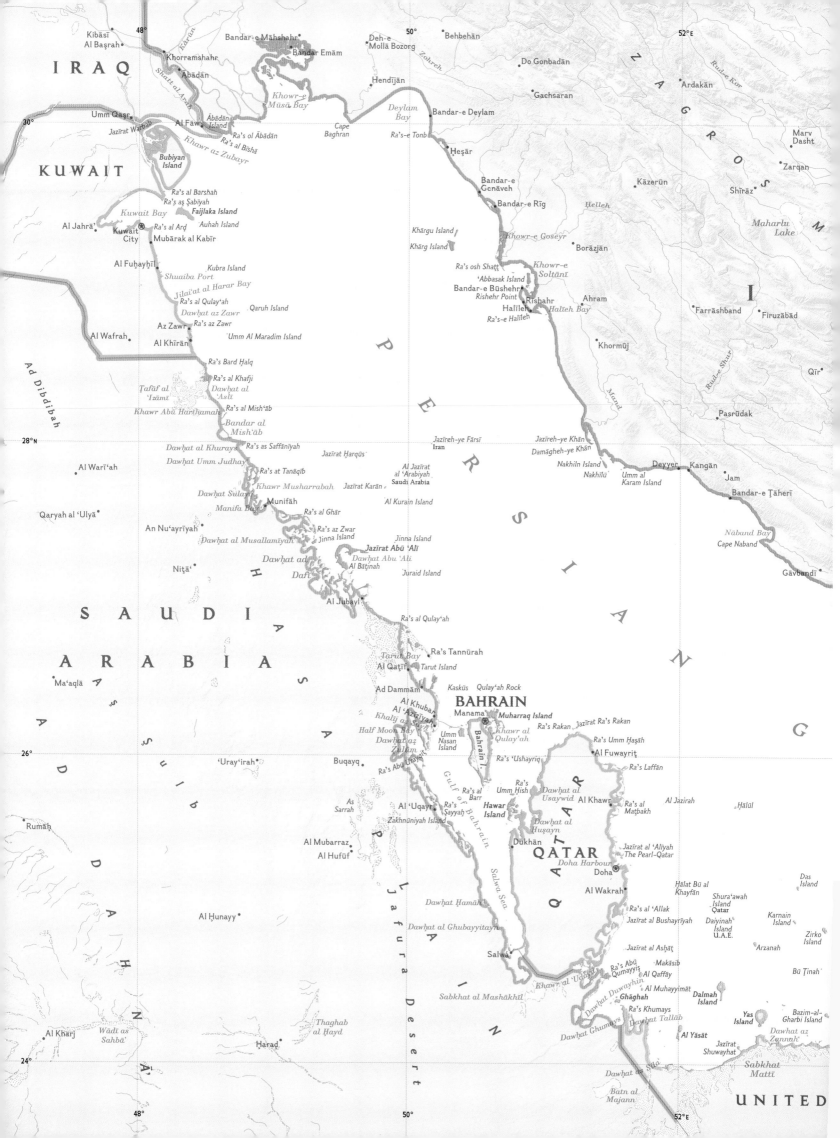

Persian Gulf

POLITICAL MAP AND DEPTH CONTOURS

Depth contours in meters

Sea Level
100
250
500
1000
1500
2000
2500
3000
3500

SCALE 1:3,000,000

KILOMETERS
0   40   80

STATUTE MILES
0   40   80

54°   56°   58°   30°

Sūriān
Marvast
Qāderābād
Sa'ādatābād
Arsanjān
Sarvestān
Fasā
Rafsanjān
Shahr-e Bābak

Tashk
Lake

Nārīz
Lake

Kavīr-e
Namak-e
Masīrjān

Ab Shūr

Sīrjān
Gaţrūyeh
Neyrīz
Estahbān

R   A   N

O   I   N
U   N   T   A   I   N   S
Jahrom
Dārāb
Jūyom
Fūrg
Shūr

Aliabad
Ḩājjīābād
Ţārom
Sa'ādatābād

Galāshgerd
Kahnūj
Bīzhanābād

28°N

Lār

Shamīl
Rudan
Manūjān

Hāmūn-e
Jaz Mūrīān

Mehrān
Bastak
Shūr

Hasan Langī
Mīnāb

Shīv
Bandar 'Abbās
Bandar-e Moqām
Ra's-e Nakhīlū
Bandargāh-e Bandar 'Abbās
Hormuz
Island
Bandar-e Tīāb

Khalīj-e
Nakhīlū
Bandar-e Khamīr
Clarence Strait
Qeshm
Larak
Island
Kūhestak
Sanderk
Ramshakī

Levan
Island
Bandar-e
Chārak
Shūrehzār-e
Mehragān
Khowr-e Bārikū
QISHM ISLAND
Sūzā
Strait of Hormuz
Sīrīk
Angohrān

Chīru Point
Charak
Bay
Ra's-e Yarīd
Ra's-e Shāvari
Bāsa 'īdū
Henjām
Island
As Salāmah wa Banātuhā
Persian Gulf's
deepest point
119 m (344 ft)

Hindarābī
Island
Khowr-e Kīsh
Ra's-e Shenās
Bandar-e
Lengeh
Ra's-e Dastakān
Ra's Musandam
M   a   k   r   a   n

Kish Island
Khalīj-e
Moghūyeh
Shīās Bay
Al Ghanam Island
Ra's ash Shaykh Mas'ūd
Ra's Khayseh
Ra's-e Shīr
Gavān-e Bālā
26°

Fārūr
Island
Tunb
Islands
Kumzār
MASANDAM PENINSULA
Khasab
Malcom Inlet
Ra's Sarkān
Bandar-e Kangān

Jazīreh-ye Banī Forūr
Ash Sha'm
Khawr al Khuwayr
Līmā
Ra's Līmā'
Damāgheh-ye Kūh
Jāsk
Gābrīk
Sūrak

G   U   L   F
Sari
Island
Abu Musa
Khawr Ḩulaylah
OMAN
Ra's al Khaymah
Ra's Ḩaffah
Jāsk
East
Bay
Ra's Jagīn

PERSIAN GULF
Historically and most commonly
known as the Persian Gulf, this
body of water is referred to by
some as the Arabian Gulf.
Jazīrat al Ḩamrā'
Jazīrat as Sīnīyah
Bay'ah
Dibba Bay
Dibā
Ra's Dibā
Damāgheh-ye Meydān

GREAT PEARL BANK
Umm al Qaywayn
Khawr al Jafrah
Dadnā
Khawr Fakkān
Madha
Oman
Khawr Fakkān

Sīr bu
Na'air Island
The Palm-Deira
The World
Ajmān
Sharjah
Al Dhaid
Nahwa
U.A.E.
Al Fujairah

The Palm-Jumeirah
Dubai
Kalbā

The Palm Jebel Ali
Ra's Ḩasyān
Jabal Alī

RUQQ AZ ZUKUM
Jazīrat aţ Ţawīlah
Shināş

Dawḩat Ḩanyūrah
GULF OF

Khawr al Ḩalij
Jazīrat Ra's Ghurāb
Jazīrat as Sa'dīyāt
Jazīrat Abū Ẓaby
Khawr Fāhid
OMAN

Al Mubarraz
Al Bahrānī Island
Abu Dhabi
Khawr
Qirqishān
Wādī Sūq
Ra's Şallān
Şuhār
Ra's al Hijārī

Khawr al Fīyay
Jazīrat
Marawwaḩ
Sabkhat as
Salamīyah
Al Buraymī
Şaḩam
Daimaniyat
Islands

Maqaishat I.
Al Ain
Al Ghuzayfah
Al Khābūrah
Suadi Point
Fahal Island

Al Mughayrā'
Ţarīf
Ḩabshān
ARAB   EMIRATES
Al Qābil
O   M   A   N
Barkā
As Sīb
Ghubbat
al Ḩayl

54°   56°   58°   24°

ARCTIC OCEAN

Arctic Ocean

I n this once fully white dome at the top of the world, cold rules. Dark for half the year, bright the rest, with never setting sun, Arctic skies stage shimmering curtain-like displays of aurora borealis. In this ice-riven place all longitudes converge; one can literally circle the world in a few steps. It is more magic still because a compass alone cannot lead you here. Instead it will send you to the magnetic pole, perhaps 1,500 kilometers away. To reach the geographic top of the world, you must determine a magnetic variation, which differs based on your location.

Ringed by the Arctic Circle some 1,700 kilometers to its south, the Arctic Ocean counts among its neighbors the northernmost parts of Canada, the United States, Greenland, Iceland, Norway, Finland, Sweden, and Russia.

While cold rules here, so does human life. For about 5,000 years the Inuit have hunted whales and seals in Canada and Greenland; the Saami and Chukchi have herded reindeer in Lapland; the Nenets have fished in Siberia. Still, the total human population of Arctic regions may be no more than four million. Villages, many nomadic, shift with the seasons, sunlight, and migration patterns of the animals they hunt or herd.

This cap of the world lies not on land but amid ice floes in the frozen Arctic Ocean. The Arctic ice cover, subfreezing water temperatures of 1.6°C, and low temperature of the surrounding air, have held explorers, scientists, and cartographers at bay. But over time an accurate picture has come into view.

Experts have determined that perhaps 66 million years ago a spreading ridge opened under the Eurasian continent. As it spewed out magma to form new ocean crust, a fragment of the continent above broke off and was propelled eastward by seafloor spreading. As it moved, it gradually sank, finally settling on the floor beneath today's North Pole. To its south lie the Alpha Cordillera and Mendeleyev Ridge, and to the north the Nansen (Gakkel) Ridge—the world ocean's deepest and slowest spreading ridge, which ends near Greenland with a string of active volcanoes.

Iconic polar bears, walruses, and seals bridge ice and sea, while the water sustains whales and the krill they consume, as well as a profusion of other life: tiny diatoms, crustaceans, polychaete worms, jellyfish, starfish, and the revered Arctic cod, a staple of the ocean's food web.

Less affected by industry, farming, and pollution than other ocean basins, the Arctic Ocean is nevertheless at risk. Planetary warming is melting the sea ice, opening the Arctic to shipping and possible offshore mining and oil drilling operations. The Arctic's vital role in regulating the temperature of the Earth is now being compromised by climate change and melting ice cover.

RIGHT: "Unicorns of the sea," narwhals dive as much as 1,000 meters into Baffin Bay's Arctic waters. PAGES 486-487: Insulated by layers of fur and fat a polar bear swims off Greenland.

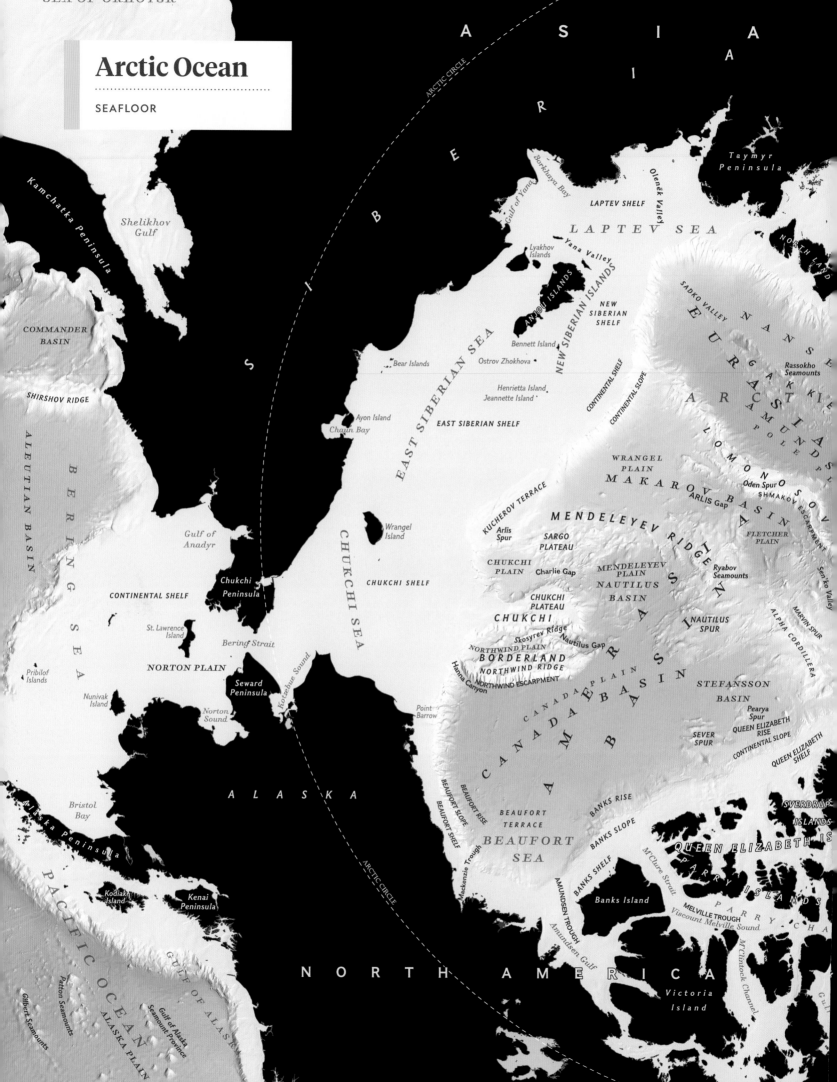

Arctic Ocean

SEAFLOOR

SEA OF OKHOTSK

A S I A

S I B E R I A

ARCTIC CIRCLE

Kamchatka Peninsula

Shelikhov Gulf

COMMANDER BASIN

SHIRSHOV RIDGE

ALEUTIAN BASIN

BERING SEA

Gulf of Anadyr

CONTINENTAL SHELF

St. Lawrence Island

Pribilof Islands

Nunivak Island

NORTON PLAIN

Norton Sound

Bristol Bay

Alaska Peninsula

PACIFIC OCEAN

Kodiak Island

Kenai Peninsula

GULF OF ALASKA

Gilbert Seamounts

Patton Seamounts

Gulf of Alaska Seamount Province

ALASKA PLAIN

ALASKA

ARCTIC CIRCLE

NORTH AMERICA

Taymyr Peninsula

Gulf of Yana

Borkhaya Bay

LAPTEV SHELF

Olenëk Valley

LAPTEV SEA

NORTH LAND

Lyakhov Islands

Yana Valley

ANJOU ISLANDS

NEW SIBERIAN ISLANDS

NEW SIBERIAN SHELF

NANSE

SADKO VALLEY

EURAS

GAKKEL

AMUND

POLE

Bennett Island

Bear Islands

Ostrov Zhokhova

CONTINENTAL SHELF

Rassokho Seamounts

EAST SIBERIAN SEA

Henrietta Island

Jeannette Island

CONTINENTAL SLOPE

CONTINENTAL SLOPE

ARCTIC

LOMONOSOV

Ayon Island

Chaun Bay

EAST SIBERIAN SHELF

WRANGEL PLAIN

MAKAROV

Oden Spur

BASIN

SHMAKOV ESCARPMENT

KUCHEROV TERRACE

MENDELEYEV RIDGE

ARLIS Gap

FLETCHER PLAIN

Sen'ko Valley

Wrangel Island

Arlis Spur

SARGO PLATEAU

MENDELEYEV PLAIN

Ryabov Seamounts

CHUKCHI SEA

CHUKCHI SHELF

CHUKCHI PLAIN

Charlie Gap

NAUTILUS BASIN

MARVIN SPUR

ALPHA CORDILLERA

CHUKCHI PLATEAU

NAUTILUS SPUR

Chukchi Peninsula

CHUKCHI

Skosyrev Ridge

Nautilus Gap

Bering Strait

BORDERLAND

Kotzebue Sound

NORTHWIND PLAIN

Seward Peninsula

NORTHWIND RIDGE

CANADA PLAIN

STEFANSSON BASIN

Hanna Canyon

NORTHWIND ESCARPMENT

CANADA

Pearya Spur

Point Barrow

BASIN

QUEEN ELIZABETH RISE

SEVER SPUR

CONTINENTAL SLOPE

QUEEN ELIZABETH SHELF

BEAUFORT SHELF

BEAUFORT SLOPE

BEAUFORT RISE

BANKS RISE

SVERDRUP ISLANDS

BEAUFORT TERRACE

BANKS SLOPE

QUEEN ELIZABETH IS

Mackenzie Trough

BEAUFORT SEA

BANKS SHELF

PARRY ISLANDS

M'Clure Strait

AMUNDSEN TROUGH

Banks Island

MELVILLE TROUGH

PARRY CHA

Amundsen Gulf

Viscount Melville Sound

M'Clintock Channel

Victoria Island

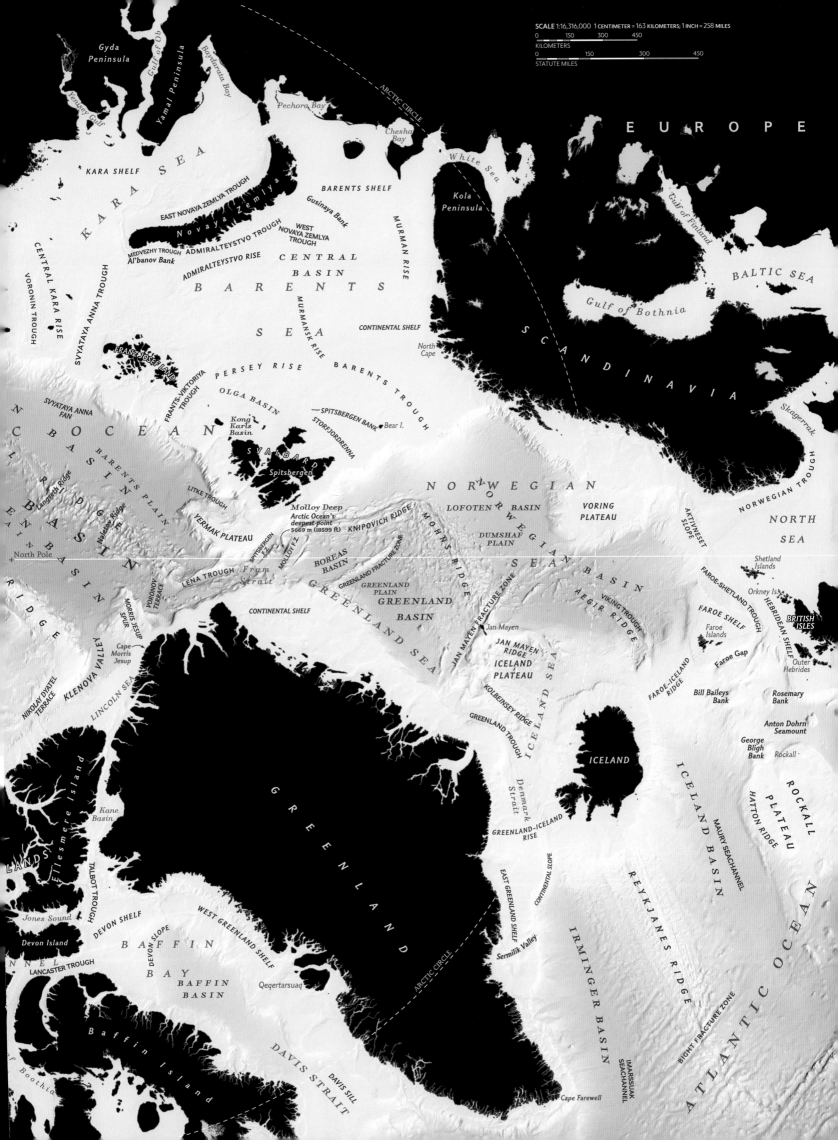

SCALE 1:16,316,000 1 CENTIMETER = 163 KILOMETERS; 1 INCH = 258 MILES

KILOMETERS 0 150 300 450

STATUTE MILES 0 150 300 450

EUROPE

Gyda Peninsula

Gulf of Ob

Yamal Peninsula

Baydarata Bay

Yenisey Gulf

KARA SHELF

KARA SEA

Pechora Bay

Chesha Bay

White Sea

Gulf of Finland

BALTIC SEA

Kola Peninsula

BARENTS SHELF

Gulf of Bothnia

SCANDINAVIA

EAST NOVAYA ZEMLYA TROUGH

Novaya Zemlya

Gusinaya Bank

WEST NOVAYA ZEMLYA TROUGH

MURMAN RISE

MEDVEZHY TROUGH ADMIRALTEYSTVO TROUGH

Al'banov Bank

ADMIRALTEYSTVO RISE

CENTRAL BASIN

B A R E N T S

ADMIRALTEYSTVO TROUGH

CENTRAL KARA RISE

VORONIN TROUGH

SVYATAYA ANNA TROUGH

FRANZ JOSEF LAND

PERSEY RISE

S E A

MURMANSK RISE

MURMANSK RISE

CONTINENTAL SHELF

North Cape

Skagerrak

SVYATAYA ANNA FAN

FRANTS-VIKTORIYA TROUGH

OLGA BASIN

BARENTS TROUGH

SPITSBERGEN BANK

Bear I.

NORTH SEA

C O C E A N

B A S I N

Kong Karls Basin

STORFJORDRENNA

NORWEGIAN TROUGH

Shetland Islands

BARENTS PLAIN

SVALBARD

Spitsbergen

N O R W E G I A N

LOFOTEN BASIN

VORING PLATEAU

AKTIVNESET SLOPE

FAROE-SHETLAND TROUGH

Orkney Is.

HEBRIDEAN SHELF

D G

Langsseth Ridge

Nafsen Ridge

LITKE TROUGH

YERMAK PLATEAU

Molloy Deep
Arctic Ocean's
deepest point
5,669 m (18,599 ft)

KNIPOVICH RIDGE

MOHNS RIDGE

DUMSHAF PLAIN

VIKING TROUGH

AEGIR RIDGE

S E A   B A S I N

FAROE SHELF

FAROE Islands

BRITISH ISLES

R I D G E

North Pole

SPITSBERGEN FZ

MOLLOY FZ

BOREAS BASIN

GREENLAND FRACTURE ZONE

FARO E-ICELAND RIDGE

Bill Baileys Bank

Outer Hebrides

B A S I N

Lena Trough

Fram Strait

GREENLAND SEA

GREENLAND PLAIN

GREENLAND BASIN

JAN MAYEN FRACTURE ZONE

Jan Mayen

Faroe Gap

Rosemary Bank

R I D G E

VORONOV TERRACE

MORRIS JESUP SPUR

CONTINENTAL SHELF

JAN MAYEN RIDGE

ICELAND PLATEAU

Anton Dohrn Seamount

NIKOLAY DYATEL TERRACE

Cape Morris Jesup

KOLBEINSEY RIDGE

ICELAND SEA

George Bligh Bank

Rockall

KLENOVA VALLEY

LINCOLN SEA

Ellesmere Island

Kane Basin

GREENLAND

G R E E N L A N D

GREENLAND TROUGH

ICELAND

ICELAND BASIN

ROCKALL PLATEAU

LANDS

TALBOT TROUGH

Denmark Strait

MAURY SEACHANNEL

HATTON RIDGE

DEVON SHELF

WEST GREENLAND SHELF

DEVON SLOPE

Jones Sound

Devon Island

BAFFIN BAY

BAFFIN BASIN

Qeqertarsuaq

GREENLAND-ICELAND RISE

EAST GREENLAND SHELF

CONTINENTAL SLOPE

Sermilik Valley

REYKJANES RIDGE

IRMINGER BASIN

A T L A N T I C   O C E A N

NNEL

LANCASTER TROUGH

ARCTIC CIRCLE

Baffin Island

DAVIS STRAIT

DAVIS SILL

BIGHT FRACTURE ZONE

f Boothia

Cape Farewell

IMARSSUAK SEACHANNEL

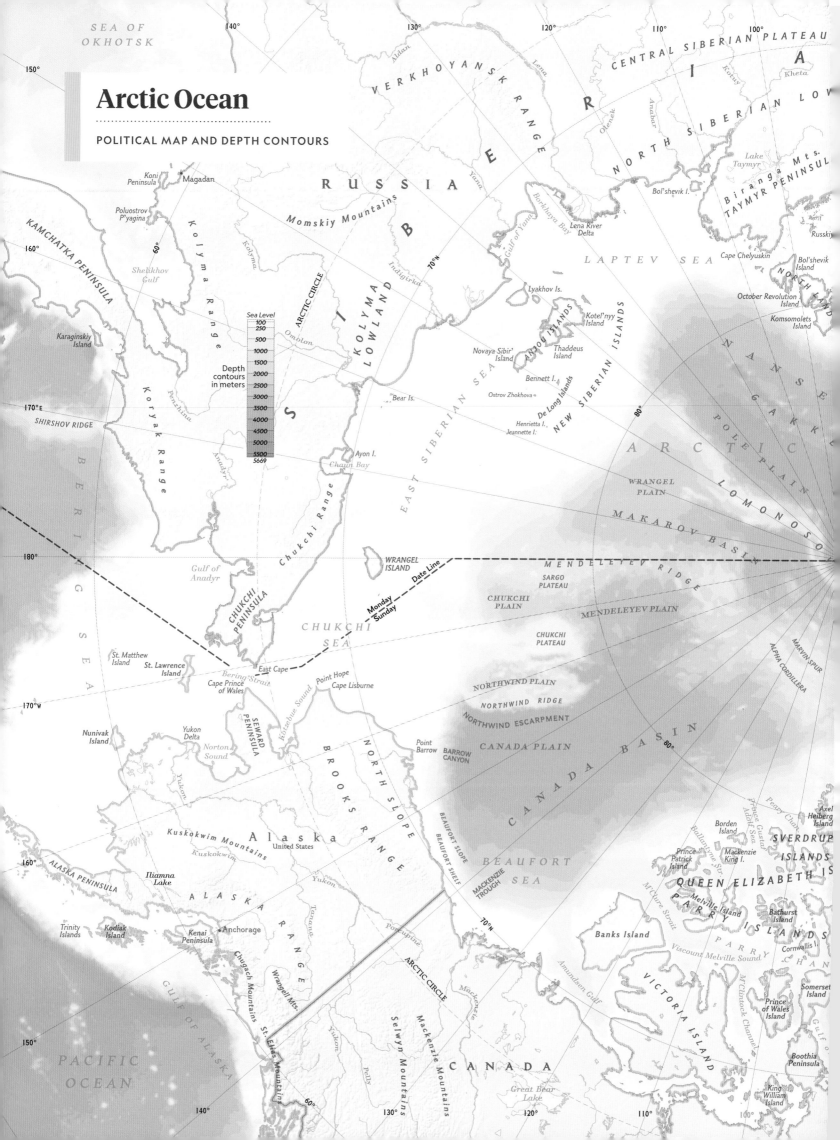

Arctic Ocean

POLITICAL MAP AND DEPTH CONTOURS

Arctic Ocean

SEA ROUTES & CONTINENTAL CLAIMS

Once impenetrable, the Arctic Ocean now is accessed by four sea routes (top), two hugging Russia's coast. With new strategic importance as climate change melts its ice and reveals trillions of dollars in resources, Arctic regions are claimed by eight encircling nations.

ASIA
Tokyo
8700 mi
Northwest Passage
New York
NORTH AMERICA
11300 mi
Panama Canal route
London
8100 mi
Northeast Passage
EUROPE
ASIA
Tokyo
AFRICA
Suez Canal route
13000 mi

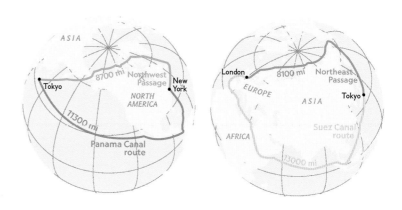

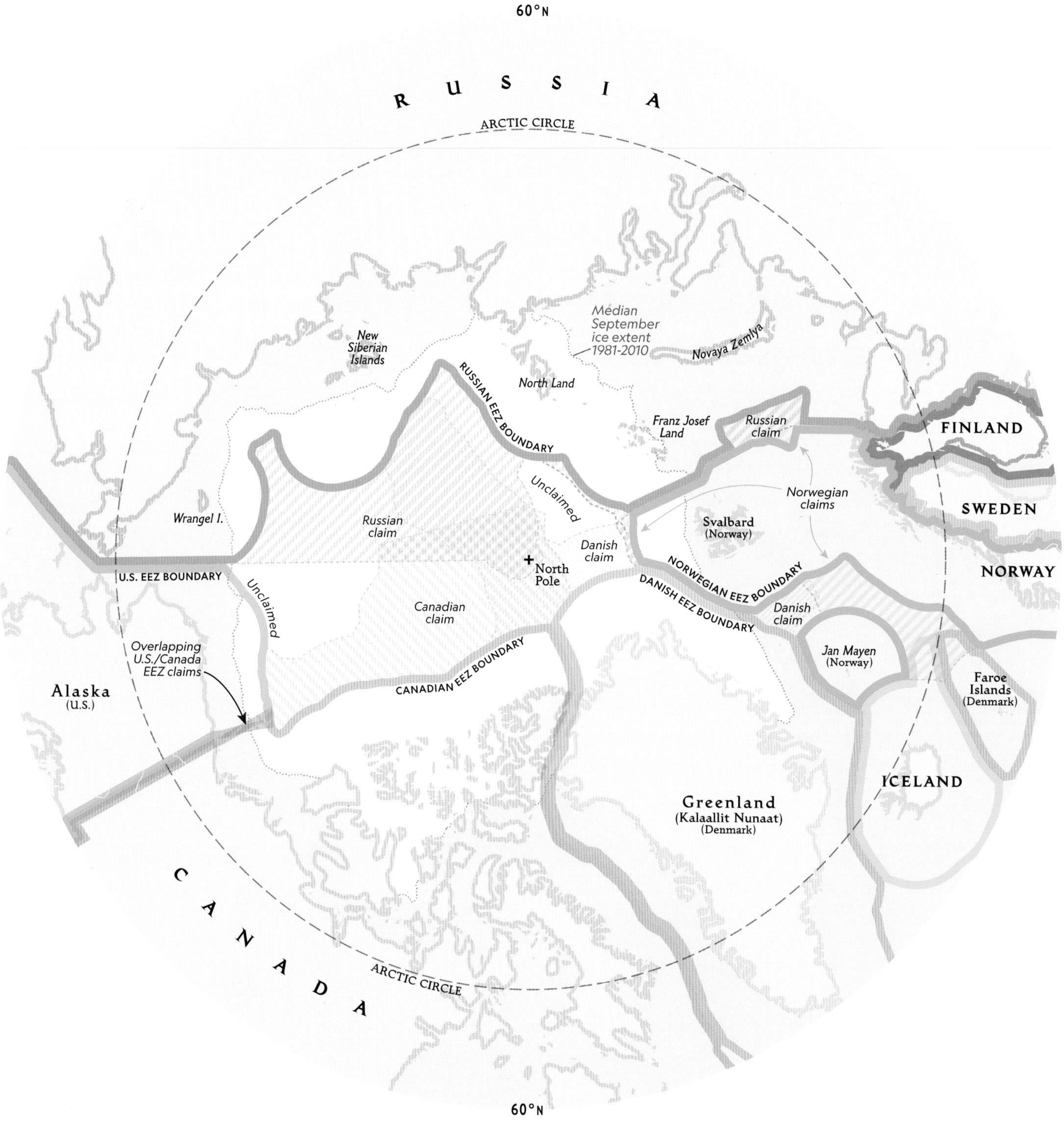

60°N
RUSSIA
ARCTIC CIRCLE
New Siberian Islands
Median September ice extent 1981-2010
Novaya Zemlya
North Land
RUSSIAN EEZ BOUNDARY
Franz Josef Land
Russian claim
FINLAND
Unclaimed
Norwegian claims
SWEDEN
Wrangel I.
Russian claim
Svalbard (Norway)
Danish claim
North Pole
NORWAY
U.S. EEZ BOUNDARY
NORWEGIAN EEZ BOUNDARY
DANISH EEZ BOUNDARY
Unclaimed
Canadian claim
Danish claim
Overlapping U.S./Canada EEZ claims
CANADIAN EEZ BOUNDARY
Jan Mayen (Norway)
Faroe Islands (Denmark)
Alaska (U.S.)
ICELAND
Greenland (Kalaallit Nunaat) (Denmark)
CANADA
ARCTIC CIRCLE
60°N

Gliding on hundreds of tiny tube feet, *Asterias* starfish thrive in an Arctic kelp forest.

I SOUTHERN OCEAN

Southern Ocean

After breaking apart from Gondwanaland 180 million years ago, Antarctica was shoved southward by seafloor spreading to rest at the bottom of Earth. Its position out of the sun's direct rays prompted the growth of the massive ice sheet some 14.2 million square kilometers wide and 3.2 kilometers thick, covering this fifth largest continent. Only about 2 percent of the land is free of ice.

The surrounding water, the Southern Ocean, covers more than 20,000 square kilometers as it rims the glacial landmass. It is the only ocean not bounded by continents but by a natural oceanic boundary, the Antarctic Convergence—the edge of a powerful, fast-moving current that forms a ring around the continent, varying between 45° S and 60° S latitude.

The seafloor holds the South Sandwich Trench, a subduction zone and the deepest part of the ocean surrounding Antarctica; the rugged Scotia Arc, a submerged mountain range that links South America's Cape Horn and the Antarctic Peninsula; and basins, channels, extensive outer banks, and the massive crater of a meteorite—the only one yet found in the deep ocean—some two million years old.

American seal hunter John Davis claimed to land on the Antarctic continent in 1821, and in 1823 James Weddell sighted the sea south of Patagonia that today bears his name. In the early 1900s British explorers Robert Falcon Scott and Ernest Shackleton set out to reach the South Pole, but it was Roald Amundsen who first arrived in 1911, planting the Norwegian flag at the geographic bottom of the globe. Today, through the 1959 Antarctic Treaty, the land is dedicated to peaceful scientific research only. Seventy stations from 20 countries dot the icescape. Except at scientific outposts, no human has ever settled on Antarctica.

While perhaps healthier than northern waters, the Southern Ocean is not an isolated system. Soon after discovery, the Antarctic continent and various subantarctic islands became destinations for whalers, sealers, and bird hunters who took so many animals for oil, meat, fur, feathers, and bone that natural populations have not recovered; protective policies did not begin until well into the 20th century. Although mammals and birds are currently protected throughout the Southern Ocean, large-scale fishing by many nations has depleted krill and various species of slow-growing, deep-sea fish including Patagonian toothfish (Chilean sea bass) and orange roughy. Plastic waste and even pesticides are reaching the Southern Ocean, and global warming is breaking up and melting Antarctic ice much faster than expected, with sea level projected to rise many meters in the coming century.

BY THE NUMBERS

Southern Ocean

....

Total Area
20,973,318 square kilometers

Total Volume
71,515,351 cubic kilometers

Greatest Depth
7,075 meters (south end of the South Sandwich Trench)

Includes 6.5 percent of Earth's water area.

Approximately 98 percent of the Antarctic continent lies under permanent ice sheets.

Geographic Boundaries
60° S to Antarctic coastline / circumpolar

Average Depth
3,270 meters

RIGHT: Vital sustenance for Antarctic animals, krill (*Euphausia superba*) in turn dine on microscopic phytoplankton. PAGES 496-497: King penguins colonize South Georgia Island.

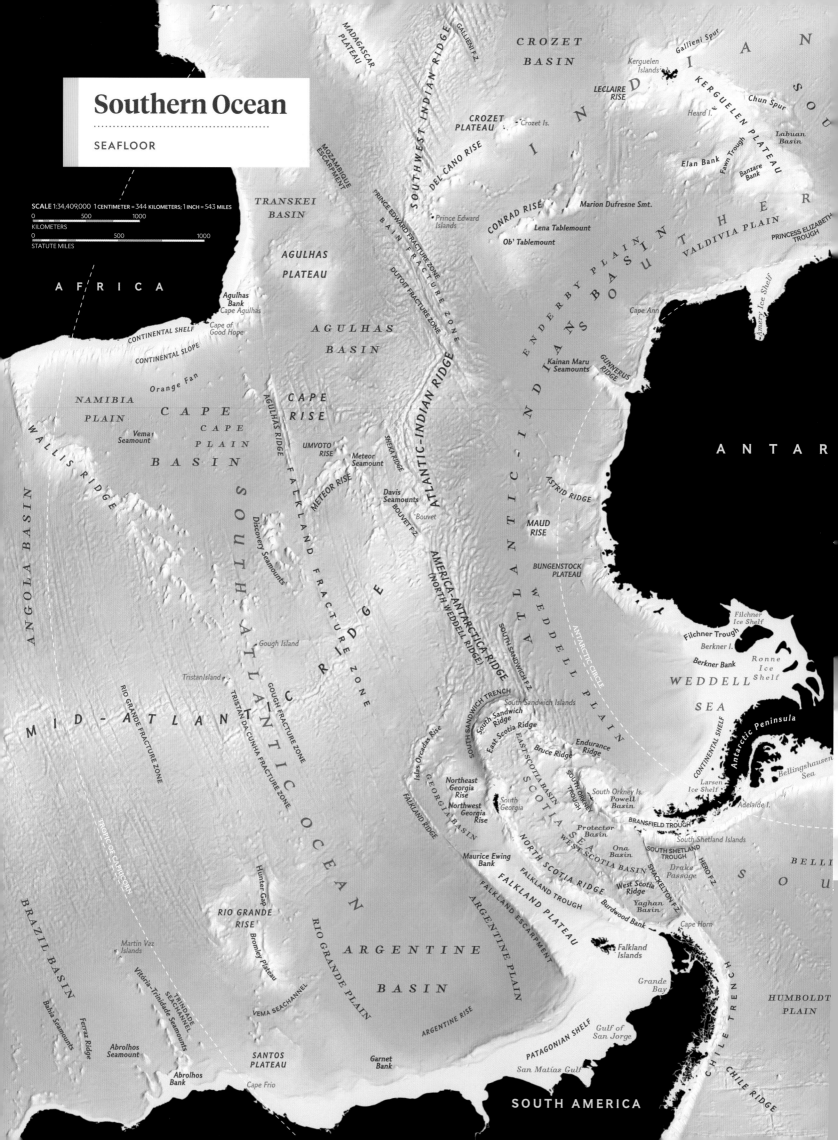

Southern Ocean

SEAFLOOR

SCALE 1:34,409,000   1 CENTIMETER = 344 KILOMETERS; 1 INCH = 543 MILES

KILOMETERS
0   500   1000

STATUTE MILES
0   500   1000

MADAGASCAR PLATEAU

CROZET BASIN

Galleni Spur

Kerguelen Islands

Chun Spur

SOUTHWEST INDIAN RIDGE

GALLIENI F.Z.

LECLAIRE RISE

CROZET PLATEAU

Crozet Is.

DEL CANO RISE

Heard I.

KERGUELEN PLATEAU

Labuan Basin

Elan Bank

Fawn Trough

Banzare Bank

INDIAN

SOU

MOZAMBIQUE ESCARPMENT

PRINCE EDWARD FRACTURE ZONE

CONRAD RISE

Marion Dufresne Smt.

Prince Edward Islands

Lena Tablemount

Ob' Tablemount

DU TOIT FRACTURE ZONE

BAIN FRACTURE ZONE

VALDIVIA PLAIN

PRINCESS ELIZABETH TROUGH

THER

TRANSKEI BASIN

AGULHAS PLATEAU

AFRICA

Agulhas Bank
Cape Agulhas

CONTINENTAL SHELF

Cape of Good Hope

CONTINENTAL SLOPE

AGULHAS BASIN

ENDERBY PLAIN

SOUTHERN

INDIANS

BOUVET

Amery Ice Shelf

Cape Ann

Kainan Maru Seamounts

GUNNERUS RIDGE

ANTAR

Orange Fan

NAMIBIA PLAIN

CAPE BASIN

CAPE PLAIN

CAPE RISE

AGULHAS RIDGE

UMVOTO RISE

SHEBA RIDGE

ATLANTIC-INDIAN RIDGE

ASTRID RIDGE

Vema Seamount

Meteor Seamount

METEOR RISE

Davis Seamounts

MAUD RISE

WALLIS RIDGE

FALKLAND FRACTURE ZONE

BOUVET F.Z.

Bouvet

BUNGENSTOCK PLATEAU

ANGOLA BASIN

SOUTH-ATLANTIC RIDGE

Discovery Seamounts

AMERICA-ANTARCTICA RIDGE (NORTH WEDDELL RIDGE)

ATLANTIC-INDIAN

WEDDELL PLAIN

ANTARCTIC CIRCLE

Filchner Ice Shelf

Filchner Trough

Berkner I.

Ronne Ice Shelf

Gough Island

GOUGH FRACTURE ZONE

Berkner Bank

WEDDELL SEA

Tristan Island

TRISTAN DA CUNHA FRACTURE ZONE

SOUTH SANDWICH TRENCH

SOUTH SANDWICH F.Z.

Antarctic Peninsula

MID-ATLAN

RIO GRANDE FRACTURE ZONE

TIC OCEAN

Islas Orcadas Rise

South Sandwich Islands

South Sandwich Ridge

East Scotia Ridge

EAST SCOTIA BASIN

Bruce Ridge

Endurance Ridge

South Orkney Is.

SOUTH ORKNEY TROUGH

Powell Basin

CONTINENTAL SHELF

Larsen Ice Shelf

Bellingshausen Sea

TROPIC OF CAPRICORN

RIO GRANDE RISE

Hunter Gap

Martin Vaz Islands

Vitória-Trindade Seamounts

Bromley Plateau

RIO GRANDE PLAIN

VEMA SEACHANNEL

ARGENTINE BASIN

Georgia Basin

Northeast Georgia Rise

Northwest Georgia Rise

FALKLAND RIDGE

South Georgia

SCOTIA SEA

NORTH SCOTIA RIDGE

WEST SCOTIA BASIN

Protector Basin

Ona Basin

SOUTH SHETLAND TROUGH

BRANSFIELD TROUGH

South Shetland Islands

West Scotia Ridge

Yaghan Basin

SHACKLETON F.Z.

Drake Passage

HERO F.Z.

Adelaide I.

SOU

BELL

Maurice Ewing Bank

FALKLAND PLATEAU

FALKLAND ESCARPMENT

ARGENTINE PLAIN

Burdwood Bank

FALKLAND TROUGH

Cape Horn

BRAZIL BASIN

Bahia Seamounts

Ferraz Ridge

Abrolhos Seamount

Abrolhos Bank

TRINDADE SEACHANNEL

ARGENTINE BASIN

SANTOS PLATEAU

Cape Frio

Garnet Bank

ARGENTINE RISE

PATAGONIAN SHELF

San Matías Gulf

Gulf of San Jorge

Grande Bay

CHILE TRENCH

CHILE RIDGE

HUMBOLDT PLAIN

SOUTH AMERICA

SOUTHEAST INDIAN RIDGE

SOUTH AUSTRALIAN BASIN

SOUTH AUSTRALIAN PLAIN

AUSTRALIA

SOUTHERN OCEAN
There is international agreement that the icy waters around Antarctica form a distinct ocean region. While there is no consensus on its name or its extent, most countries call it the Southern Ocean and use 60° south latitude as an approximation of its northern limit.

Great Barrier Reef

Queensland Plateau

CORAL SEA BASIN

O C E A N

S O U T H   I N D I A N   B A S I N

I N D I A N - A N T A R C T I C   R I D G E

Australian-Antarctic
Discordance

Marion Plateau

C O R A L   S E A

C O R A L   S E A

CATO TROUGH

Bruce Spur

CONTINENTAL SLOPE

Britannia Tablemounts

Kenn Plateau

DAVIS SEA

Cape Poinsett

Tasmania

Bass Canyon

TASMAN PLAIN

Tasmantid Seamounts

DAMPIER RIDGE

Fairway Plateau

ANTARCTIC CIRCLE

SAINT VINCENT FRACTURE ZONE

GEORGE V FRACTURE ZONE

GAMBIER FRACTURE ZONE

TASMAN ESCARPMENT

SOUTH TASMAN RISE

TASMAN F.Z.

L'Atalante Valley

EAST TASMAN PLATEAU

North Tasman Seamounts

Gascoyne Smt.

Lord Howe I.
Ball's Pyramid

Lord Howe Seamounts

New Caledonia

NEW CALEDONIA TROUGH

TASMAN BASIN

T A S M A N   S E A

L O R D   H O W E   R I S E

NORFOLK RIDGE

A N T A R C T I C A

Macquarie I.

MACQUARIE RIDGE

Dolphin Spur
Bellona Valley

NORFOLK TROUGH

West Norfolk Ridge

Reinga Basin

South Norfolk Basin

North Norfolk Basin

NEW HEBRIDES TRENCH

HJORT TRENCH

MACQUIRE TRENCH

EMERALD BASIN

AUKLAND ESCARPMENT

Auckland Is.

AUKLAND ISLANDS SHELF

South Island

CHALLENGER PLATEAU

Egmont Terrace

Three Kings Ridge

BALLENY TROUGH

BALLENY F.Z.

Balleny Is.

BALLENY BASIN

CAMPBELL PLATEAU

CAMPBELL ISLAND SHELF

NEW ZEALAND

North Island

SOUTH FIJI BASIN

Drygalski Basin

Ross I.

Mawson Bank

Cape Adare

Colville Basin

Hikurangi Terrace

COLVILLE RIDGE

South Pole

ROSS SEA

Iselin Bank

CAMPBELL ESCARPMENT

SUBANTARCTIC SLOPE

Antipodes Is.

BOUNTY TROUGH

CHATHAM RISE

SOUTH CHATHAM SLOPE

NORTH CHATHAM SLOPE

HAVRE TROUGH

Roosevelt I.

Little America Basin

Bounty Plateau

Bollons Seamount

Chatham Is.

Hatherton Seamounts

KERMADEC RIDGE

Kermedic Is.

Tonga Ridge

Siple I.

Carney I.

PITMAN FRACTURE ZONE

A N T A R C T I C   R I D G E

ANTIPODES FRACTURE ZONE

S O U T H W E S T

KERMADEC TRENCH

Tonga Islands

TONGA TRENCH

Amundsen Sea

Marie Byrd Seamounts

Amundsen Ridges

ANTARCTIC CIRCLE

P A C I F I C -

Thurston I.

O C E A N

UDINTSEV FRACTURE ZONE

P A C I F I C   O C E A N

LOUISVILLE RIDGE

Peter I Island

A M U N D S E N   P L A I N

S O U T H E R N

MARLIN RISE

AGUILA FRACTURE ZONE

De Gerlache Seamounts

NGSHAUSEN PLAIN

T

SOUTHEAST PACIFIC BASIN

ELTANIN FRACTURE ZONE

THARP FRACTURE ZONE

HEEZEN FRACTURE ZONE

EAST PACIFIC RISE

S O U T H   P A C I F I C   B A S I N

COOK ISLANDS

Maria Islands

San Martin Seamounts

MENARD FRACTURE ZONE

Henry Trough

Austral Seamounts

AUSTRAL ISLANDS

SOCIETY RIDGE

AGASSIZ FRACTURE ZONE

ADVENTURE TROUGH

Neilson Reef

Rapa

Marotiri

President Thiers Bank

Tahiti

SOCIETY ISLANDS

TROPIC OF CAPRICORN

Foundation Seamounts

RESOLUTION F.Z.

TUAMOTU RIDGE

TUAMOTU ARCHIPELAGO

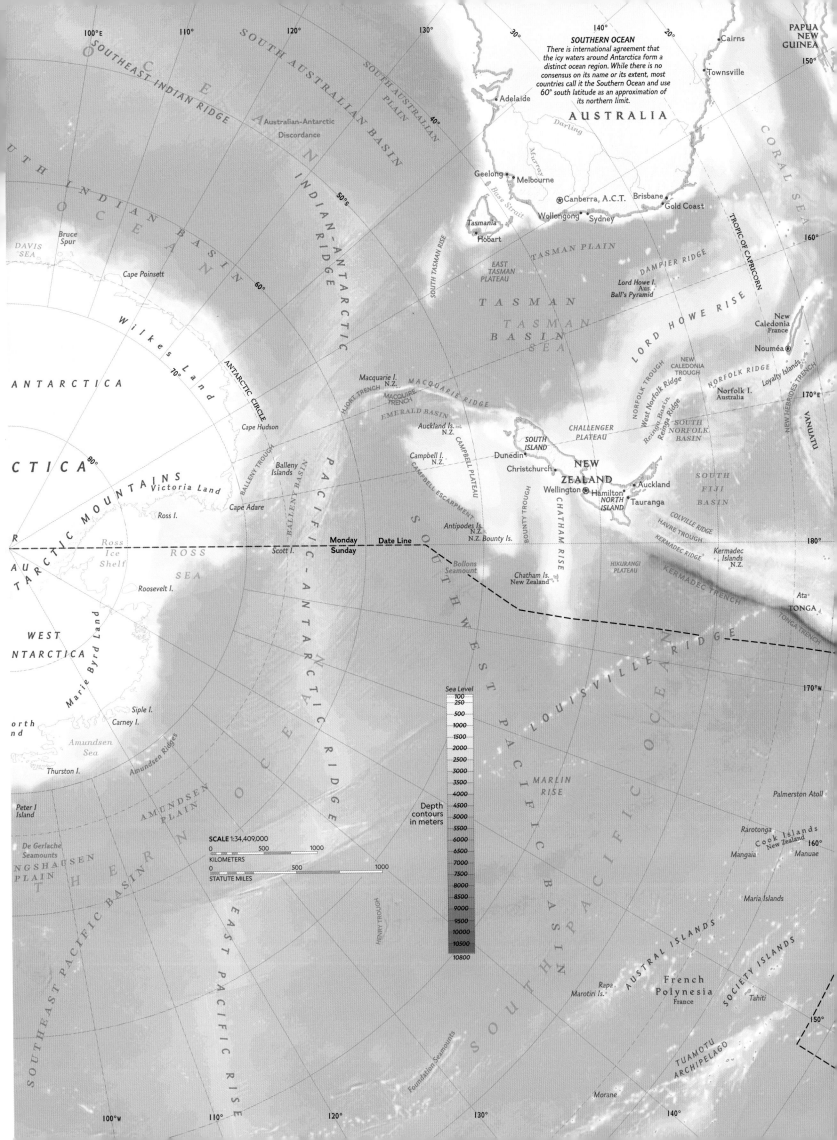

100°E · 110° · 120° · 130° · 30° · 140° · 20° · 150°

PAPUA NEW GUINEA

SOUTHERN OCEAN
There is international agreement that the icy waters around Antarctica form a distinct ocean region. While there is no consensus on its name or its extent, most countries call it the Southern Ocean and use 60° south latitude as an approximation of its northern limit.

• Cairns
• Townsville

OCEAN

SOUTHEAST INDIAN RIDGE

SOUTH AUSTRALIAN BASIN

SOUTH AUSTRALIAN PLAIN

Australian-Antarctic Discordance

40°

AUSTRALIA
Darling
Murray

• Adelaide

SOUTH INDIAN BASIN

INDIAN-ANTARCTIC RIDGE

50°S

• Geelong • Melbourne
⊗ Canberra, A.C.T.
Wollongong • Sydney • Brisbane
• Gold Coast

TROPIC OF CAPRICORN

CORAL SEA

160°

DAVIS SEA
Bruce Spur

Cape Poinsett

Wilkes Land

60°

Bass Strait

Tasmania
• Hobart

EAST TASMAN PLATEAU

TASMAN PLAIN

DAMPIER RIDGE

Lord Howe I. Aus.
Ball's Pyramid

LORD HOWE RISE

New Caledonia France

Nouméa •

ANTARCTICA

70°

ANTARCTIC CIRCLE

Cape Hudson

Macquarie I. N.Z.

MACQUARIE RIDGE

NORFOLK TROUGH
NEW CALEDONIA TROUGH

West Norfolk Ridge
Reinga Basin
Reinga Ridge

NORFOLK RIDGE

Loyalty Islands

Norfolk I. Australia

170°E

VANUATU

NEW HEBRIDES TRENCH

CTICA

CTICA

80°

ANTARCTIC MOUNTAINS

Victoria Land

Ross I.

Cape Adare

HJORT TRENCH
MACQUARIE TRENCH

EMERALD BASIN

Auckland Is. N.Z.

BALLENY TROUGH

Balleny Islands

BALLENY BASIN

Campbell I. N.Z.

CAMPBELL PLATEAU

CAMPBELL ESCARPMENT

SOUTH ISLAND

Dunedin
Christchurch •

CHALLENGER PLATEAU

NEW ZEALAND
Wellington ⊗ Hamilton •
NORTH ISLAND

• Auckland
• Tauranga

SOUTH FIJI BASIN

R

Ross Ice Shelf

ROSS SEA

Scott I.

Monday
Sunday

Date Line

SOUTH

PACIFIC-ANTARCTIC RIDGE

Antipodes Is. N.Z.
N.Z. Bounty Is.

BOUNTY TROUGH

CHATHAM RISE

COLVILLE RIDGE

HAVRE TROUGH

KERMADEC RIDGE

180°

AU

ARCTICA

Roosevelt I.

Bollons Seamount

Chatham Is. New Zealand

HIKURANGI PLATEAU

KERMADEC TRENCH

Kermadec Islands N.Z.

Ata •
TONGA

WEST NTARCTICA

Marie Byrd Land

Siple I.
Carney I.

Amundsen Ridges

SOUTH WEST PACIFIC OCEAN

LOUISVILLE RIDGE

TONGA TRENCH

170°W

orth nd

Amundsen Sea

Thurston I.

AMUNDSEN PLAIN

AMUNDSEN OCEAN

Sea Level
100
250
500
1000
1500
2000
2500
3000
3500
4000
4500
5000
5500
6000
6500
7000
7500
8000
8500
9000
9500
10000
10500
10800

Depth contours in meters

MARLIN RISE

Palmerston Atoll

Peter I Island

De Gerlache Seamounts

SCALE 1:34,409,000
0   500   1000
KILOMETERS
0   500   1000
STATUTE MILES

SOUTH PACIFIC BASIN

Rarotonga
Cook Islands New Zealand

160°

NGSHAUSEN PLAIN

T

Mangaia

Manuae

SOUTHEAST PACIFIC BASIN

EAST PACIFIC RISE

HENRY TROUGH

Maria Islands

Foundation Seamounts

AUSTRAL ISLANDS

Rapa
Marotiri Is.

French Polynesia France

SOCIETY ISLANDS

Tahiti

TUAMOTU ARCHIPELAGO

150°

Morane

100°W · 110° · 120° · 130° · 140°

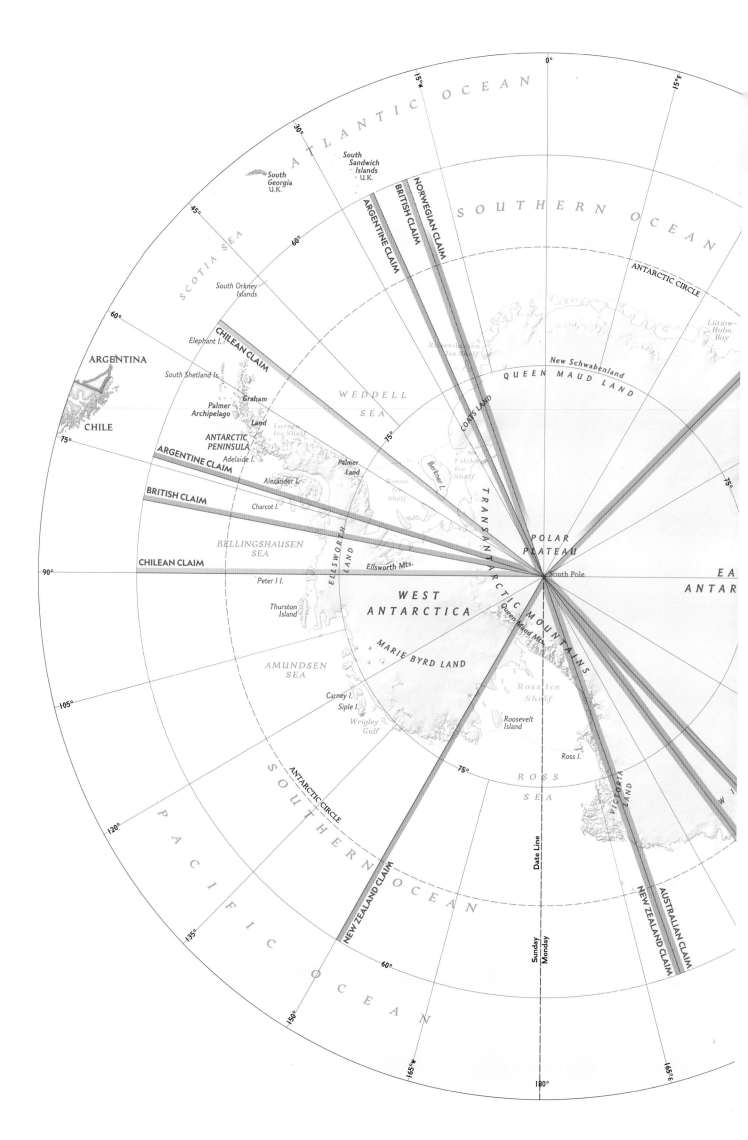

ATLANTIC OCEAN

South
Georgia
U.K.

South
Sandwich
Islands
U.K.

BRITISH CLAIM

ARGENTINE CLAIM

NORWEGIAN CLAIM

SOUTHERN OCEAN

ANTARCTIC CIRCLE

Lützow-
Holm
Bay

SCOTIA SEA

South Orkney
Islands

Riiser-larsen
Ice Shelf

New Schwabenland

QUEEN MAUD LAND

ARGENTINA

Elephant I.

CHILEAN CLAIM

South Shetland Is.

Palmer
Archipelago

Graham

CHILE

Land

ANTARCTIC
PENINSULA

Larsen
Ice Shelf

WEDDELL

SEA

COATS LAND

Berkner I.

Filchner
Ice
Shelf

ARGENTINE CLAIM

Adelaide I.

Palmer
Land

Ronne
Ice
Shelf

BRITISH CLAIM

Alexander I.

Charcot I.

ANTARCTIC
CIRCLE

POLAR
PLATEAU

EA
ANTAR

CHILEAN CLAIM

BELLINGSHAUSEN
SEA

Ellsworth Mts.

ELLSWORTH
LAND

WEST
ANTARCTICA

TRANSANTARCTIC MOUNTAINS

South Pole

75°

Peter I I.

Thurston
Island

Queen Maud Mts.

Carney I.
Siple I.

MARIE BYRD LAND

Ross Ice
Shelf

Roosevelt
Island

AMUNDSEN
SEA

Wrigley
Gulf

75°

Ross I.

ROSS

SEA

VICTORIA LAND

W

ANTARCTIC CIRCLE

Date Line

SOUTHERN OCEAN

NEW ZEALAND CLAIM

Sunday
Monday

NEW ZEALAND CLAIM

AUSTRALIAN CLAIM

PACIFIC

OCEAN

Southern Ocean

....

ANTARCTIC TREATY MAP

LEFT: Owned by no country, Antarctica supports 70 scientific research stations from 20 nations. ABOVE: A bright orange stomach and frilled arms distinguish this *Diplulmaris antarctica* jellyfish.

ACKNOWLEDGMENTS

Keeping up with the state of the ocean has proved to be a team sport. It has taken the combined talent, dedication, scholarship, and good humor of numerous individuals over many months to make this opus possible. First among those to be recognized are senior editors Susan Hitchcock and Allyson Johnson, whose equanimity and personal commitment kept the project on an even keel throughout. Thanks, too, to editorial director and publisher Lisa Thomas and general manager Hector Sierra, who consistently supported what proved to be an unexpectedly ambitious project, and to the vital endorsement of National Geographic's leaders, Jean Case, Gary Knell, and Jill Tiefenthaler. Thanks are due to Daneen Goodwin and Ann Day for their commitment to this book's global outreach. We want to salute photography editor Matt Propert and the gifted contributing photographers, who have made possible a uniquely 21st-century vision of the ocean and its life, much of it discovered in the past decade. David Whitmore's compelling design is the ideal showcase for their imagery and the ocean story. Debbie Gibbons and her National Geographic maps team embody the spirit of exploration through their beautiful and timely cartography and diagrams, enhancing the comprehensive ocean maps published in *National Geographic Ocean: An Illustrated Atlas* in 2008.

To Mission Blue executive director Laura Cassiani and the entire Mission Blue team, especially chair of the Hope Spots Council Dan Laffoley, Vice-Chair, IUCN World Commission on Marine Protected Areas. Thank you to Craig McLean at NOAA for first believing in this project and committing his support in 2015. To Phil Stephenson for his unending dedication to and support of National Geographic's ocean initiatives, including *National Geographic Kids Extreme Ocean,* and the Pristine Seas initiative, and to this tangible tribute for the ocean. To Francesco Raeli and Caterina Boitani at Rolex for their interest and support of this book, of Mission Blue, and Sylvia Earle's message for the ocean. And we thank Thane Ritchie for his far-reaching vision to make the ocean accessible by supporting this book and future related projects through film, social media, Hope Spot support, and more, and for Robert Bigelow's engagement. To Patty Elkus and Wolcott Henry, ocean champions. To Ellen Prager, for sharing her deep knowledge of the ocean, and to Liz Taylor, DOER visionary, our north star. We join the ocean in offering all of you our deepest thanks.

A southern stingray glides in a glassy sea near Grand Cayman's Stingray City.

ABOUT THE AUTHORS

SYLVIA A. EARLE

Sylvia Alice Earle, National Geographic Society explorer since 1998, is founder of Mission Blue, founder of Deep Ocean Exploration and Research, a founding Ocean Elder, founding IUCN Patron, Council Chair of Harte Research Institute, adviser to Earth Observatory Singapore at NTU, and former chief scientist of NOAA. Author of 240 publications, leader of more than 100 expeditions, with years of research at sea and thousands of hours underwater, she has a B.S. from Florida State University, M.A. and Ph.D. from Duke University, and 33 honorary doctorates. She researches the ecology and conservation of marine ecosystems and development of technologies for access to the deep sea. A speaker in more than 100 countries; subject of the Netflix film *Mission Blue; Time* magazine's first "Hero for the Planet"; a Library of Congress Living Legend; and recipient of Carl Sagan, Lewis Thomas, Walter Cronkite, and Rachel Carson honors, she holds awards for global leadership from the United Nations, Tällberg Foundation, On Cue, and the World Affairs Council. Her other 150 honors include the Netherlands' Order of the Golden Ark, Princess of Asturias Award for Concord (Peace), TED Prize, Explorers Club Medal, Royal Geographic Society's Patron's Medal, and the National Geographic Hubbard Medal.

BARBARA BROWNELL GROGAN
co-author

Former editor in chief for National Geographic Books, Barbara Brownell Grogan has worked with Sylvia Earle on six National Geographic publications, including *The World Is Blue* and *Ocean: An Illustrated Atlas*. As principal for Rivanna Publishing Ventures and editor in chief for Potomac Global Media, she envisions and packages books on science, nature, history, and culture for all ages. Her expertise as a science writer includes exploring facets of nature's medicine and integrative health, on which topic she has authored five books.

LINDA K. GLOVER
technical consultant

Technical consultant Linda K. Glover is an expert in ocean science and policy. She has assisted in including global seafloor maps in Google Earth and published several ocean books including *Defying Ocean's End* and *National Geographic Ocean: An Illustrated Atlas* with Sylvia Earle, and she serves on nonprofit boards for ocean conservation and science. She is known for translating complex scientific information into language accessible to the general public.

INDEX

ILLUSTRATIONS & MAP CREDITS

Cover, Christian Vizl; 1, David Liittschwager/National Geographic Image Collection (NGIC); 2-3, Greg Lecoeur; 4-5, Thomas P. Peschak/NGIC; 6-7, David Liittschwager/NGIC; 8, Kip Evans Photography; 9 (UP), James D. Watt/BluePlanetArchive.com; 9 (LO LE), George W. Bush Presidential Library and Museum; 9 (LO RT), Brian Skerry/NGIC; 10, Flip Nicklin/Minden Pictures; 11, Blue Planet Archive/Steven Kovacs/Alamy Stock Photo; 12, Kip Evans Photography; 13, Bruce Shafer—ScubaShafer.com; 15, David Wrobel/BluePlanetArchive.com; 16-17, Thomas P. Peschak/NGIC; 18-19, Keith Ladzinski/NGIC; 20-21, NASA/Bill Anders; 22-3, @CJKale; 24, Michael Melford; 25, Dr. Neil Overy/Science Source; 26, David Shale/naturepl.com; 27, Alex Mustard/NPL/Alamy Stock Photo; 28 (UP), World Ocean Floor Panorama, Bruce C. Heezen and Marie Tharp, 1977. Copyright by Marie Tharp 1977/2003. Reproduced by permission of Marie Tharp Maps LLC and Lamont Doherty Earth Observatory; 28 (LO), Emory Kristof/NGIC; 29, Granger, All rights reserved; 31 (LE), Kevin Schafer/Minden Pictures; 31 (RT), ullstein bild via Getty Images; 32, NOAA Pacific Marine Environmental Laboratory/NSF/Science Source; 33, BluePlanetArchive; 34, Norbert Wu/Minden Pictures; 35 (UP), Diva Amon and Craig Smith, ABYSSLINE Project; 35 (LO), Joel Sartore/NG Photo Ark, photographed at Gulf Specimen Marine Lab; 36-7, Thomas P. Peschak/NGIC; 38, Doug Perrine/naturepl.com; 39, Sigma Sreedharan; 41 (LE), Thomas P. Peschak/NGIC; 41 (RT), Reynold Riksa Dewantara; 42, Sean Gallup/Getty Images; 43 (UP), World History Archive/Alamy Stock Photo; 43 (LO), Anton Balazh/Shutterstock; 44, Brian Skerry/NGIC; 45, Flip Nicklin/NGIC; 46, UVic Photo Services; 47 (UP), Kip Evans Photography; 47 (LO), Enric Sala/NG Explorer Programs; 48-9, Monica Bertolazzi/Getty Images; 50-51, Laurent Ballesta; 52, International Ocean Discovery Program; 53 (LE), Kiyoshi Ota/Bloomberg via Getty Images; 53 (RT), Fritz Goro/The LIFE Picture Collection via Getty Images; 54-5, Jad Davenport; 56-7, Dmitry Marchenko; 59, Brian Christensen/Stocktrek Images/Science Source; 60-61, Jason Edwards; 62-3, Thomas P. Peschak; 64, Andrea Marshall; 64-5, Paul Nicklen/NGIC; 66, University of Georgia Marketing & Communications. All rights reserved; 67 (BOTH), Frans Lanting; 69, Brook Peterson; 70, NOAA/Science Source; 72, Paul Nicklen/NGIC; 72-3, MichaelAW.com; 74-5, Brett Monroe Garner/Getty Images; 75 (ALL), David Liittschwager; 76-7 (BOTH), David Liittschwager/NGIC; 78-9, Planet Observer/UIG/Alamy Stock Photo; 79, Jennifer Idol/Stocktrek Images/Science Source; 80, Dan Tchernov, MKMS University of Haifa & Project CETI; 81, David Gruber and John Sparks; 82-3, Paul Nicklen/NGIC; 84, Paul Nicklen/NGIC; 85, courtesy of the Monterey Bay Aquarium Research Institute ©2004 MBARI; 86, Rebecca Hale, NG Staff; 87, wildestanimal/Shutterstock; 88-9, Reiko Takahashi; 90, Ocean Exploration Trust; 91 (UP), Paula Joy Cartwright; 91 (LO), Natural History Museum, London, UK ©Natural History Museum, London/Bridgeman Images; 92-3, Erik Aeder; 94-5, David Doubilet/NGIC; 96, L. Barry Hetherington; 97, Chris Bickford; 98, Encyclopaedia Britannica/UIG Via Getty Images; 98-9, Mauricio Handler; 99, NOAA; 101, Robbie George; 102-103,

Matt Propert; 104, Chris McGrath/Getty Images; 106-107, Paul Souders/Worldfoto/Minden Pictures; 109 (UP), @buck_taylor_; 109 (LO), OET/SRF/NOAA; 110, Bruce Miller/Alamy Stock Photo; 111, Joel Sartore; 112-13, ©Craig Foster, Sea Change Project; 114-15, Paul Nicklen/NGIC; 117, Michael Melford; 118-19, ©BenThouard.com; 120 (UP), Karsten Petersen, www.global-mariner.com; 120 (LO), Universal History Archive/UIG/Shutterstock; 121, Andrew H. Walker/Getty Images for CHANEL; 122-3 (UP), JIJI PRESS/AFP via Getty Images; 122-3 (LO), John Stanmeyer/NGIC; 123, Mike Theiss; 124, Thomas P. Peschak; 124-5, BluePlanet Archive/Richard Herrmann; 126, BluePlanetArchive/David B. Fleetham; 127, Mark Thiessen, NG Staff; 128, Bettmann/Getty Images; 129 (LE), NGIC, courtesy Cmdr. Edward Peary Stafford; 129 (RT), Robert E. Peary Collection/NGIC; 130-31, Bruce Shafer—ScubaShafer.com; 132-3, Andy Mann; 134-5, SCOTLAND: The Big Picture/NPL/Minden Pictures; 136-7, Paul A. Zahl/NGIC; 138, Jason Edwards; 139, Fred Bavendam/Minden Pictures; 141 (UP), Chris Newbert/Minden Pictures; 141 (LO), Steven Kovacs/BluePlanetArchive.com; 142-3, David Doubilet/NGIC; 144, Zack Bolton, NGS; 145 (UP), Bruno Guenard/Biosphoto/Minden Pictures; 145 (LO), D.P. Wilson/FLPA/Minden Pictures; 146-7, Brian Skerry/NGIC; 148-9, David Wrobel/BluePlanetArchive.com; 150, Paul Nicklen; 151, Chris Newbert/Minden Pictures; 152, © Woods Hole Oceanographic Institution (WHOI); 153, Dray van Beeck/NiS/Minden Pictures; 154-5, Chris Newbert/Minden Pictures; 156-7, David Liittschwager/NGIC; 157, Frans Lanting; 158, © WHOI; 159 (UP), Wolfgang Baumeister/Science Source; 159 (LO), Michael Abbey/Science Source; 161, Martin Shields/Science Source; 162, Claire Ting/Science Source; 162-3, M.I. Walker/Science Source; 164, Fred Bavendam/Minden Pictures; 164-5, Franco Banfi/NPL/Minden Pictures; 166-7, Leslie Newman & Andrew Flowers/Science Source; 167, Mathieu Foulquie/Biosphoto/Minden Pictures; 169, Norbert Wu/Minden Pictures; 170, Pascal Kobeh/NPL/Minden Pictures; 171, Brent Hedges/NPL/Minden Pictures; 172 (UP), Thomas Smoyer; 172 (LO), Gary Bell/Oceanwide/Minden Pictures; 173-5 (ALL), Edith Widder, Ph.D., ORCA; 176, David Liittschwager/NGIC; 177 (LE—UP to LO): Power and Syred/Science Source, Wim Von Egmond/Science Source, bombloombom/Imazins/Getty Images, Biophoto Associates/Science Source; 177 (RT—UP to LO): Jerome Pickett-Heaps/Science Source, Eric V. Grave/Science Source, Sylvia Earle, Deborah Maxemow/Getty Images; 178 (LE—UP to LO): Sylvia Earle, Gerd Guenther/Science Source, Frans Lanting, The Natural History Museum, London/Science Source; 178 (RT—UP to LO): Damocean/Getty Images, Flip Nicklin/Minden Pictures, Frans Lanting, Steve Gschmeissner/Science Source; 179 (LE—UP to LO): Shane Gross/NPL/Minden Pictures, MichaelAW.com, Paul J. Fusco/Science Source, Claudio Contreras/NPL/Minden Pictures; 179 (RT—UP to LO): Sylvia Earle, David M. Dennis/Animals Animals—All rights reserved, Michael Melford, Juergen Freund/Alamy Stock Photo; 180 (LE—UP to LO): Birgitte Wilms/Minden Pictures, Franco Banfi/NPL/Minden Pictures, Pete Oxford/Minden Pictures, L. Newman & A. Flowers/Science Source; 180 (RT—UP to LO): Kenneth M. Highfill/Science Source,

370-71, Karim Iliya; 372, Courtesy Ameer Abdulla/NG Explorer Programs; 373, Frans Lanting; 374-5, BluePlanetArchive/Andrey Nekrasov/imageBROKER; 376, Carrie Vonderhaar/Ocean Futures Society; 377, Esther Horvath; 378, James Nesterwitz/Alamy Stock Photo; 379 (UP), Chip Somodevilla/Getty Images; 379 (LO), Dan Costa; 380, Larry Busacca/Getty Images; 381 (UP), Millard H. Sharp/Science Source; 381 (LO), Paul Nicklen; 382, Alexander Gerst/ESA via Getty Images; 383 (LE), NASA; 383 (RT), Lockheed Martin (CC BY 2.0—https://creativecommons .org/licenses/by/2.0/legalcode); 384-5, Nadia Aly; 386-7, Thomas P. Peschak; 388, Nancy R. Schiff/Getty Images; 389 (UP), Paul Nicklen/NGIC; 389 (LO), Colors and shapes of underwater world/Getty Images; 390, Guido Mocafico, courtesy Hamiltons Gallery, London; 391, Jordi Chias/naturepl.com; 392-3, David Doubilet/NGIC; 394, Bill Curtsinger; 394-5, Hiroya Minakuchi/Minden Pictures; 396 (UP), Michael Melford/NGIC; 396 (LO), Roland Seitre/Minden Pictures; 397 (UP), Steve McCurry; 397 (LO), Brian Skerry/NGIC; 398-9, Giordano Cipriani/Getty Images; 400, Cole Sartore/NGIC; 401 (BOTH), Joel Sartore/NG Photo Ark, photographed at Pure Aquariums; 402, Magnus Lundgren/Wild Wonders of China/naturepl.com; 403, Dr. Peter M. Forster/Getty Images; 404-405, Wil Meinderts/Buiten-Beeld/Alamy Stock Photo; 406-407, Thomas P. Peschak/NGIC; 408, Alex Mustard/naturepl.com; 409, Courtesy Schmidt Ocean Institute; 410, Laura Cavanaugh/FilmMagic/Getty Images; 411, Paul Nicklen; 412-13, Laurent Ballesta; 413, Parker Amstutz/4ocean; 414, Courtesy of the IF/THEN® Collection. © 2020 Orange Capital Media. All rights reserved; 415 (BOTH), Enric Sala/NGIC; 416-17, Janine Marx; 418, Greg Lecoeur; 419 (LE), Sylvia Earle; 419 (RT), David Doubilet/NGIC; 420-21, David Doubilet/NGIC; 422-3, David Doubilet/NGIC; 424, Walter Meayers Edwards/NGIC; 425, Private Collection/Photo © Photo Josse/Bridgeman Images; 426, © Natural History Museum, London, UK/Bridgeman Images; 426-7, British Library, London, UK ©British Library Board. All Rights Reserved/Bridgeman Images; 434-5, Solvin Zankl/naturepl.com; 437, Jim Richardson/NGIC; 443, Alex Postigo; 447, David Moynahan Photography; 451, Helmut Corneli/Alamy Stock Photo; 454-5, Blue Planet Archive JMI/Alamy Stock Photo; 457, Brook Peterson/Stocktrek Images/Getty Images; 463, Alex Mustard/Science Source; 467, YinYang/Getty Images; 470-71, Sakis Papadopoulos/robertharding.com; 473, Pascal Kobeh/naturepl.com; 478-9, Alex Mustard/naturepl.com; 483, Anton Balazh/Shutterstock, with elements from NASA; 486-7, Andy Mann; 489, Paul Nicklen; 495, Wild Wonders of Europe/Magnus Lundgren/naturepl.com; 496-7, Frans Lanting; 499, Flip Nicklin/Minden Pictures; 505, Jordi Chias/naturepl.com; 506, Christian Vizl/TandemStock.com; 519, Al Giddings.

Maps and graphics credits

Pages 30, 40, 58, 61, 68, 71, 107, 114, 140: Chuck Carter, Eagre Games Inc.

Page 23, Tibor G. Tóth; 24, Ron Blakey, © Colorado Plateau Geosystems, Inc. cpgeosystems.com; 24-5, Source: International Commission on Stratigraphy. International Chronostratigraphic Chart, v2014/10. stratigraphy.org; 87, Stefan Fichtel.

Sources: C. W. Clark, Cornell Lab of Orinthology; Brandon Southall, University of California, Santa Cruz; Kathleen Vigness-Raposa, Marine Acoustics, Inc.; 100, Source: National Geographic Society; Lunar Reconnaissance Orbiter (LRO); NASA; 107, Source: Toby Garfield, NOAA, 111, Source: NASA/Goddard Space Flight Center Scientific Visualization Studio, https://svs .gsfc.nasa.gov/3881; 116, Sources: Bruno Tremblay, McGill University; Patricia DeRepentigny, University of Colorado Boulder; Robert Newton, Columbia University; Climatic Research Unit, University of East Anglia; 127, Sources: Scott Benson, Southwest Fisheries Science Center, NOAA; State of the World's Sea Turtles (SWOT), OBIS-SEAMAP; 245, Source: Brian Lapointe, Florida Atlantic University; 307, The Ocean Health Index is developed by the National Center for Ecological Analysis and Synthesis (NCEAS) and Conservation International; 333, Data: Benjamin Halpern and others, National Center for Ecological Analysis and Synthesis, University of California, Santa Barbara; 339, Data: D. A. Kroodsma, J. Mayorga, T. Hochberg, N. A. Miller, K. Boerder, F. Ferretti, A. Wilson, B. Bergman, T. D. White, B. A. Block, P. Woods, B. Sullivan, C. Costello, and B. Worm. "Tracking the global footprint of fisheries." *Science* 359.6378 (2018); 342, Data: Andrés Cózar Cabañas, University of Cádiz; Laurent Lebreton, Ocean Cleanup Foundation; Rachel W. Obbard, Dartmouth College; Alan J. Jamieson, Newcastle University; 345, Data: Flanders Marine Institute (2019). Maritime Boundaries Geodatabase: Maritime Boundaries and Exclusive Economic Zones (200NM), version 11. Available online at marineregions .org/; doi.org/10.14284/386; 350-51, Data: Marine Protected Areas: Marine Conservation Institute. (2020). MPAtlas. Seattle, WA. www.mpatlas.org [Accessed Sept 2020]; UNEP-WCMC and IUCN (2020), Protected Planet: The World Database on Protected Areas (WDPA) [On-line], Cambridge, UK: UNEP-WCMC and IUCN. Available at: www.protected-planet.net. Mission Blue Hope Spots: Mission Blue https://mission-blu.org/hope-spots/. Pristine Seas: Enric Sala, National Geographic Society; 369, Data: Verisk Maplecroft's Climate Change Vulnerability Index; 379, Data: Felix Landerer, NASA/JPL; M. Perrette et al., 2013; Organization for Economic Co-operation and Development; 432-3, Data adapted from International Hydrographic Organization Publication S-23 Limits of Oceans and Seas Draft 4th Edition 2002 incorporating National Geographic Map Policy; 494, Source: National Geospatial-Intelligence Agency; U.S. Coast Guard Office of Waterways and Ocean Policy; International Boundaries Research Unit.

All ocean floor maps (428-9, 438-9, 458-9, 474-5, 490-91, 500-501): Relief art: Tibor G. Tóth

All ocean political maps (430-31, 440-41, 444-5, 448-9, 452-3, 460-61, 464-5, 468-9, 476-7, 481, 492-3, 502-503): Data: UNESCO's Intergovernmental Oceanographic Commission (IOC), International Hydrographic Organization (IHO), General Bathymetric Chart of the Oceans (GEBCO), GEBCO's Sub-Committee on Undersea Feature Names (SCUFN), Earth Reference Seamount Catalog: earthref.org/SBN, Australian Offshore Mineral Locations map 2006, Geoscience Australia, The Scientific Committee on Antarctic Research (SCAR), International Council for Science

ROLEX

n the 1950s, Rolex, the Swiss watchmaker, became involved in marine exploration in the course of improving its technology for divers' watches and in later years has developed an intense interest in protecting the oceans.

Rolex has been involved with Sylvia Earle since 1970 and has admired her efforts through Mission Blue to protect ocean biodiversity through the creation of a global network of marine-protected Hope Spots. For this reason, Rolex supports Mission Blue as a pillar of the company's Perpetual Planet initiative, which was launched to assist key individuals and organizations who are finding solutions for preserving the environment and improving human well-being.

As well as Mission Blue, Perpetual Planet encompasses the Rolex Awards for Enterprise, a program that supports individuals with projects that preserve natural and cultural heritage, advance scientific knowledge, and improve the quality of life. The initiative also includes an enhanced partnership with National Geographic to study the impact of climate change.

THE NATIONAL GEOGRAPHIC SOCIETY INVESTS IN BOLD PEOPLE WITH TRANSFORMATIVE IDEAS WHO ILLUMINATE AND PROTECT THE WONDER OF OUR WORLD.

Since 1888, the National Geographic Society has driven impact by identifying and investing in a global community of Explorers: leading change makers in science, education, storytelling, conservation, and technology. National Geographic Explorers help bring our mission to life by defining some of the most critical challenges of our time, uncovering new knowledge, advancing new solutions, and inspiring transformative change in our world.

One of the most revered members of this community is Explorer at Large Sylvia Earle, world-renowned oceanographer, who has worked to protect Earth's blue heart and all its treasures. Throughout her career, Earle has led more than 100 oceanic expeditions and has logged over 7,000 hours underwater.

To learn more about the Explorers we invest in—like Earle—and the efforts we support, visit *natgeo.org*.

From a submerged platform off Oahu, Hawaii, in 1979, divers release the *Jim* suit with Sylvia Earle inside and the submersible *Star II*, to initiate a dive to the seafloor.

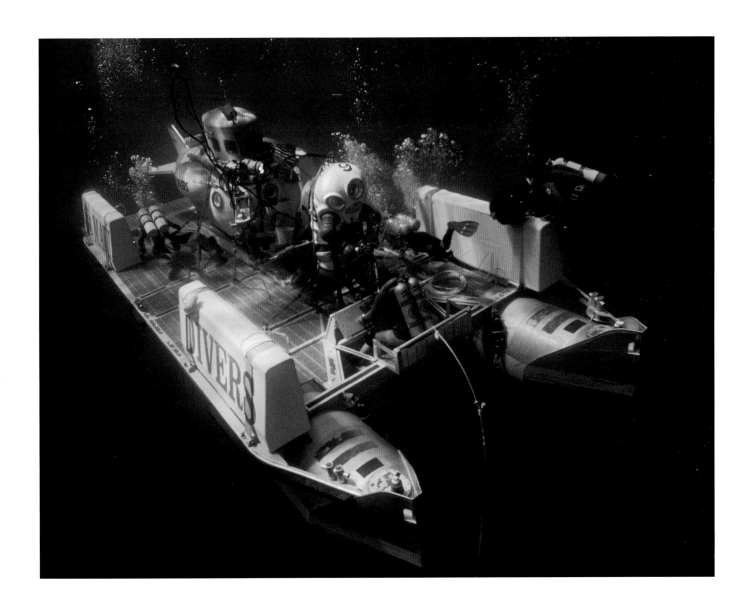